STUDY GUIDE

AND

SOLUTIONS MANUAL

ESSENTIAL iGENETICS

Bruce A. Chase

University of Nebraska at Omaha

Benjamin Cummings

San Francisco • Boston • New York
Cape Town • Hong Kong • London • Madrid • Mexico City
Montreal • Munich • Paris • Singapore • Syndey • Tokyo • Toronto

Publishing Assistant: Michael McArdle
Compositor: TBH Typecast, Inc.
Manufacturing Buyer: Pam Augspurger
Marketing Manager: Josh Frost

ISBN 0-8053-4707-0

2 3 4 5 6 7 8 9 10 — DPC — 05 04 03 02
www.aw.com/bc

Benjamin Cummings

CONTENTS

1

GENETICS: AN INTRODUCTION

CHAPTER OUTLINE

BUILDING CONCEPT MAPS USING GENETICS TERMS

Every chapter of the text introduces new terms used in the "language of genetics." Like new words of a foreign language, each term has a precise meaning and contextual use. To help develop your ability to "speak" the language of genetics, each chapter of this guide will provide lists of key terms, symbols, and concepts. After reading a chapter in the text, review the lists. Without consulting the text and in your own words, write a brief, precise definition of each term. Check your definition against that in the text. Then visualize the relationships between the terms by constructing a concept map using the terms in each list.

What is a concept map? A concept map is a visual image that depicts the relationships between ideas. Terms are arranged on a page, and related terms are connected with lines. A word or phrase that explains the relationship of the concepts is placed next to the lines. Different concept maps can be constructed from the same list of terms, depending on how the relationships between the concepts are viewed. Here is a sample concept map illustrating relationships between ancestral canines and modern dogs.

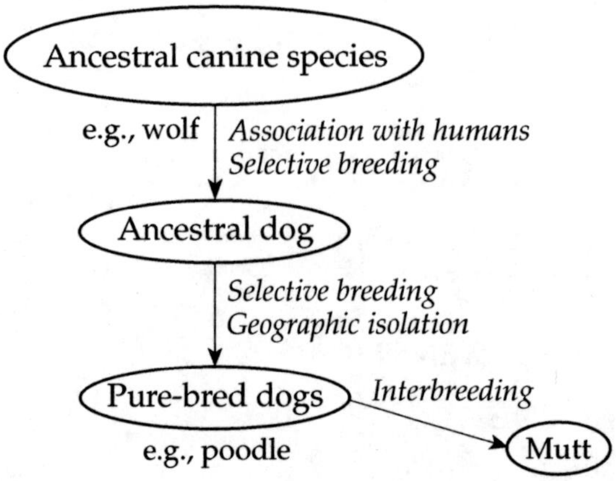

Here are some hints* to help you build concept maps using the provided lists of terms:

1. Write each term on a small piece of paper. In some lists, specific examples of a term will be given. For example, in this chapter, one list contains "genetic material," "DNA," and "RNA." DNA and RNA could be considered examples of genetic material. Keep the examples separate, perhaps even using different colored paper for them.
2. Arrange the terms on a larger sheet of paper, with the broadest or most abstract ideas at the top and the most specific ideas at the bottom.
3. Arrange the concepts so that two related ideas are placed near (e.g., under) each other.
4. Draw lines between related concepts. Rearrange the pieces of paper as you see fit.
5. On the connecting lines, write words or phrases that explain the relationship between the concepts. Be as precise and succinct as you can. This is challenging, so be patient.
6. Put examples under the concepts they belong with. Connect examples to a concept using an explanatory phrase—"e.g." or "specifically."
7. Copy the results onto a single sheet of paper. Draw circles around the concepts, but not the examples.

While concept maps can be generated with pencil and paper, computer tools for concept mapping also exist. The site *http://users.edte.utwente.nl/lanzing/cm_home.htm* provides a useful list of links.

REVIEW OF KEY TERMS, SYMBOLS, AND CONCEPTS

In your own words, write a brief, precise definition of each term in the groups on the the next two pages. Check your definitions using the text. Then develop a concept map using the terms in each list.

THINKING ANALYTICALLY

Learning genetics is more than just learning a set of facts. Geneticists use analytical reasoning to incorporate new results into established conceptual frameworks. Analytical reasoning is essential to use new data to expand and sometimes refute established concepts.

*Adapted from *http://www.dc.peachnet.edu/~shale/humanities/composition/handouts/concept.html.*

1	2	3	4
genome	gene	genetics	eukaryote
eukaryotes	gene expression	genome project	diploid
nucleus	enzyme	genetic map	haploid
prokaryotes	polypeptide	gene locus	gemete
bacteria	protein	mutation	zygote
eubacteria	one gene–one enzyme hypothesis	recombination	homologue
archaebacteria	one gene–one polypeptide hypothesis	selection	homologous chromosome
chromosome	operon	hypothetico-deductive investigation	nonhomologous chromosome
gene	messenger RNA	transmission genetics	sex chromosome
allele	transfer RNA	molecular genetics	autosome
genotype	ribosomal RNA	population genetics	centromere
phenotype	small nuclear RNA	quantitative genetics	p arm
homozygous	ribosomes	recombinant DNA technology	q arm
heterozygous	nucleolus	PCR	metacentric chromosome
dominant	transcription	basic research	submetacentric chromosome
recessive	translation	applied research	acrocentric chromosome
genetic material	genetic code	model organisms	telocentric chromosome
DNA	codon	clone	karyotype
RNA	amino acid	DNA typing	ideogram
nucleotide	RNA polymerase	genetic database	band
ribose	centrioles	BLAST	G banding
deoxyribose	endoplasmic reticulum	GenBank	
adenine (A)	mitochondria	PubMed	
guanine (G)	chloroplast	OMIM	
cytosine (C)		NCBI	
thymine (T)			
uracil (U)			

5	6	7	8
mitosis	meiosis I, II	meiosis	meiosis
karyokinesis	prophase I, II	sexual reproduction	meiospore
cytokinesis	metaphase I, II	spermatogenesis	gametogenesis
cell cycle	anaphase I, II	oogenesis	sporogenesis
M, G_1, S, G_2	telophase I, II	spermatozoa	gametophyte
sister chromatid	leptonema	primary spermatogonia	sporophyte
chromatid	zygonema	primary spermatocytes	stamen
daughter chromosome	pachynema	meiocyte	pistil
prophase	diplonema	secondary spermatocyte	alternation of generations
prometaphase	diakinesis		ovule
metaphase	pairing		pollen
	crossing-over		

(continued)

5	6	7	8
anaphase	synapsis	spermatid	anther
telophase	synaptonemal complex	primary oogonia	gametophyte generation
mitotic spindle	bivalent	secondary oogonia	sporophyte generation
microtubule	tetrad	primary oocyte	spore
tubulin	dyad	unequal cytokinesis	
centriole	chiasma, chiasmata	secondary oocyte	
kinetochore microtubule	genetic recombination	first polar body	
metaphase plate	recombinant chromosome	second polar body	
scaffold	pseudoautosomal region	ovum	
disjunction			
cell plate			

Following each chapter is a set of questions and problems. The questions ask you to recall and integrate information, while the problems require you to apply analytical reasoning. Solving problems using analytical reasoning is essential to learning genetics. It reinforces what you have read and promotes an understanding of the relationships between terms, facts, and concepts.

Unfortunately, learning to reason analytically is not always straightforward. It also takes time. The time will be well spent, however, as you will grow tremendously in your reasoning ability. Here are some pointers that will be helpful to keep in mind as you work on problems.

1. Read the problem straight through without pausing. Get the sense of what it is about.
2. Then read it again, slowly and critically. This time,
   a. Jot down pertinent information.
   b. Scan the problem carefully, analyzing the *terms* it uses. Often, clues to an answer can be found in the way the problem is stated. (Geneticists use very precise language!)
3. If you are unsure how to continue, ask yourself, "What are the options here?"
   a. Write down all the options you perceive.
   b. Carefully analyze whether a particular option will provide a solution.
4. When you arrive at a solution, read the problem again, making sure that your analysis is consistent with *all* the data in the problem. Try to see if there is a more general principle behind the solution or, for that matter, a clearer or more straightforward solution.
5. If you get stuck (everyone does!), DON'T GO IMMEDIATELY TO THE ANSWER KEY. This only gives an *illusion* of solving a problem. On an exam, and in real life, there is no answer key. Don't be tempted "to just know the answer." Go on to another problem. Come back to this one later. Your mind can use the time to sort through loose ends.
6. If you cannot solve the problem after coming back to it a second time, read through the answer key. Then close the answer key and try the problem once more. The next day, try it a third time without the answer key.

Applying Rules of Probability

To solve some problems in this and subsequent chapters, you will need to apply rules of probability. Two rules are commonly employed in genetics:

Product rule: The probability of two independent events occurring simultaneously is the product of each of their individual probabilities.

Sum rule: The probability of either one of two independent mutually exclusive events occurring is the sum of their individual probabilities.

To apply these rules, first consider the possible outcomes of a situation. Then identify the mutually exclusive independent events leading to these outcomes. Apply the rules after rephrasing the question. Suppose there are two mutually exclusive independent events, A and B. If you can rephrase the question to ask for the chance that *both A and B* occur (together), multiply the individual probabilities: $P(\text{A and B}) = P(\text{A}) \times P(\text{B})$. If you can rephrase the question statement to ask for the chance that *either A or B* occurs (separately), add the individual probabilities: $P(\text{A or B}) = P(\text{A}) + P(\text{B})$.

QUESTIONS FOR PRACTICE

Multiple Choice Questions

1. Transmission genetics is primarily concerned with
   a. the distribution and behavior of genes in populations
   b. the passing of genes from generation to generation and their recombination
   c. the structure and function of genes at the molecular level
   d. the means by which mutations are retained in nature
2. Which one of the following is not a criterion to choose an organism for genetic experimentation?
   a. The organism should be able to be used in applied research.
   b. The organism should be easy to handle.
   c. The organism should exhibit genetic variation.
   d. The organism should have a relatively short life cycle.
3. Which one of the following eukaryotic cell structures does not contain DNA?
   a. a nucleus
   b. a mitochondrion
   c. the endoplasmic reticulum
   d. a chloroplast
4. Which one of the following is not an accurate description of a chromosome?
   a. It is a colored body localized in the nucleus.
   b. It is a protein and nucleic acid complex.
   c. It is the cellular structure that contains the genetic material.
   d. In eukaryotes, it is composed of many DNA molecules attached end to end.
5. Which one of the following is not true?
   a. Genes encode proteins.
   b. Genes are proteins.
   c. Genes are transcribed into RNA molecules.
   d. mRNA molecules are translated into proteins on ribosomes.
6. A eukaryotic organism is one that
   a. is multicellular
   b. has a plasma membrane
   c. has genetic material encapsulated in a nuclear membrane
   d. all of the above

7. A centriole is an organelle that is
   a. present in the center of a cell's cytoplasm
   b. composed of microtubules and important for organizing the spindle fibers
   c. surrounded by a membrane
   d. part of a chromosome
8. The rough endoplasmic reticulum is
   a. an intracellular double-membrane system to which ribosomes are attached
   b. an intracellular membrane that is studded with microtubular structures
   c. a membranous structure found within mitochondria
   d. only found in prokaryotic cells
9. Which one of the following is available at the NCBI website and is a search tool that is useful to compare a nucleotide sequence with other nucleotide sequences in a database?
   a. PubMed
   b. BLAST
   c. OMIM
   d. GenBank
   e. PCR
10. Which one of the following can be translated?
    a. rRNA
    b. tRNA
    c. snRNA
    d. mRNA
    e. DNA
11. What two events occur in leptonema of prophase I?
    a. synapsis and crossing-over
    b. terminalization and disjunction
    c. synaptonemal complex formation and crossing-over
    d. pairing and chiasma formation
    e. pairing and the beginning of crossing-over
12. Which one of the following is *not* true?
    a. Crossing-over begins in leptonema.
    b. Chiasmata are visible manifestations of crossing-over.
    c. Chiasmata are visible in pachynema.
    d. Chiasmata are visible before terminalization.
13. Which one of the following statements is true about mitosis?
    a. The nucleolus disappears during metaphase.
    b. Nuclear membranes re-form at the end of telophase.
    c. Centromeres are aligned on an equatorial plane at prophase.
    d. The nucleolus re-forms at the end of anaphase.
14. Sister chromatids are
    a. synonymous with homologous chromosomes
    b. present only in meiosis and not in mitosis
    c. identical products of chromosome duplication held together by a replicated but unseparated centromere
    d. visible in interphase just after S phase

Answers: 1b, 2a, 3c, 4d, 5b, 6c, 7b, 8a, 9b, 10d, 11e, 12c, 13b, 14c

Thought Questions

1. Why is it difficult to draw a sharp boundary between molecular genetics, transmission genetics, population genetics, and quantitative genetics?
2. Geneticists use the hypothetico-deductive method of investigation. How does this rational approach still allow research projects to go in exciting, unpredictable directions?
3. How are basic and applied research interrelated in the following research areas?
   a. studies of mutations that affect the production of alcohol (ethanol) in yeast
   b. recombinant DNA technology
   c. studies of mutations that result in pesticide resistance in *Drosophila melanogaster*
   d. studies of mutations affecting the timing of seed ripening in *Arabidopsis*
   e. studies of transposons in corn
4. What features make an organism well suited for genetic experimentation?
5. What is the relationship between a genotype and a phenotype?
6. What is a gene?
7. When during meiosis does the chromosomal behavior underlying Mendel's laws occur?
8. What different types of analyses can be made using genetic databases?
9. How are genetic maps different from ideograms? How is each used?
10. What hopes and concerns are associated with the field of genomics?

Thought Questions for Media

After reviewing the media at the *Essential iGenetics* website, try to answer these questions.

1. How are centromeres, centrioles, and centrosomes different in terms of their cellular location and function during mitosis?
2. At what point in the cell cycle are chromosomes replicated? At what point are replicated chromosomes so extended that individual chromatids cannot be seen? When can they first be visualized? When are they maximally condensed?
3. When do centromeres replicate? When do replicated centromeres split apart?
4. During anaphase, chromosomes move to opposite poles and polar and kinetochore microtubules change length. Which type of microtubule shortens, and which type of microtubule elongates?
5. During anaphase, why do chromosomes appear V-shaped, I-shaped, or J-shaped?
6. How do the chromosomes of daughter cells produced by mitosis compare with the chromosomes of the parental cell?

SOLUTIONS TO TEXT PROBLEMS

1.1 Interphase is a period corresponding to the cell cycle phases of

a. mitosis
b. S
c. $G_1 + S + G_2$
d. $G_1 + S + G_2 + M$

Answer: c

1.2 Chromatids joined together by a centromere are called

a. sister chromatids
b. homologues
c. alleles
d. bivalents (tetrads)

Answer: a

1.3 Mitosis and meiosis always differ in regard to the presence of

a. chromatids
b. homologues
c. bivalents
d. centromeres
e. spindles

Answer: c

1.4 State whether each of the following statements is true or false. Explain your choice.

a. The chromosomes in a somatic cell of any organism are all morphologically alike.
b. During mitosis the chromosomes divide and the resulting sister chromatids separate at anaphase, ending up in two nuclei, each of which has the same number of chromosomes as the parental cell.
c. At zygonema, any chromosome can synapse with any other chromosome in the same cell.

Answer:

a. False. Chromosomes can have different sizes and can have their centromeres positioned differently. For many organisms, this statement will be false. However, in a diploid cell, a pair of homologous chromosomes will be morphologically alike.
b. True
c. False. In meiosis I, only homologous chromosomes will synapse together.

1.5 For each mitotic event described in the following table, write the name of the event in the blank provided in front of the description. Then put the events in the correct order (sequence); start by placing a 1 next to the description of interphase and continue through 6, which should correspond to the last event in the sequence.

Name of Event		Order of Event
____________	The cytoplasm divides and the cell contents are separated into two separate cells.	______
____________	Chromosomes become aligned along the equatorial plane of the cell.	______

(continued)

Name of Event		Order of Event
	Chromosome replication occurs.	
	The migration of the daughter chromosomes to the two poles is complete.	
	Replicated chromosomes begin to condense and become visible under the microscope.	
	Sister chromatids begin to separate and migrate toward opposite poles of the cell.	

Answer: Try to visualize the events of mitosis dynamically as you solve this problem.

Name of Event		Order of Event
cytokinesis	The cytoplasm divides and the cell contents are separated into two separate cells.	6
metaphase	Chromosomes become aligned along the equatorial plane of the cell.	3
interphase	Chromosome replication occurs.	1
telophase	The migration of the daughter chromosomes to the two poles is complete.	5
prophase	Replicated chromosomes begin to condense and become visible under the microscope.	2
anaphase	Sister chromatids begin to separate and migrate toward opposite poles of the cell.	4

1.6 Determine whether the answer to these statements is yes or no. Explain the reasons for your answer.

a. Can meiosis occur in haploid species?
b. Can meiosis occur in a haploid individual?

Answer:

a. Yes. If a sexual mating system exists in the species, two haploid cells can fuse to produce a diploid cell that can then go through meiosis to produce haploid progeny. The fungi *Neurospora crassa* and *Saccharomyces cerevisiae* exemplify this life cycle.

b. No. Meiosis only can occur in a diploid cell. A haploid individual cannot form a diploid cell, so meiosis cannot occur.

1.7 Select the correct answer. The general life cycle of a eukaryotic organism has the sequence

a. N → meiosis → 2N→ fertilization → N

b. 2N → meiosis → N → fertilization → 2N

c. N → mitosis → 2N → fertilization → N

d. 2N → mitosis → N → fertilization → 2N

Answer: b. Only diploid (2N) cells can undergo meiosis, and haploid cells (N) fuse at fertilization, regenerating the diploid state.

1.8 Which statement is true?

a. Gametes are 2N; zygotes are N.

b. Gametes and zygotes are 2N.

c. The number of chromosomes can be the same in gamete cells and in somatic cells.

d. The zygotic and somatic chromosome numbers cannot be the same.

e. Haploid organisms have haploid zygotes.

Answer: c. In a haploid organism, gametes and somatic cells are both N. This provides an example of how (c) is true.

1.9 All of the following happen in prophase I of meiosis except

a. chromosome condensation

b. pairing of homologues

c. chiasma formation

d. terminalization

e. segregation

Answer: e

1.10 Give the name of the stage of mitosis or meiosis at which each of the following events occurs:

a. Chromosomes are located in a plane at the center of the spindle.

b. The chromosomes move away from the spindle equator to the poles.

Answer:

a. Metaphase: metaphase in mitosis, metaphase I and metaphase II in meiosis

b. Anaphase: anaphase in mitosis, anaphase I and anaphase II in meiosis

1.11 Given the diploid, meiotic mother cell shown in the following figure, diagram the chromosomes as they would appear

a. in late pachynema

b. in a nucleus at prophase of the second meiotic division

c. in the first polar body resulting from oogenesis in an animal

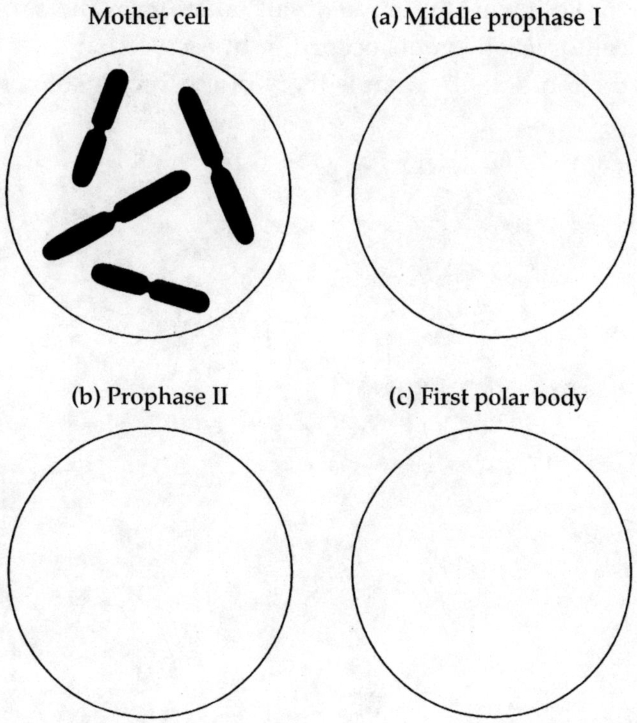
Mother cell
(a) Middle prophase I
(b) Prophase II
(c) First polar body

Answer:

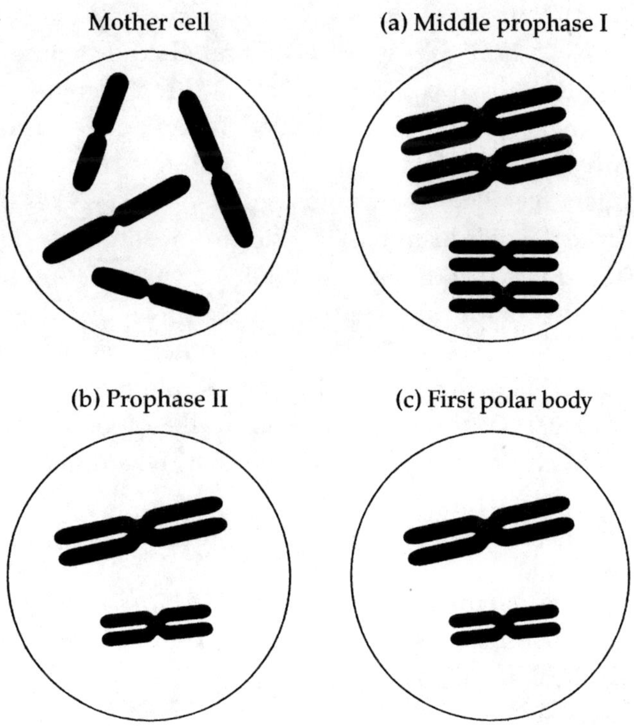
Mother cell
(a) Middle prophase I
(b) Prophase II
(c) First polar body

1.12 The cells in the following figure were all taken from the same individual (a mammal). Identify the cell division events occurring in each cell and explain your reasoning. What is the sex of the individual? What is the diploid chromosome number?

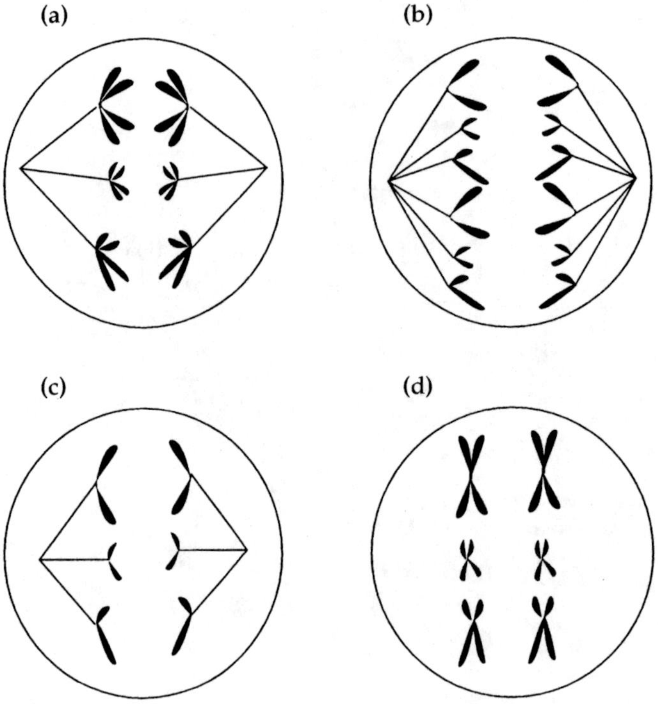

Answer: To reconcile how all the cells illustrated could come from one individual, consider that the cells shown could come from either somatic or germ-line cells. Cell (a) shows three pairs of previously synapsed homologues disjoining and must therefore illustrate anaphase I of meiosis. That three pairs of chromosomes are present indicates that the organism has 2N = 6 chromosomes, so N = 3. Cell (b) shows the disjoining of chromatids. Since the organism has 2N = 6 chromosomes, and the daughter cells that will form as a result of this cell division will have 6 chromosomes, (b) must illustrate part of mitosis, specifically anaphase. Figure (c) also shows the disjoining of chromatids. Since the daughter cells will receive three chromosomes, this must be anaphase II of meiosis. Figure (d) shows the pairing of homologues and therefore illustrates metaphase I of meiosis. As the individual has three pairs of identically appearing chromosomes, there are two identical sex (= X) chromosomes, indicating that the animal is female.

1.13 Does mitosis or meiosis have greater significance in the study of heredity? Explain your choice.

Answer: Meiosis has greater significance. While mitosis generates progeny cells that are genetically identical to a parent cell, meiosis generates gametes that are genetically diverse. Gamete diversity is obtained when nonparental combinations of genes are obtained in two ways: the random assortment of maternal and paternal chromosomes at anaphase I of meiosis I and crossing-over events in prophase I of meiosis I. When gametes from different parents fuse at fertilization, more diversity is obtained. Hence, meiosis provides a means for genetic variation.

1.14 Consider a diploid organism that has three pairs of chromosomes. Assume that the organism receives chromosomes A, B, and C from the female parent and A′, B′, and C′ from the male parent. To answer the following questions, assume that no crossing-over occurs.

a. What proportion of the gametes of this organism would be expected to contain all the chromosomes of maternal origin?
b. What proportion of the gametes would be expected to contain some chromosomes from both maternal and paternal origin?

Answer:

a. The chance that a gamete would have a particular maternal chromosome is ½. Apply the product rule to determine the chance of obtaining a gamete with all three maternal chromosomes: $P(\text{A } \textit{and} \text{ B } \textit{and} \text{ C}) = P(\text{A}) \times P(\text{B}) \times P(\text{C}) = (1/2)^3 = 1/8$.
b. There are two approaches to this question.
    1. List all of the possible options to obtain gametes that satisfy the specified condition —that some maternal and paternal chromosomes are present. Then use the product rule to determine the probability of each, and then, using the sum rule, determine the proportion of the gametes having maternal and paternal chromosomes. Gametes that satisfy the condition that both maternal and paternal chromosomes are present are ABC′, AB′C, A′BC, AB′C′, A′BC′, and A′B′C. For each one of these six gamete types, the chance of obtaining it is ⅛. This is based on applying the product rule and logic similar to that used in part (a) of this question. Apply the sum rule to determine the chance of obtaining one of these types of gametes: $P(\text{ABC}' \textit{ or } \text{AB}'\text{C} \textit{ or } \text{A}'\text{BC} \textit{ or } \text{AB}'\text{C}' \textit{ or } \text{A}'\text{BC}' \textit{ or } \text{A}'\text{B}'\text{C}) = 1/8 + 1/8 + 1/8 + 1/8 + 1/8 + 1/8 = 6/8 = 3/4$.
    2. Realize that the set of gametes with some maternal and paternal chromosomes is composed of all gametes *except* those that have only maternal or only paternal chromosomes. That is, *P*(gamete with *both* maternal *and* paternal chromosomes) = 1 – *P*(gamete with *only* maternal *or* only paternal chromosomes). From part (a), the chance of a gamete having chromosomes from only one parent is ⅛. Using the sum rule, *P*(gamete with both maternal and paternal chromosomes) = 1 – (⅛ + ⅛) = ¾.

1.15 Normal diploid cells of a theoretical mammal are examined cytologically at the mitotic metaphase stage for their chromosome complement. One short chromosome, two medium-length chromosomes, and three long chromosomes are present. Explain how the cells might have such a set of chromosomes.

Answer: Since the cells are normal and the cell is diploid, the chromosomes should exist in pairs. The cell could have come from a male mammal with one medium-length pair, one long-length pair, and a heteromorphic sex chromosome (X, Y) pair.

1.16 Is the following statement true or false? Explain your answer. "Meiotic chromosomes can be seen after appropriate staining in nuclei from rapidly dividing skin cells."

Answer: False. Skin cells consist entirely of mitotic, not meiotic, cells.

1.17 Is the following statement true or false? Explain your answer. "All the sperm from one human male are genetically identical."

Answer: False. Genetic diversity in the male's sperm is achieved during meiosis. There is crossing-over between nonsister chromatids as well as independent assortment of the male's maternal and paternal chromosomes during spermatogenesis.

1.18 The horse has a diploid set of 64 chromosomes, and a donkey has a diploid set of 62 chromosomes. Mules (viable but usually sterile progeny) are produced when a male donkey is mated to a female horse. How many chromosomes will a mule cell contain?

Answer: A female horse has ova with 32 chromosomes, while a male donkey has sperm with 31 chromosomes. A mule has $32 + 31 = 63$ chromosomes.

1.19 The red fox has 17 pairs of large, long chromosomes. The arctic fox has 26 pairs of shorter, smaller chromosomes.

a. What do you expect to be the chromosome number in somatic tissues of a hybrid between these two foxes?

b. The first meiotic division in the hybrid fox shows a mixture of paired and single chromosomes. Why do you suppose this occurs? Can you suggest a possible relationship between this fact and the observed sterility of the hybrid?

Answer:

a. The red fox will have gametes with 17 chromosomes, and the arctic fox will have gametes with 26 chromosomes. The hybrid will have $17 + 26 = 43$ chromosomes. Once the fertilized egg is formed, mitotic division will produce somatic cells with 43 chromosomes.

b. Similar chromosomes pair in meiosis. The pairing pattern seen in the hybrid indicates that some of the chromosomes in the arctic and red foxes share evolutionary similarity, while others do not. Unpaired chromosomes will not segregate in an orderly manner, giving rise to unbalanced meiotic products with either extra or missing chromosomes. This can lead to sterility for two reasons: First, meiotic products that are missing chromosomes may not have genes necessary to form gametes. Second, even if gametes are able to form, a zygote generated from them will not have the chromosome set from either the hybrid, the red, or the arctic fox. The zygote will be an aneuploid with missing or extra genes, causing it to be inviable.

1.20 At the time of synapsis preceding the reduction division in meiosis, the homologous chromosomes align in pairs, and one member of each pair passes to each of the daughter nuclei. In an animal with five pairs of chromosomes, assume that chromosomes 1, 2, 3, 4, and 5 have come from the father, and 1′, 2′, 3′, 4′, and 5′ have come from the mother. Assuming no crossing-over, in what proportion of the gametes of this animal will all the paternal chromosomes be present together?

Answer: The chance of getting a particular paternal chromosome is ½. Using the product rule and logic similar to that in the answer presented in problem 1.14a, the chance of getting all five paternal chromosomes is $(1/2)^5 = 1/32$.

2

MENDELIAN GENETICS

CHAPTER OUTLINE

REVIEW OF KEY TERMS, SYMBOLS, AND CONCEPTS

In your own words, write a brief, precise definition of each term in the groups on the following page. Check your definitions using the text. Then develop a concept map using the terms in each list.

THINKING ANALYTICALLY

There are different levels of understanding of Mendelian genetics. You can certainly obtain some understanding of this material from a careful reading and outlining of the chapter, especially if you pay attention to detail and the logic of Mendel's experiments. Indeed, understanding why Mendel's approach gave insights into the principles of inheritance, when those for thousands of years before him failed, is a substantial accomplishment. However, it is through solving problems

1	2
hereditary trait	probability
gene, particulate factor	product rule
locus, loci	sum rule
allele, allelomorph	principle of segregation
genotype	principle of independent assortment
phenotype	branch diagram
dominant	chi-square (χ^2)test
recessive	goodness-of-fit test
heterozygote, heterozygous	null hypothesis
homozygote, homozygous	expected values
self-fertilization, selfing	*P*
cross-fertilization, cross	significance
true-breeding, pure-breeding	degrees of freedom
P, F_1, F_2, F_3 generation, cross	pedigree analysis
monohybrid, dihybrid, reciprocal cross	proband, propositus, proposita
gamete, zygote	1:1, 3:1, all : none ratios
hybrid, dihybrid, trihybrid individual, cross	wild-type allele
testcross	rare homozygote
haploid, diploid	*AA, A–, Aa, aa*
gene segregation	loss-of-function, gain-of-function mutation
principle of segregation	recessive, dominant trait
principle of independent assortment	recessive, dominant pedigree

that you will develop a practical, useful understanding of this material. Solving the problems in this chapter will not only reinforce what you have read, but help you develop an understanding of the relationships between its terms, ideas, and facts.

With some effort, you can discover one of the multiple "right" approaches to solving a Mendelian genetics problem. Pursuing any number of "right" approaches will reinforce your understanding of Mendelian genetic principles and help you gain a deeper understanding of relationships you read about in the chapter. Be warned, however, that there is also a "wrong" approach. It is easy to describe. In the wrong approach, you read through the problem and then see if you can understand an answer presented in this guide or another key. This is about as beneficial as reading today's weather forecast after you have already stepped out into the rain. You have learned that it's raining, but you're already wet. What use is your new knowledge? What will you do tomorrow? When you read a solution to a problem, you have not yourself devised the solution, so you have not developed the skills to analyze the situation, apply your knowledge of Mendelian principles, and identify a path that leads to a solution. You may understand the answer, but you won't have developed any of the skills to solve a similar problem. On an exam, you will be stuck, and the purpose of doing the problems—to gain a deeper understanding of Mendelian principles—will have been thwarted.

What, then, is a "right" approach? It's not the same thing as finding *the* correct answer, usually because there is more than one approach to solving the problem. Here are some pointers, some repeated from Chapter 1 of this guide, that are worth emphasizing:

1. Read the problem straight through without pausing. Get the sense of what it is about.
2. Then read it again, slowly and critically. This time,
   a. Jot down pertinent information.
   b. Assign descriptive gene symbols. Use a dash (e.g., – – or *A*–) when unsure of a genotype.
   c. Scan the problem from start to finish and then from the end to the beginning, and carefully analyze the *terms* used in the problem. Often, clues to the answer can be found in the way the problem is stated. (Geneticists use very precise language!)
3. If you are unsure how to continue, ask yourself, "What are the options here?"
   a. Write down all the options you perceive.
   b. Carefully analyze whether a particular option will provide a solution.
4. When you think you have a solution, read through the problem once more and make sure that your analysis is consistent with *all* the data in the problem. Try to see if there is a more general principle behind the solution or, for that matter, a clearer or more straightforward solution.
5. If you do get stuck (and everyone does!), DO NOT GO IMMEDIATELY TO THE ANSWER KEY. This will only give you the *illusion* of having solved the problem. On an exam, or in real life, there won't be an answer key available. Learn to control the temptation "to just know the answer" and go on to another problem. Come back to this one another time, perhaps the next day. Your mind may be able to use the time to sort through loose ends.
6. If you cannot solve the problem after coming back to it a second time, read through the answer key. Then close the answer key and try the problem once more. The next day, try it a third time without the answer key.

QUESTIONS FOR PRACTICE

Multiple Choice Questions

1. A dominant gene is one that
   a. suppresses the expression of genes at all loci
   b. masks the expression of neighboring gene loci
   c. masks the expression of a recessive allele
   d. masks the expression of all the foregoing
2. A dihybrid is an individual that
   a. is heterozygous for every gene
   b. is heterozygous for two genes under study
   c. is the result of a testcross
   d. is used for a testcross
3. The cross of an uncertain genotype with a homozygous recessive genotype at the same locus is a
   a. pure-breeding cross
   b. monohybrid cross
   c. testcross
   d. dihybrid cross
4. The genotypic ratio of the progeny of a monohybrid cross is typically
   a. 1:2:1
   b. 9:3:3:1
   c. 27:9:9:9:3:3:3:1
   d. 3:1

5. A typical phenotypic ratio of a dihybrid cross with dominant and recessive alleles is
   a. 9:1
   b. 1:2:1
   c. 3:1
   d. 9:3:3:1
6. Pedigrees showing rare recessive traits
   a. have about half of the progeny affected when one parent is affected
   b. have heterozygotes that are phenotypically affected
   c. have about ¾ of the progeny affected when both parents are affected
   d. often skip a generation
7. In a chi-square test, a *P* value equal to 0.04 tells one that
   a. there is a 4 percent chance the hypothesis is correct
   b. there is a 4 percent chance the hypothesis is incorrect
   c. if the experiment were repeated, chance deviations from the expected values as large as those observed would be seen only 4 percent of the time
   d. if the experiment were repeated, chance deviations from the expected values as large as those observed would be seen at least 96 percent of the time
8. In a chi-square test, a *P* value equal to 0.04 indicates that
   a. the hypothesis is unlikely to be true
   b. the hypothesis is false and must be rejected
   c. the hypothesis is true
   d. the hypothesis is likely to be true
9. Gain-of-function mutations are associated with all of the following except
   a. dominant traits
   b. new properties in a gene
   c. phenotypes in heterozygous individuals
   d. a decrease in the normal activity of a gene
10. Two parents affected with a genetic disease have five children, all of whom are affected. Which one statement must be true?
    a. Both parents must have at least one mutant allele; the trait might be dominant or recessive.
    b. Both parents must be homozygous for a recessive trait.
    c. At least one parent must be homozygous.
    d. At least one parent must be homozygous for a dominant trait.
11. How many different genotypes are obtainable from the cross *Aa Bb* × *Aa Bb?*
    a. 3
    b. 4
    c. 9
    d. 16
12. What is the chance of obtaining a heterozygous individual from a testcross?
    a. 0%
    b. 50%
    c. at least 50%
    d. 100%

Answers: 1c, 2b, 3c, 4a, 5d, 6d, 7c, 8a, 9d, 10a, 11c, 12c

Thought Questions

1. When are Mendel's principles realized during meiosis?
2. How might the discovery of chromosomes toward the end of the nineteenth century have led to the rediscovery of Mendel's work?
3. Over a period of many years, numerous attempts were made to understand how physical traits are passed from one generation to the next. Gregor Mendel was first to make a breakthrough. How do you account for his success, in light of years of failure before him?
4. a. Clearly distinguish between an allele, a gene, and a locus.
   b. Can you see an analogy between a gene and an allele and
      - a digit and an index finger?
      - a canine and a German shepherd?
      - a dime and a mint 1954 dime?

      Where do these analogies break down?
5. Why might the term *degrees of freedom* be so named, and why (for the problems in this chapter) is it equal to $n - 1$?
6. Why are the frequencies of harmful recessive mutant alleles usually higher than the frequencies of harmful dominant mutations, even though individuals with the recessive trait are rare?
7. Construct a decision-making tree (a flowchart) that can be used to decide if a pedigree shows a dominant or recessive trait.
8. Why is it important to know if a trait is common or rare when analyzing a pedigree?

Thought Questions for Media

After reviewing the media at the *Essential iGenetics* website, try to answer these questions.

1. What is the difference between a locus, a gene, and an allele? Can these terms be used interchangeably?
2. What is the difference between a phenotypic ratio and a genotypic ratio? Can these ever be the same in the offspring of a cross?
3. When a true-breeding plant is self-fertilized, why aren't plants with different traits ever produced? If two true-breeding plants are crossed, why do all of the offspring have just one phenotype?
4. In a cross between two true-breeding plants differing in two characteristics, will the F_1 offspring always exhibit both traits of just one parent? Will they always be dihybrids?
5. What does it mean for "genes on different chromosomes to behave independently in the production of gametes"?
6. How many types of gametes are produced by a monohybrid? By a dihybrid?
7. How many different zygote genotypes can be produced in a monohybrid cross? In a dihybrid cross?
8. How many different zygote phenotypes can be produced in a monohybrid cross? In a dihybrid cross?
9. Why doesn't a recessive trait ever appear in a heterozygote? How do we know that the recessive allele is still present in a heterozygote?

1. Two tagged tribbles wander out of the iActivity tribble laboratory and show up in your bed. One is tagged "male *bb ss* (solid yellow)," and one is tagged "female *Bb Ss* (spotted brown)." You recall that spotted (*S*) is dominant to solid (*s*) and brown (*B*) is dominant to yellow (*b*) and that tribbles reproduce quickly. You seize this opportunity to make a (small) fortune selling spotted yellow tribbles. What crosses would you do to develop a true-breeding, spotted yellow strain? Since it is important to keep the population of non–spotted yellow tribbles to a minimum, you want to be sure that the animals chosen for the strain are true-breeding. How can you do this, since they cannot be selfed?
2. As you read the morning paper at breakfast, you notice that a (tiny) reward has been posted by the iActivity tribble laboratory for three missing tribbles. All the ad says is that one is a heterozygote and two are homozygotes. Midway through breakfast, three untagged tribbles waddle across your plate. One is a solid brown male, one is a solid brown female, and the third is a spotted brown male. Assuming that these animals were not just stained with coffee, could they be the missing tribbles? If so, how would you determine, using crosses with just these animals, which animal is heterozygous, which animals are homozygous, and if the homozygotes are both homozygous dominant, both homozygous recessive, or one is homozygous dominant and one is homozygous recessive?

SOLUTIONS TO TEXT PROBLEMS

2.1 In tomatoes, red fruit color is dominant to yellow. Suppose a tomato plant homozygous for red is crossed with one homozygous for yellow. Determine the appearance of (a) the F_1 tomatoes, (b) the F_2 tomatoes, (c) the offspring of a cross of the F_1 back to the red parent, and (d) the offspring of a cross of the F_1 back to the yellow parent.

Answer: Consider two ways to solve this problem.

1. First, notice that this is a situation akin to Mendel's crosses. In such crosses, dominant traits mask the appearance of recessive traits, and when a true-breeding dominant plant is crossed to a true-breeding recessive plant, all the F_1 progeny show the dominant trait. Here the dominant trait is red, the recessive trait is yellow, and the parents are true-breeding, since they are homozygous.
   (a) Just as in Mendel's crosses, the F_1 plants will be heterozygous and all show the dominant, red trait.
   (b) The F_2 plants that result from the cross of two F_1 heterozygotes will also show the same ratios seen by Mendel, that is, 3 dominant : 1 recessive, or 3 red : 1 yellow.
   (c) When an F_1 plant is crossed back to the red parent, a heterozygous plant is crossed to a true-breeding dominant plant. Here, each offspring receives a dominant allele from the red parent, and so all the progeny must show the dominant, red phenotype.
   (d) When an F_1 plant is crossed back to the yellow parent, a heterozygous plant is crossed back to a true-breeding recessive plant. This is akin to one of Mendel's testcrosses, and so the progeny should show a 1:1 ratio of red:yellow plants.
2. Assign *R* as the allele symbol for dominant red color and *r* as the allele symbol for recessive yellow color.

(a) Then the initial cross between two homozygous plants can be depicted as $RR \times rr$, and the F_1 progeny obtain an R allele from the red parent and an r allele from the yellow parent. Therefore, they are all heterozygotes and are Rr. As the R (red) allele is dominant to the r (yellow) allele, the Rr progeny are red.
(b) The F_2 are obtained from $Rr \times Rr$ and so will be composed of 1 RR : 2 Rr : 1 rr types of progeny. There will be 3 red (RR or Rr) : 1 yellow (rr).
(c) The F_1 crossed back to the red parent can be depicted as $Rr \times RR$. All the progeny will obtain an R allele from the red parent; the progeny can be written as R– (either RR or Rr), and all will be red.
(d) The F_1 crossed back to the yellow parent can be depicted as $Rr \times rr$. There will be two equally frequent types of progeny, Rr and rr. Thus, half the progeny will be red and half yellow.

2.2 In maize, a dominant allele A is necessary for seed color, as opposed to colorless (a). Another gene has a recessive allele wx that results in waxy starch, as opposed to normal starch (Wx). The two genes segregate independently. Give phenotypes and relative frequencies for offspring resulting when a plant of genetic constitution $Aa\ WxWx$ is testcrossed.

Answer: A plant that is genotypically $Aa\ WxWx$ has two types of gametes: $A\ Wx$ and $a\ Wx$ (notice that Wx is a symbol for one allele, not two). In a testcross, this plant is crossed to one homozygous for recessive alleles at both the color gene and the waxy gene, $aa\ wxwx$. The gametes of this plant are all $a\ wx$. This cross can be illustrated in the following Punnett square:

		Gametes of $Aa\ WxWx$	
		$A\ Wx$	$a\ Wx$
Gametes of $aa\ wxwx$	$a\ wx$	$Aa\ Wxwx$	$aa\ Wxwx$

Progeny will be of two equally frequent genotypes: $Aa\ Wxwx$ and $aa\ Wxwx$. Thus, half will be colored (Aa), half will be colorless (aa), and all will have normal starch ($Wxwx$).

2.3 F_2 plants segregate ¾ colored : ¼ colorless. If a colored plant is picked at random and selfed, what is the probability that both colored and colorless plants will be seen among a large number of its progeny?

Answer: In the F_2, there is a 3:1 colored:colorless phenotypic ratio. This reflects a 1 CC : 2 Cc : 1 cc genotypic ratio, where colored (C) is dominant to colorless (c). Thus, there are two types of colored plants (CC, Cc) that are present in a 1 CC : 2 Cc ratio. If one is picked at random, there is a ⅓ chance of picking a homozygous CC plant and a ⅔ chance of picking a heterozygous Cc plant. If selfed, the CC plant will only produce CC (colored) plants, while the Cc plant will produce both colored and colorless plants (in a 3:1 ratio). Thus, to see more than one type of plant in the progeny, a Cc plant must be chosen initially. The chance of this is ⅔. (Notice that the question asks for the chance that a particular colored plant will have two types of progeny and *not* for the ratio of progeny types when two types are seen.)

2.4 In guinea pigs, rough coat (*R*) is dominant over smooth coat (*r*). A rough-coated guinea pig is bred to a smooth one, giving eight rough and seven smooth progeny in the F_1 generation.

a. What are the genotypes of the parents and their offspring?
b. If one of the rough F_1 animals is mated to its rough parent, what progeny would you expect?

Answer:

a. Since rough (*R*) is dominant over smooth (*r*), a rough parent is *RR* or *Rr* (i.e., *R*–). A cross between a rough and smooth guinea pig is *R*– × *rr*. If the cross is *RR* × *rr*, then all the progeny will be *Rr* and be rough. If the cross is *Rr* × *rr*, then half the progeny will be *Rr* (rough) and half will be *rr* (smooth). The 8 rough : 7 smooth progeny ratio approximates a 1:1 *Rr*:*rr* ratio, and so the initial cross must have been *Rr* × *rr*. The rough progeny must be *Rr*, while the smooth progeny must be *rr*.
b. If one of the rough F_1 animals is mated back to its rough parent, the cross is *Rr* × *Rr*. This cross would produce both rough (*R*–) and smooth (*rr*) progeny in a 3:1 ratio.

2.5 In cattle, the polled (hornless) condition (*P*) is dominant over the horned (*p*) phenotype. A particular polled bull is bred to three cows. Cow A, which is horned, produces a horned calf; polled cow B produces a horned calf; and horned cow C produces a polled calf. What are the genotypes of the bull and the three cows, and what phenotypic ratios do you expect in the offspring of these three matings?

Answer: First depict the phenotypes as genotypes: A polled animal exhibiting the dominant, hornless condition can be depicted as *P*–, while a horned animal exhibits the recessive condition and must be *pp*. Therefore, the three crosses and their progeny can be depicted as

cow		×	**polled bull**	**progeny**	
phenotype	**genotype**		**genotype**	**phenotype**	**genotype**
A: horned	*pp*	×	*P*–	horned	*pp*
B: polled	*P*–	×	*P*–	horned	*pp*
C: horned	*pp*	×	*P*–	polled	*P*–

Now follow how each set of parents contributed alleles to their progeny. In the crosses with cows A and B, a horned, *pp* offspring can only be obtained if each parent contributes a recessive *p* allele to the progeny. Therefore, both the polled bull and cow B must be heterozygous: each is *Pp*. With this information, both the crosses with cows A and C appear to be testcrosses (*Pp* × *pp*) and will produce 1:1 phenotypic ratios of polled (*Pp*) and horned (*pp*) progeny. The cross with cow B is a cross between two heterozygotes (*Pp* × *Pp*) and will produce a 3:1 phenotypic ratio of polled to horned progeny.

2.6 In the jimsonweed, purple flowers are dominant to white. Self-fertilization of a particular purple-flowered jimsonweed produces 28 purple-flowered and 10 white-flowered progeny. What proportion of the purple-flowered progeny will breed true?

Answer: The ratio of 28 purple to 10 white plants is close to 3:1. A 3:1 ratio means that the purple plant that was initially selfed must have been heterozygous, or *Pp*. A cross of *Pp* × *Pp* would yield progeny with a genotypic ratio of 1 *PP* : 2 *Pp* : 1 *pp*. Thus, among the purple progeny, there would be ⅓ *PP* homozygotes and ⅔ *Pp* heterozygotes. Since the homozygous *PP* plants are the only purple plants that will breed true, only ⅓ of the purple-flowered progeny will breed true.

2.7 Two black female mice are crossed with the same brown male. In a number of litters, female X produced 9 blacks and 7 browns, and female Y produced 14 blacks. What is the mechanism of inheritance of black and brown coat color in mice? What are the genotypes of the parents?

Answer: Notice that the two crosses give very different results. Use both sets of results to answer the question. In the cross of black female X with the brown male, the 9 black and 7 brown progeny approximate a 1:1 phenotypic ratio. This suggests that this cross is similar to a testcross and might be depicted as *Bb* × *bb*. However, it is not clear from this cross alone which color trait is dominant. This question can be answered by considering the cross of black female Y with the brown male, where only black progeny are seen. This is like a cross between two true-breeding individuals where black is dominant to brown, *BB* × *bb*. Thus, the brown male is homozygous recessive (*bb*), female X is heterozygous (*Bb*), and female Y is homozygous dominant (*BB*).

2.8 Bean plants may have different symptoms when infected with a virus. Some show local lesions that do not seriously harm the plant; others show general systemic infection. The following genetic analysis was made:

P local lesions × systemic infection
F_1 all local lesions
F_2 785 local lesions : 269 systemic infection

What is probably the genetic basis of this difference in beans? Assign gene symbols to all the genotypes occurring in the genetic analysis. Design a testcross to verify your assumptions.

Answer: In this question, note that the trait that is being examined is the *response* of plants to a virus. One envisions that crosses are performed between plants, and then, instead of looking at flower color or seed shape to score a trait, plants are individually tested for their response to viral infection. A cross of a plant showing only local lesions to one showing systemic lesions gives only plants that show local lesions, indicating that a local lesion response might be considered dominant to a systemic lesion response. If the allele for the local lesion response is depicted as *L* and the allele for the systemic lesion response as *l*, the parental generation can be written as *LL* × *ll*. Then the F_1 would be all *Ll* and the F_2 would be 1 *LL* : 2 *Ll* : 1 *ll*. This fits with the observed numbers of 785 local lesion and 269 systemic lesion (3 *L–* : 1 *ll*) plants.

To test the hypothesis that the local lesion response is dominant to the systemic lesion response, testcross the F_2 plants to true-breeding plants that show a systemic lesion response (i.e., are *ll*). The progeny of this testcross can then be assayed for their response to viral infection so that the genotype of the F_2 can be determined. One would expect ⅔ of those F_2 plants showing a local lesion response to be heterozygotes, i.e., *Ll*. When they are testcrossed, half of their progeny should show local responses and half systemic responses. The remaining $\frac{1}{3}$ will be *LL* and give only progeny with local responses.

2.9 A normal *Drosophila* (fruit fly) has both brown and scarlet pigment granules in the eyes, which appear red as a result. Brown (*bw*) is a recessive allele on chromosome 2 that, in the homozygous condition, results in the absence of scarlet granules (so that the eyes appear brown). Scarlet (*st*) is a recessive allele on chromosome 3 that, when homozygous, results in scarlet eyes because of the absence of brown pigment. Any fly homozygous for recessive brown and recessive scarlet alleles produces no eye pigment and has white eyes. The following results were obtained from crosses:

P brown-eyed fly × scarlet-eyed fly
F_1 red eyes (both brown and scarlet pigment present)
F_2 $\frac{9}{16}$ red : $\frac{3}{16}$ scarlet : $\frac{3}{16}$ brown : $\frac{1}{16}$ white

a. Assign genotypes to the P and F_1 generations.
b. Design a testcross to verify the F_1 genotype, and predict the results.

Answer: Here, two independently assorting genes affecting eye pigment are being followed. Use the symbols suggested in the problem to depict alleles at each locus. For the locus controlling scarlet granules, use *bw* to represent the recessive allele and *Bw* to represent the dominant allele. For the locus controlling brown granules, use *st* to represent the recessive allele and *St* to represent the dominant allele. When writing the genotypes of the flies being crossed, be certain to write the alleles present at both genes.

a. Our first task is to assign genotypes to the P and F_1 generations. Brown-eyed flies lack scarlet granules, so they must be *bwbw* (no scarlet) and *St–*; these flies do make brown granules. Note that they can either be *bwbw StSt* or *bwbw Stst* and that we cannot tell their exact genotype yet. Using similar reasoning, scarlet-eyed flies must be *Bw– stst*. We can determine the exact genotype of these P generation flies by considering what progeny they produce. Since each parent can be either of two genotypes, four different crosses are possible. These are illustrated in the following branch diagrams:

(i) brown eyes × scarlet eyes
bwbw StSt × *BwBw stst*

bwbw × *BwBw* *Stst* × *stst*

BwBw —— *Stst* → *Bwbw Stst*
(scarlet granules) (brown granules) (all red eyes)

(ii) brown eyes × scarlet eyes
bwbw Stst × *BwBw stst*

bwbw × *BwBw* *Stst* × *stst*

Bwbw (scarlet granules)
→ $\frac{1}{2}$ *Stst* (brown granules) → $\frac{1}{2}$ *Bwbw Stst* (red eyes)
→ $\frac{1}{2}$ *stst* (no brown granules) → $\frac{1}{2}$ *Bwbw stst* (scarlet eyes)

(iii) brown eyes × scarlet eyes
bwbw StSt × *Bwbw stst*

bwbw × *BwBw*	*StSt* × *stst*	
$\frac{1}{2}$ *bwBw* (scarlet granules)	— *Stst* (brown granules)	→ $\frac{1}{2}$ *bwBw Stst* (red eyes)
$\frac{1}{2}$ *bwbw* (no scarlet granules)	— *Stst* (brown granules)	→ $\frac{1}{2}$ *bwbw Stst* (brown eyes)

(iv) brown eyes × scarlet eyes
bwbw Stst × *Bwbw stst*

bwbw × *Bwbw*	*Stst* × *stst*	
$\frac{1}{2}$ *bwBw* (scarlet granules)	$\frac{1}{2}$ *Stst* (brown granules)	→ $\frac{1}{4}$ *bwBw Stst* (red eyes)
	$\frac{1}{2}$ *stst* (no brown granules)	→ $\frac{1}{4}$ *bwBw stst* (scarlet eyes)
$\frac{1}{2}$ *bwbw* (no scarlet granules)	$\frac{1}{2}$ *Stst* (brown granules)	→ $\frac{1}{4}$ *bwbw Stst* (brown eyes)
	$\frac{1}{2}$ *stst* (no brown granules)	→ $\frac{1}{4}$ *bwbw stst* (white eyes)

Notice that even though the parents in each of the above crosses look the same, their progeny are not. The only way all red-eyed progeny can be obtained is if the cross is that shown in (i). To see if the observed F_2 is consistent with this view, use a branch diagram:

red eyes × red eyes
Bwbw Stst × *Bwbw stst*

Bwbw × *Bwbw*	*Stst* × *Stst*	
$\frac{3}{4}$ *Bw–* (scarlet granules)	$\frac{3}{4}$ *St–* (brown granules)	→ $\frac{9}{16}$ *Bw– St–* (red eyes)
	$\frac{1}{4}$ *stst* (no brown granules)	→ $\frac{3}{16}$ *Bw– stst* (scarlet eyes)
$\frac{1}{4}$ *bwbw* (no scarlet granules)	$\frac{3}{4}$ *St–* (brown granules)	→ $\frac{3}{16}$ *bwbw St–* (brown eyes)
	$\frac{1}{4}$ *stst* (no brown granules)	→ $\frac{1}{16}$ *bwbw stst* (white eyes)

b. To verify the F_1 genotype, cross the red-eyed F_1 to doubly homozygous recessive white-eyed flies of the genotype *bwbw stst*. One expects to see a 1:1:1:1 ratio of red : brown : scarlet : white flies, as shown in the following branch diagram.

red eyes × white eyes
Bwbw Stst × *bwbw stst*

Bwbw × *bwbw*	*Stst* × *stst*	
½ *Bwbw* (scarlet granules)	½ *Stst* (brown granules) →	¼ *Bwbw Stst* (red eyes)
	½ *stst* (no brown granules) →	¼ *Bwbw stst* (scarlet eyes)
½ *bwbw* (no scarlet granules)	½ *Stst* (brown granules) →	¼ *bwbw Stst* (brown eyes)
	½ *stst* (no brown granules) →	¼ *bwbw stst* (white eyes)

2.10 Grey seed color (*G*) in garden peas is dominant to white seed color (*g*). In the following crosses, the indicated parents with known phenotypes but unknown genotypes produced the listed progeny. Give the possible genotypes of each female parent based on the segregation data.

Parents	Progeny		Female Parent
Female × Male	**Grey**	**White**	**Genotype**
grey × white	81	82	?
grey × grey	118	39	?
grey × white	74	0	?
grey × grey	90	0	?

Answer: First notice that white seeds are always *gg* while grey seeds are either *GG* or *Gg*. Thus, one can assign partial genotypes (either *G*– or *gg*) based on the phenotypes of the parents. The problem is to determine whether a grey parent is *GG* or *Gg*. In monohybrid crosses, there are three kinds of phenotypic ratios that can be generated. The cross *Gg* × *Gg* gives a 3 grey (*G*–) : 1 white (*gg*) ratio, the cross *Gg* × *gg* gives a 1 grey (*Gg*) : 1 white (*gg*) ratio, and the crosses *GG* × *gg* and *GG* × *G*– give all grey (*G*–) progeny. As the first cross of grey × white gives a 1:1 ratio, it must be *Gg* × *gg*. As the second cross of grey × grey gives a 3:1 ratio, it must be *Gg* × *Gg*. As the third cross of grey × white gives all grey, it must be *GG* × *gg*. In the fourth cross, only grey progeny are produced. The genotype of one of the parents must be *GG*, but the genotype of the other can only be specified as *G*–. One has:

Parents	Progeny		Female Parent
Female × Male	**Grey**	**White**	**Genotype**
grey × white	81	82	*Gg*
grey × grey	118	39	*Gg*
grey × white	74	0	*GG*
grey × grey	90	0	*GG* or *G*–

2.11 Fur color in the babbit, a furry little animal and popular pet, is determined by a pair of alleles, *B* and *b*. *BB* and *Bb* babbits are black, and *bb* babbits are white. A farmer wants to breed babbits for sale. True-breeding white (*bb*) female babbits breed poorly. The farmer purchases a pair of black babbits, and these mate and produce six black and two white offspring. The farmer immediately sells his white babbits and then consults with you for a breeding strategy to produce more white babbits.

a. If he performed random crosses between pairs of F_1 black babbits, what proportion of the F_2 progeny would be white?

b. If he crossed an F_1 male to the parental female, what is the probability that this cross will produce white progeny?

c. What would be the farmer's best strategy to maximize the production of white babbits?

Answer: As the farmer starts with a pair of black babbits, he must be starting with animals that are either *BB* or *Bb*. Since his pair is not true-breeding and indeed gives a 3:1 ratio of black to white progeny, he must have two heterozygotes, or *Bb* babbits. Therefore, one expects to find a 1 *BB* : 2 *Bb* genotypic ratio among the black F_1 progeny that were not sold.

a. Consider that for a white babbit to be obtained in a cross between a pair of F_1 babbits, both babbits must be *Bb* in genotype, and a *bb* offspring must be produced by these parents. Now ascertain the chance of these events occurring. If one picks randomly among the black F_1 progeny, there will be a ⅓ chance of picking a *BB* individual and a ⅔ chance of picking a *Bb* individual. The chance of two *Bb* parents giving a *bb* offspring is ¼. Using the product rule, one has

$$\begin{aligned} P\text{ (white offspring)} &= P(\text{both } F_1 \text{ babbits are } Bb \text{ and a } bb \text{ offspring is produced}) \\ &= P(\text{both } F_1 \text{ babbits are } Bb) \times P(bb \text{ offspring}) \\ &= (2/3 \times 2/3) \times (1/4) \\ &= 1/9 \end{aligned}$$

b. If he crosses an F_1 male (*Bb* or *BB*) to the parental female (*Bb*), two types of crosses are possible. The crosses and probabilities are

(i) *Bb* (F_1 male) × *Bb* (parental female) $P = 2/3 \times 1 = 2/3$

(ii) *BB* (F_1 male) × *Bb* (parental female) $P = 1/3 \times 1 = 1/3$

Here, only the first cross can produce white progeny, ¼ of the time. Using the product rule, the chance that this strategy will yield white progeny is

$P = 2/3$ (chance of *Bb* × *Bb* cross) × ¼ (chance of *bb* offspring) = ⅙

c. As neither of the strategies in (b) work very well, evaluate other potential strategies (use trial and error). One that works better in the long run, but is more work initially, is as follows. Re-mate the initial two black babbits (which we have determined to be *Bb*) to obtain a white male offspring: $P = 1/4$ (white *bb*) × ½ (male) = ⅛. Retain this male (note that the fertility of white males is not affected) and breed it back to its mother. This cross would be *Bb* × *bb* and give ½ white (*bb*) and ½ black (*Bb*) offspring. The progeny of this cross could be used to develop a "breeding colony" consisting of black (*Bb*) females and white (*bb*) males. These would consistently produce half white and half black offspring.

2.12 In the jimsonweed, purple flower (*P*) is dominant to white (*p*), and spiny pods (*S*) are dominant to smooth (*s*). In a cross between a jimsonweed homozygous for white flowers and spiny pods and one homozygous for purple flowers and smooth pods, determine the phenotype of (a) the F_1 generation; (b) the F_2 generation; (c) the progeny of a cross of the F_1 back to the white spiny parent; and (d) the progeny of a cross of the F_1 back to the purple smooth parent.

Answer: Since you are told that each parent is homozygous, the parental genotypes can be determined from their phenotypes. One has a white (*pp*) spiny (*SS*) plant crossed to a purple (*PP*) smooth (*ss*) plant. This initial information can be used with branch diagrams to solve each part of the problem as follows:

P: white spiny × purple smooth
pp SS × *PP ss*

Gametes: *p S* *P s*
F_1: *Pp Ss* (purple spiny)
$F_1 \times F_1$: *Pp Ss* × *Pp Ss*

Pp × *Pp* *Ss* × *Ss*

¾ *P–* < ¾ *S–* ⟶ 9⁄16 *P– S–* (purple spiny)
¼ *ss* ⟶ 3⁄16 *P– ss* (purple smooth)

¼ *pp* < ¾ *S–* ⟶ 3⁄16 *pp S–* (white spiny)
¼ *ss* ⟶ 1⁄16 *pp ss* (white smooth)

F_1 × white spiny: *Pp Ss* × *pp SS*

Pp × *pp* *Ss SS*

½ *Pp* — *S–* ⟶ ½ *Pp S–* (purple spiny)
½ *pp* — *S–* ⟶ ½ *pp S–* (white spiny)

F_1 × purple smooth: *Pp Ss* × *PP ss*

Pp × *PP* *Ss* × *ss*

P– < ½ *Ss* ⟶ ½ *P– Ss* (purple spiny)
½ *ss* ⟶ ½ *P– ss* (purple smooth)

2.13 Cleopatra is normally a very refined cat. When she finds even a small amount of catnip, however, she purrs madly, rolls around in the catnip, becomes exceedingly playful, and appears intoxicated. Cleopatra and Antony, who walks past catnip with an air of indifference, have produced five kittens who respond to catnip just like Cleopatra. When the kittens mature, two of them mate and produce four kittens that respond to catnip and one that does not. When another of Cleopatra's daughters mates with Augustus (a nonrelative), who behaves just like Antony, three catnip-sensitive and two catnip-insensitive kittens are produced. Propose a hypothesis for the inheritance of catnip sensitivity that explains these data.

Answer: Try fitting the data to a model in which catnip sensitivity/insensitivity is controlled by a pair of alleles at one gene. Hypothesize that since sensitivity is seen in all of the progeny of the initial mating between catnip-sensitive Cleopatra and catnip-insensitive Antony, sensitivity is dominant. Let *S* represent the sensitive allele and *s* the insensitive allele. Then the initial cross would have been *S*– × *ss*, and the progeny are *Ss*. If two of the *Ss* kittens mate, one would expect 3 *Ss* (sensitive) : 1 *ss* (insensitive) kittens. In the mating with Augustus, the cross would be *Ss* × *ss* and should give a 1 *Ss* (sensitive) : 1 *ss* (insensitive) progeny ratio. The observed progeny ratios are not far off from these expectations.

An alternative hypothesis is that sensitivity (*s*) is recessive and insensitivity (*S*) is dominant. For Antony and Cleopatra to have sensitive (*ss*) offspring, they would need to be *Ss* and *ss*, respectively. When two of their *ss* progeny mate, only sensitive, *ss* offspring should be produced. Since this is not observed, this hypothesis does not explain the data.

2.14 Using the information in problem 2.12, what progeny would you expect from the following jimsonweed crosses? You are encouraged to use the branch diagram approach.

a. *PP ss* × *pp SS*
b. *Pp SS* × *pp ss*
c. *Pp Ss* × *Pp SS*
d. *Pp Ss* × *Pp Ss*
e. *Pp Ss* × *Pp ss*
f. *Pp Ss* × *pp ss*

Answer: Using branch diagrams, one has:

a. *PP ss* × *pp SS*

PP × *pp*   *ss* × *SS*
Pp —— *Ss* ⟶ all *Pp Ss* (purple spiny)

b. *Pp SS* × *pp ss*

Pp × *pp*   *SS* × *ss*
½ *Pp* —— *Ss* ⟶ ½ *Pp Ss* (purple spiny)
½ *pp* —— *Ss* ⟶ ½ *pp Ss* (white spiny)

c. *Pp Ss* × *Pp SS*

Pp × *Pp*   *Ss* × *SS*
¾ *P*– —— *S*– ⟶ ¾ *P*– *S*– (purple spiny)
¼ *pp* —— *S*– ⟶ ¼ *pp S*– (white spiny)

d. *Pp Ss* × *Pp Ss*

Pp × *Pp*   *Ss* × *Ss*

¾ *P*– < ¾ *S*– ⟶ 9/16 *P*– *S*– (purple spiny)
¾ *P*– < ¼ *ss* ⟶ 3/16 *P*– *ss* (purple smooth)
¼ *pp* < ¾ *S*– ⟶ 3/16 *pp S*– (white spiny)
¼ *pp* < ¼ *ss* ⟶ 1/16 *pp ss* (white smooth)

e. *Pp Ss* × *Pp ss*

Pp × *Pp*   *Ss* × *ss*

¾ *P–*
- ½ *Ss* ⟶ ⅜ *P– Ss* (purple spiny)
- ½ *ss* ⟶ ⅜ *P– ss* (purple smooth)

¼ *pp*
- ½ *Ss* ⟶ ⅛ *pp Ss* (white spiny)
- ½ *ss* ⟶ ⅛ *pp ss* (white smooth)

f. *Pp Ss* × *Pp ss*

Pp × *pp*   *Ss* × *ss*

½ *Pp*
- ½ *Ss* ⟶ ¼ *Pp Ss* (purple spiny)
- ½ *ss* ⟶ ¼ *Pp ss* (purple smooth)

½ *pp*
- ½ *Ss* ⟶ ¼ *pp Ss* (white spiny)
- ½ *ss* ⟶ ¼ *pp ss* (white smooth)

2.15 In summer squash, white fruit (*W*) is dominant over yellow (*w*), and disk-shaped fruit (*D*) is dominant over sphere-shaped fruit (*d*). The following problems give the appearances of the parents and their progeny. Determine the genotypes of the parents in each case.

a. White, disk × yellow, sphere gives ½ white, disk and ½ white, sphere.

b. White, sphere × white, sphere gives ¾ white, sphere and ¼ yellow, sphere.

c. Yellow, disk × white, sphere gives all white, disk progeny.

d. White, disk × yellow, sphere gives ¼ white, disk; ¼ white, sphere; ¼ yellow, disk; and ¼ yellow, sphere.

e. White, disk × white, sphere gives ⅜ white, disk; ⅜ white, sphere; ⅛ yellow, disk; and ⅛ yellow, sphere.

Answer: First define symbols for the alleles specifying each trait. Then use information about which allele is dominant or recessive to make initial assignments of possible genotypes. For example, if a plant is white, it must have a dominant *W* allele, but as it may be either *WW* or *Ww*, it would be initially noted as *W–*. If a plant is yellow, it has to be *ww*. Then, by considering just one pair of allelic traits at a time and recalling the Mendelian progeny ratios you have seen [a 1:1 ratio follows from a testcross (*Aa* × *aa*), an all-to-none ratio follows if at least one parent is homozygous dominant (*AA* × *A–* or *aa*), and a 3:1 ratio follows from a monohybrid cross (*Aa* × *Aa*)], you can ascertain whether a parental *W–* plant is *Ww* or *WW*.

Cross	Parents		Progeny
a.	white, disk × yellow, sphere *WW Dd* *ww dd*		½ white, disk; ½ yellow, sphere
b.	white, sphere × white, disk *Ww dd* *Ww dd*		¾ white, sphere; ¼ yellow, sphere
c.	yellow, disk × white, sphere *ww DD* *WW dd*		all white, disk
d.	white, disk × yellow, sphere *Ww Dd* *ww dd*		¼ white, disk; ¼ white, sphere; ¼ yellow, disk; ¼ yellow, sphere
e.	white, disk × white, disk *Ww Dd* *Ww dd*		⅜ white, disk; ⅜ white, sphere; ⅛ yellow, disk; ⅛ yellow, sphere

2.16 Genes *a*, *b*, and *c* assort independently and are recessive to their respective alleles *A*, *B*, and *C*. Two triply heterozygous (*Aa Bb Cc*) individuals are crossed.

a. What is the probability that a given offspring will be phenotypically *A B C*, that is, will exhibit all three dominant traits?
b. What is the probability that a given offspring will be genotypically homozygous for all three dominant alleles?

Answer: First consider just one pair of alleles, say *A* and *a*. Since the cross is *Aa* × *Aa*, the progeny will be 1 *AA* : 2 *Aa* : 1 *aa*, or 3 *A*– : 1 *aa*.

a. Since the three genes assort independently, one can use the product rule to determine the chance of obtaining an *A*– *B*– *C*– offspring.

   P = (chance of *A*–)(chance of *B*–)(chance of *C*–)

   $P = (3/4)(3/4)(3/4)$

   $P = 27/64$

b. Since the chance of obtaining an *AA* offspring will be ¼, and the three genes assort independently, use the product rule to determine the chance of obtaining an *AA BB CC* offspring.

   P = (chance of *AA*)(chance of *BB*)(chance of *CC*)

   $P = (1/4)(1/4)(1/4)$

   $P = 1/64$

You can also solve this problem by setting up a branch diagram or, much more laboriously, by setting up a Punnett square.

2.17 In garden peas, tall stem (*T*) is dominant over short stem (*t*), green pods (*G*) are dominant over yellow pods (*g*), and smooth seeds (*S*) are dominant over wrinkled seeds (*s*). Suppose a homozygous short, green, wrinkled pea plant is crossed with a homozygous tall, yellow, smooth one.

a. What will be the appearance of the F_1 generation?
b. If the F_1 plants are interbred, what will be the appearance of the F_2 generation?
c. What will be the appearance of the offspring of a cross of the F_1 back to its short, green, wrinkled parent?
d. What will be the appearance of the offspring of a cross of the F_1 back to its tall, yellow, smooth parent?

Answer: First note that the problem asks only for the *appearance* of the offspring of the various crosses and not the genotypes. Then write out the genotype of the parental cross: *tt GG ss* × *TT gg SS*.

a. For each pair of traits, a homozygous dominant plant is crossed to a homozygous recessive plant. Thus, all the F_1 progeny will show the dominant trait and be tall, green, and smooth (*Tt Gg Ss*).
b. The F_2 results from selfing the F_1, i.e., *Tt Gg Ss* × *Tt Gg Ss*. The appearance of the F_2 is most readily determined by employing a branch diagram.

Tt × *Tt*	*Gg* × *Gg*	*Ss* × *Ss*	
¾ *T–*	¾ *G–*	¾ *S–*	→ 27/64 *T– G– S–* (tall, green, smooth)
		¼ *ss*	→ 9/64 *T– G– ss* (tall, green, wrinkled)
	¼ *gg*	¾ *S–*	→ 9/64 *T– gg S–* (tall, yellow, smooth)
		¼ *ss*	→ 3/64 *T– gg ss* (tall, yellow, wrinkled)
¼ *tt*	¾ *G–*	¾ *S–*	→ 9/64 *tt G– S–* (short, green, smooth)
		¼ *ss*	→ 3/64 *tt G– ss* (short, green, wrinkled)
	¼ *gg*	¾ *S–*	→ 3/64 *tt gg S–* (short, yellow, smooth)
		¼ *ss*	→ 1/64 *tt gg ss* (short, yellow, wrinkled)

c. The branch diagram of the cross *Tt Gg Ss* × *tt GG ss* is:

Tt × *tt*	*Gg* × *GG*	*Ss* × *ss*	
½ *Tt*	all *G–*	½ *Ss*	→ ¼ *Tt G– Ss* (tall, green, smooth)
		½ *ss*	→ ¼ *Tt G– ss* (tall, green, wrinkled)
½ *tt*	all *G–*	½ *Ss*	→ ¼ *tt G– Ss* (tall, green, smooth)
		½ *ss*	→ ¼ *tt G– ss* (short, green, wrinkled)

d. The branch diagram of the cross *Tt Gg Ss* × *TT gg SS* is:

Tt × *Tt*	*Gg* × *Gg*	*Ss* × *Ss*	
3/4 *T–*	3/4 *G–*	3/4 *S–*	→ 27/64 *T– G– S–* (tall, green, smooth)
		1/4 *ss*	→ 9/64 *T– G– ss* (tall, green, wrinkled)
	1/4 *gg*	3/4 *S–*	→ 9/64 *T– gg S–* (tall, yellow, smooth)
		1/4 *ss*	→ 3/64 *T– gg ss* (tall, yellow, wrinkled)
1/4 *tt*	3/4 *G–*	3/4 *S–*	→ 9/64 *tt G– S–* (short, green, smooth)
		1/4 *ss*	→ 3/64 *tt G– ss* (short, green, wrinkled)
	1/4 *gg*	3/4 *S–*	→ 3/64 *tt gg S–* (short, yellow, smooth)
		1/4 *ss*	→ 1/64 *tt gg ss* (short, yellow, wrinkled)

2.18 *C/c, O/o,* and *I/i* are three independently segregating pairs of alleles in chickens. *C* and *O* are dominant alleles, both of which are necessary for pigmentation. *I* is a dominant inhibitor of pigmentation. Individuals of genotype *cc* or *oo* or *Ii* or *II* are white, regardless of what other genes they possess. Assume that White Leghorns are *CC OO II*, White Wyandottes are *cc OO ii*, and White Silkies are *CC oo ii*. What types of offspring (white or pigmented) are possible, and what is the probability of each, from the following crosses?

a. White Silkie × White Wyandotte
b. White Leghorn × White Wyandotte
c. (Wyandotte–Silkie F_1) × White Silkie

Answer: For an individual to be pigmented, they must have one of each of the dominant *C* and *O* alleles and not have any of the dominant *I* alleles. That is, they must be *C– O– ii*.

a. White Silkie (*CC oo ii*) × White Wyandotte (*cc OO ii*). These are two true-breeding strains at each of three genes and differ from each other at two loci, the *C/c* and *O/o* loci. The progeny will be dihybrids, are all *Cc Oo ii*, and are all pigmented.

b. White Leghorn (*CC OO II*) × White Wyandotte (*cc OO ii*). These two are also true-breeding at each of three genes and differ from each other at two loci, the *C/c* and *I/i* loci. Their progeny will be dihybrids, are all *Cc Oo Ii*, and are all white (because of *Ii*).

c. Wyandotte–Silkie F_1 (*Cc Oo ii*) × White Silkie (*CC oo ii*). As both parents are *ii*, all the progeny will be as well, and so if they are *C– O–*, they will be pigmented. The proportions of white and pigmented progeny can be determined by using a branch diagram or considering each locus separately as follows: At the *C/c* locus, the cross is *Cc* × *CC*, so all the progeny will be *C–*. At the *O* locus, the cross is *Oo* × *oo*, a testcross, so half the progeny will be *Oo* and half will be *oo*. Thus, half of the progeny will be *C– Oo ii* and be pigmented, and half will be *C– oo ii* and be white.

2.19 Two homozygous strains of corn are hybridized. They are distinguished by six different pairs of genes, all of which assort independently and produce an independent phenotypic effect. The F_1 hybrid is selfed to give an F_2.

a. What is the number of possible genotypes in the F_2?
b. How many of these genotypes will be homozygous at all six gene loci?
c. If all gene pairs act in a dominant-recessive fashion, what proportion of the F_2 will be homozygous for all dominants?
d. What proportion of the F_2 will show all dominant phenotypes?

Answer: Although this problem can be solved using a branch diagram, the branch diagram analysis of six genes is tedious. Therefore, approach it in a different manner. A general relationship exists between the number of possible genotypes and phenotypes. At any one locus having alleles that act in a dominant-recessive fashion, there are two possible phenotypes (*A*– and *aa*) and three possible genotypes (*AA*, *Aa*, and *aa*). When several independently assorting loci are considered, the number of possibilities grows according to the number of combinations of alleles possible. For two loci, independent assortment of the alleles at each locus allows for $2 \times 2 = 4$ phenotypes (two possible phenotypes at each locus assorted with either of two possible phenotypes at a second locus) and 3×3 genotypes (three possible genotypes at one locus assorted with any of three possible genotypes at a second locus). For three loci, independent assortment of the alleles at each locus allows for $2 \times 2 \times 2 = 8$ phenotypes and $3 \times 3 \times 3 = 27$ genotypes. For n loci, the relationship becomes 2^n possible phenotypes and 3^n possible genotypes.

To determine the frequency of a particular phenotypic or genotypic class, consider that at one locus, an F_1 hybrid cross gives ¾ phenotypically dominant and ¼ phenotypically recessive progeny and ¼ homozygous dominant, ½ heterozygous and ¼ homozygous recessive progeny. The fraction of progeny of a particular genotypic or phenotypic class can be determined by considering the combinations that are possible. For example, for two independently assorting loci *A/a* and *B/b*, ¾ of the progeny of a cross of *Aa Bb* × *Aa Bb* will show the *A*– phenotype and ¾ of the progeny will show the *B*– phenotype. When independent assortment of these two loci is considered, $\frac{3}{4} \times \frac{3}{4} = \frac{9}{16}$ of the progeny will show both the *A*– and the *B*– phenotypes. (Multiply the individual probabilities using the product rule.) Thus, when n loci are involved, $(\frac{3}{4})^n$ of the progeny will show all dominant phenotypes, $(\frac{1}{4})^n$ of the progeny will show all recessive phenotypes, $(\frac{1}{4})^n$ of the progeny will be homozygous dominant at all loci, $(\frac{1}{4})^n$ of the progeny will be homozygous recessive at all loci, and $(\frac{1}{2})^n$ of the progeny will be heterozygous at all loci.

a. For six loci, $3^6 = 729$ possible genotypes will be seen.
b. Notice that the question asks for how many of the *genotypes* will be homozygous. For just one locus, two (*AA*, *aa*) of three possible (*AA*, *Aa*, *aa*) genotypes are homozygous. Thus, for six loci, $2^6 = 64$ genotypes will be homozygous for all loci. Note that included in these genotypes are those that are homozygous dominant at each locus, those that are homozygous recessive at each locus, and those that are homozygous dominant at one or more loci and homozygous recessive at the other loci.
c. In a monohybrid cross, ¼ of the progeny will be homozygous dominant. For the F_2 in this cross, $(\frac{1}{4})^6 = 1/4{,}096$ of the progeny will be homozygous dominant. Notice that this fraction of the progeny derives from a single genotype.
d. In a monohybrid cross, ¾ of the progeny show a dominant phenotype. For the F_2 in this cross, $(\frac{3}{4})^6 = 729/4{,}096 = 17\%$ of the progeny will show all six dominant phenotypes.

2.20 The coat color of mice is controlled by several genes. The agouti pattern, characterized by a yellow band of pigment near the tip of the hairs, is produced by the dominant allele *A*; homozygous *aa* mice do not have the band and are nonagouti. The dominant allele *B* determines black hairs, and the recessive allele *b* determines brown. Homozygous c^hc^h individuals allow pigments to be deposited only at the extremities (e.g., feet, nose, and ears) in a pattern called Himalayan. The genotype *C*– allows pigment to be distributed over the entire body.

a. If a true-breeding black mouse is crossed with a true-breeding brown, agouti, Himalayan mouse, what will be the phenotypes of the F_1 and F_2 generation?

b. What proportion of the black agouti F_2 will be of genotype *Aa BB* Cc^h?

c. What proportion of the Himalayan mice in the F_2 generation are expected to show brown pigment?

d. What proportion of all agoutis in the F_2 generation are expected to show black pigment?

Answer:

a. The cross is *aa BB CC* × *AA bb* c^hc^h. Since the parents are true-breeding at each of three different loci, the F_1 must be a trihybrid, or *Aa Bb* Cc^h. The F_1 will be agouti with black hairs and pigmented over the entire body. In the F_2, a branch diagram will identify eight possible phenotypic classes: agouti, black, full pigmentation (*A*–*B*– *C*–); nonagouti, black, full (*aa B*– *C*–); agouti, brown, full (*A*– *bb C*–); nonagouti, brown, full (*aa bb C*–); agouti, black, Himalayan (*A*– *B*– c^hc^h); nonagouti, black, Himalayan (*aa B*– c^hc^h); agouti, brown, Himalayan (*A*– *bb* c^hc^h); and nonagouti, brown, Himalayan (*aa bb* c^hc^h).

b. F_2 animals that are black and agouti (and either Himalayan or have full-body pigmentation) have the genotypes *A*– *B*– *C*– and *A*– *B*– c^hc^h. Among the *A*– animals, ⅔ are *Aa*. Among the *B*– animals, ⅓ are *BB*. Among the *C*– and c^hc^h animals, ½ are Cc^h. Thus, among all of the black agouti F_2, ⅔ × ⅓ × ½ = ⅑ are *Aa BB* Cc^h. If one limits the animals under consideration to those that are black agouti, *and not Himalayan* (i.e., *A*– *B*– *C*–), then ⅔ of the *C*– animals are Cc^h, and the proportion of *Aa BB* Cc^h animals is ⅔ × ⅓ × ⅔ = 4/27.

c. From the cross *Aa Bb* Cc^h × *Aa Bb* Cc^h, ¼ of the progeny will be *bb* and show brown pigment. This will be the case regardless of whether the animals are pigmented over their entire body or are Himalayan. Thus, ¼ of the Himalayan mice will show brown pigment. (Be careful not to misread this question: It does *not* ask what proportion of the progeny are both brown *and* Himalayan.)

d. From the cross *Aa Bb* Cc^h × *Aa Bb* Cc^h, ¾ of the progeny will be *B*– and show black pigment. This will be the case regardless of whether the animals are agouti or nonagouti. Thus, ¾ of the agouti mice will show black pigment. (Be careful not to misread this question: It does *not* ask what proportion of the progeny are both black *and* agouti.)

2.21 In cocker spaniels, solid coat color is dominant over spotted coat. Suppose a true-breeding, solid-colored dog is crossed with a spotted dog, and the F_1 dogs are interbred.

a. What is the probability that the first puppy born will have a spotted coat?

b. What is the probability that if four puppies are born, all of them will have solid coats?

Answer: First, assign symbols to the alleles and write down the cross. Let solid color be *S* and spotted be *s*. Then, if a true-breeding, solid-colored dog is bred to a spotted dog, the cross is $SS \times ss$, the F_1 are all Ss, and interbreeding the F_1 will give $\frac{3}{4}$ $S-$ (solid) and $\frac{1}{4}$ ss (spotted) progeny.

a. The chance that the first puppy born is spotted is just the chance of getting an *ss* offspring from an $Ss \times Ss$ cross, or $\frac{1}{4}$.

b. The chance of getting four puppies all having solid coats is the chance of getting an $S-$ offspring the first time *and* the second time *and* the third time *and* the fourth time. Apply the product rule to get $P = \frac{3}{4} \times \frac{3}{4} \times \frac{3}{4} \times \frac{3}{4} = \frac{81}{256}$.

2.22 In the F_2 generation of his cross of red-flowered × white-flowered *Pisum,* Mendel obtained 705 plants with red flowers and 224 with white.

a. Is this result consistent with his hypothesis of factor segregation, from which a 3:1 ratio would be predicted?

b. In how many similar experiments would a deviation as great as or greater than this one be expected? (Calculate χ^2 and obtain the approximate value of P from Table 2.5.)

Answer:

a. It is important to *think through* this problem and not just plug numbers into a chi-square formula. Here we want to test the hypothesis of factor segregation. The hypothesis states that each allele in a monohybrid will segregate into gametes independently, so that zygotes have a half-chance of obtaining either allele from a heterozygous parent. Drawing a Punnett square shows that a 3:1 ratio of progeny phenotypes would be expected from an $F_1 \times F_1$ cross. Mendel observed 705 red and 224 white plants in the F_2, a ratio of 3.14:1. Thus, this ratio seems consistent with his hypothesis.

b. To test *how* significant this result is, use the χ^2 test to determine how frequently these types of numbers would be obtained in similar experiments.

Class	Observed	Expected	d	d^2	d^2/e
red	705	697	8	64	0.09
white	224	232	–8	64	0.28
Total	929	929	0	–	$\chi^2 = 0.37$

$\chi^2 = 0.37$; df = 1; $0.50 < P < 0.70$

According to the chi-square test, then, Mendel's result is consistent with this hypothesis. More specifically, in approximately 60 percent of similar experiments, one would expect a deviation (i.e., a value of χ^2) as great as or greater than this one. One therefore does not reject the hypothesis.

2.23 In tomatoes, cut leaf and potato leaf are alternative characters, with cut (C) dominant to potato (c). Purple stem and green stem are another pair of alternative characters, with purple (P) dominant to green (p). A true-breeding cut green tomato plant is crossed with a true-breeding potato purple plant, and the F_1 plants are allowed to interbreed. The 320 F_2 plants were phenotypically 189 cut purple, 67 cut green, 50 potato purple; and 14 potato green. Propose a hypothesis to explain the data, and use the χ^2 test to test the hypothesis.

Answer: First, use the symbols defined in the problem to write the genotypes of the parents and expected offspring if the two loci assort independently. One has

$$\begin{aligned} \text{P:}&\ \ CC\,pp \times cc\,PP \\ F_1 \times F_1\text{:}&\ \ Cc\,Pp \times Cc\,Pp \\ F_2\text{:}&\ \ 9\,C\text{–}\,P\text{–} : 3\,C\text{–}\,pp : 3\,cc\,P\text{–} : 1\,cc\,pp \end{aligned}$$

(9 cut purple : 3 cut green : 3 potato purple : 1 potato green)

If the two loci assort independently, then one expects to see a 9:3:3:1 phenotypic ratio in the F_2. Therefore, test the hypothesis of independent assortment using the χ^2 test.

Class	Observed	Expected	d	d^2	d^2/e
cut, purple	189	280	9	81	0.45
cut, green	67	60	7	49	0.81
potato, purple	50	60	–10	100	1.66
potato, green	14	20	–6	36	1.80
Total	320	320	0	–	$\chi^2 = 4.72$

$\chi^2 = 4.72$; df = 3; $0.10 < P < 0.20$

Therefore, in experiments similar to this one, a deviation at least as great as that observed here would be seen about 20 percent of the time. Thus, the hypothesis of two independently assorting genes is accepted as being possible.

2.24 The simple case of just two mating types (male and female) is by no means the only sexual system known. The ciliated protozoan *Paramecium bursaria* has a system of four mating types, controlled by two genes (*A* and *B*). Each gene has a dominant and a recessive allele. The four mating types are expressed according to the following scheme:

Genotype	Mating Type
AA BB	A
Aa BB	A
AA Bb	A
Aa Bb	A
AA bb	D
Aa bb	D
aa BB	B
aa Bb	B
aa bb	C

It is clear, therefore, that some of the mating types result from more than one possible genotype. We have four strains of known mating type—"A," "B," "C," and "D"—but unknown genotype. The following crosses were made with the indicated results:

Cross	Mating Type of Progeny *A*	*B*	*C*	*D*
"*A*" × "*B*"	24	21	14	18
"*A*" × "*C*"	56	76	55	41
"*A*" × "*D*"	44	11	19	33
"*B*" × "*C*"	0	40	38	0
"*B*" × "*D*"	6	8	14	10
"*C*" × "*D*"	0	0	45	45

Assign genotypes to "*A*," "*B*," "*C*," and "*D*."

Answer: First summarize what you know about the mating types:

"*A*" = *A*– *B*–   "*B*" = *aa B*–   "*C*" = *aa bb*   "*D*" = *A*– *bb*

Then rewrite the table in terms of what is known.

Cross	Mating Type of Progeny *A* *A*– *B*–	*B* *aa B*–	*C* *aa bb*	*D* *A*–*bb*
"*A*" × "*B*" *A*– *B*– × *aa B*–	24	21	14	18
"*A*" × "*C*" *A*– *B*– × *aa bb*	56	76	55	41
"*A*" × "*D*" *A*– *B*– × *A*– *bb*	44	11	19	33
"*B*" × "*C*" *aa*– *B*– × *aa bb*	0	40	38	0
"*B*" × "*D*" *aa*– *B*– × *A*– *bb*	6	8	14	10
"*C*" × "*D*" *aa bb* × *A*– *bb*	0	0	45	45

Notice that a cross involving mating type C is similar to a testcross, as C is *aa bb*. The progeny ratios in crosses involving "*C*" will therefore reflect the gametes from the non-"*C*" parent. In the cross of "*B*" × "*C*," both the parents are *aa*, and the 1 B : 1 C (1 *aa B*– : 1 *aa bb*) ratio in the progeny reflects the fact that "*B*" is *Bb*. Therefore, "B" is *aa Bb*. In the cross "*C*" × "*D*," both parents are *bb*, and the 1 C : 1 D progeny ratio indicates that "*D*" is *Aa*. "*D*" is therefore *Aa bb*. To determine the genotype of "*A*," consider that the cross of "*A*" × "*C*" produces progeny of all four mating types. This would be expected only if "*A*" were *Aa Bb*.

2.25 In bees, males (drones) develop from unfertilized eggs and are haploid. Females (workers and queens) are diploid and come from fertilized eggs. *W* (black eyes) is dominant over *w* (white eyes). Workers of genotype *RR* or *Rr* use wax to seal crevices in the hive; *rr* workers use resin instead. A *Ww Rr* queen founds a colony after being fertilized by a black-eyed drone bearing the *r* allele.

a. What will be the appearance and behavior of workers in the new hive, and what are their relative frequencies?

b. Give the genotypes of male offspring, with relative frequencies.

c. Fertilization normally takes place in the air during a nuptial flight, and any bee unable to fly would effectively be rendered sterile. Suppose that a recessive mutation, *c*, occurs spontaneously in a sperm that fertilizes a normal egg and that the effect of the mutant gene is to cripple the wings of any adult not bearing the normal allele *C*. The fertilized egg develops into a normal queen named Madonna. What is the probability that wingless males will be *found in a hive* founded two generations later by one of Madonna's granddaughters?

d. By one of Madonna's great-great-granddaughters?

Answer: The initial cross is *Ww Rr* × *W r*. Note that since males are haploid, they have only one set of chromosomes. As a consequence, their gametes are only of one kind (e.g., the *W r* male will have only *W r* gametes) and their phenotype can be used to directly infer their genotype.

a. The progeny females will be ½ *W– Rr* (black-eyed, wax sealers) and ½ *W– rr* (black-eyed, resin sealers).

b. As males arise from unfertilized eggs, they receive chromosomes only from their mother. The progeny males will be ¼ *W R* (black-eyed, wax sealers), ¼ *W r* (black-eyed, resin sealers), ¼ *w r* (white-eyed, resin sealers), and ¼ *w R* (white-eyed, wax sealers).

c. The egg fertilized by the mutation-bearing sperm results in a *Cc* female. Since fertilization occurs in flight, only males that are *C* contribute genes to the next generation. Hence, the first generation arises from the cross *Cc* × *C*. There is a ½ chance of obtaining daughters that are *Cc*. Such a daughter can only be fertilized by a *C* male, so the chance of her having a *Cc* daughter is also ½. The chance of having a *Cc* granddaughter is thus ½ × ½ = ¼. Such granddaughters will have wingless males if they are prolific. Thus, the probability that wingless males will be found in a hive founded two generations later by one of Madonna's granddaughters is the probability that the granddaughter is *Cc*, or ¼.

d. $(\frac{1}{2})^4 = \frac{1}{16}$

2.26 Consider the following pedigree, in which the allele responsible for the trait (*a*) is recessive to the normal allele (*A*).

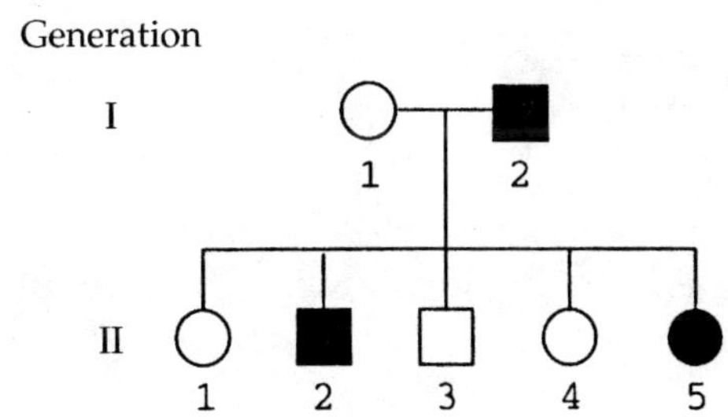

a. What is the genotype of the mother?
b. What is the genotype of the father?
c. What are the genotypes of the children?
d. Given the mechanism of inheritance involved, does the ratio of children with the trait to children without the trait match what would be expected?

Answer:

a. If the allele responsible for the trait is recessive to the normal allele, then to have affected offspring in generation II, the mother (individual I-1) must be heterozygous (i.e., be *Aa*). If she were homozygous for the normal allele (i.e., *AA*), all the offspring would be *A*– and be normal. If she were heterozygous, the cross in the first generation would be *Aa* × *aa*, and one would expect about 50 percent affected and 50 percent normal offspring. This is close to what is observed, and so the mother is *Aa*.
b. As the father is affected with a recessive trait, he must be *aa*.
c. Since the father is homozygous for the recessive allele, all the children must have inherited an abnormal allele from their father. If they inherited a normal, dominant allele from their mother, they will not be affected (*Aa:* II-1, II-3, II-4), while if they inherited the abnormal, recessive allele from their mother, they will be affected (*aa:* II-2, II-5).
d. Given that the parents are *Aa* × *aa*, one expects a 1:1 ratio of affected to unaffected children. For the five offspring shown, one sees a ratio of 2:3, which, given the number of offspring, approximates a 1:1 ratio. Remember that each birth is independent of the others, and while, if many progeny are seen, one should see about a 1:1 ratio, for small numbers of progeny, the ratios may be significantly off from what is expected. Indeed, even if the couple were to have more unaffected children, the pedigree would still be consistent with the inheritance of a recessive trait. (Can you show that if the pedigree is taken on its own, *without any information about whether the trait is dominant or recessive,* the pedigree is also consistent with the inheritance of a dominant trait?)

2.27 For the following pedigrees A and B, indicate whether the trait involved in each case could be (a) recessive or (b) dominant. Explain your answer.

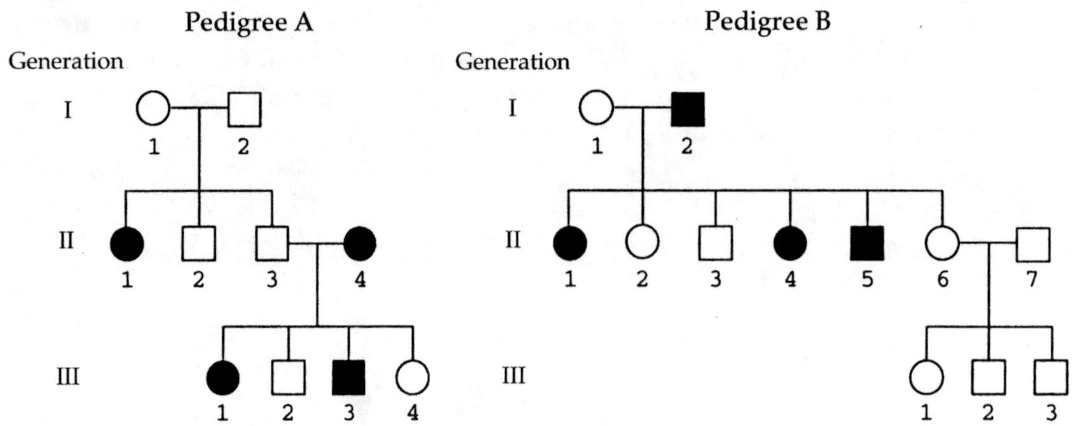

Answer: For pedigree A, parents not showing the trait in generation I have a daughter that exhibits the trait. This is typical of a cross of two heterozygotes giving a homozygous recessive, affected individual: $Aa \times Aa$ (two normal parents) give 3 $A-$ (normal) : 1 aa (affected). If this were the case, individuals II-2 and II-3 could be carriers (i.e., be Aa). If individual II-3 is a carrier and his spouse is homozygous recessive (aa), generation III would arise from the cross $Aa \times aa$, and the 1:1 ratio of affected to unaffected individuals seen would be expected. Thus, pedigree A is consistent with the trait being recessive.

In pedigree B, the trait is present in half of the individuals in generations I and II, but in none of those of generation III. However, the parents of generation III do not themselves exhibit the trait. This is consistent with the trait being dominant. With this view, individual I-1 would be aa and individual I-2 would be Aa. Half of their progeny would be expected to show the trait, as is observed. Furthermore, unaffected parents should have unaffected offspring, as is seen in generation III. Notice, however, that the pedigree is also consistent with a recessive trait that is not rare. In this case, individual I-1 would be heterozygous, and the cross in generation I would be $Aa \times aa$, giving the 50 percent affected progeny seen in generation II. Although individual II-6 would have to be Aa, if her spouse was AA, all the children would be normal.

3

CHROMOSOMAL BASIS OF INHERITANCE, SEX LINKAGE, AND SEX DETERMINATION

CHAPTER OUTLINE

CHROMOSOME THEORY OF INHERITANCE
- Sex Chromosomes
- Sex Linkage
- Nondisjunction of X Chromosomes

SEX DETERMINATION
- Genotypic Sex Determination Systems
- Environmental Sex Determination Systems

ANALYSIS OF SEX-LINKED TRAITS IN HUMANS
- X-Linked Recessive Inheritance
- X-Linked Dominant Inheritance
- Y-Linked Inheritance

REVIEW OF KEY TERMS, SYMBOLS, AND CONCEPTS

In your own words, write a brief, precise definition of each term in the groups on the following page. Check your definitions using the text. Then develop a concept map using the terms in each list.

THINKING ANALYTICALLY

The problems on Mendelian genetics in this chapter are more complicated than those in Chapter 2 and present more of a challenge to your analytical reasoning skills. Review the "Thinking Analytically" hints in Chapter 2 of this guide, and approach the problems analytically and systematically. After you have arrived at a solution, always go back and test it against the data presented in the problem. Try to avoid falling into a trap: Don't look at the answer presented here before you have attempted to solve the problem independently. Being able to understand the answer in this guide is not equivalent to being able to solve the problem independently. You won't have the guide in an exam. By trying to solve problems independently, you will improve your analytical thinking. This will enable you to do better on an exam.

1	2	3
chromosome theory	genotypic sex determination	wild type
sex chromosome	environmental sex determination	mutant allele
X chromosome	Y chromosome sex determination	Medelian symbolism
Y chromosome	X chromosome–autosome balance system	*Drosophila* symbolism
homogametic sex	disjunction, nondisjunction	A, a, AA, Aa, aa, aY, AY
heterogametic sex	segregation, chromosome pairing	a^+, a, a^+Y, A^+Y
sex linkage	primary nondisjunction	
X linkage	secondary nondisjunction	
hemizygous	X chromosome nondisjunction	
homozygous	aneuploidy	
crisscross inheritance	karyotype	
reciprocal cross	Turner, Klinefelter, XYY syndromes	
X-linked recessive trait	dosage compensation	
X-linked dominant trait	Barr body	
Y-linked trait	Lyon hypothesis, lyonization	
holandric trait	X inactivation center	
	XIC, Xic, XIST, Xist	
	testis-determining factor gene	
	sex reversal individual	
	sex-determining region Y (*SRY*)	
	SRY, Sry genes	
	transgene, transgenic organism	
	hermaphroditic	
	monoecious, dioecious	
	mating type	

Take care to represent crosses accurately with appropriate symbols. While you can often use Mendelian or normal/abnormal notation (see Box 3.1) equally well, become fluent with both. Geneticists use each type of notation. As you symbolize traits in crosses, consistently use one type of notation, be careful to use a single type of symbol for each gene (e.g., g and g^+ or g and G for green and yellow seed color, not g and Y), and clearly distinguish between X-linked and autosomal traits, as these follow different inheritance patterns. For example, when assigning members of a pedigree symbols for genetic traits, use a notation that reflects whether inheritance is autosomal or X-linked. This will make it easier for you to spot patterns as well as infer expected types of offspring. Use X^aX^a for a female with an X-linked recessive trait, X^AY or $X^{a+}Y$ for a male with an X-linked dominant trait, X^aY for a male with a recessive trait, and *Aa* or a^+a for an individual with an autosomal trait.

The proof that genes lie on chromosomes came from relating abnormal patterns of trait inheritance to abnormal patterns of chromosome inheritance. Understanding how chromosome behavior during meiosis relates to the segregation and independent assortment of traits in genetic crosses is therefore central to this material. Review the material in Chapter 1 on chromosome pairing and segregation during meiosis and relate it to how autosomal alleles in a heterozygote segregate during the formation of gametes. Then diagram what happens if the alleles are on an X chromosome and there is X chromosome nondisjunction. Are the normal crisscross inheritance patterns altered? Are the results of testcrosses with the progeny altered?

QUESTIONS FOR PRACTICE

Multiple Choice Questions

1. The chromosome theory of inheritance holds that
   a. chromosomes are inherited
   b. the chromosomes contain the hereditary material
   c. the genes are inherited
   d. the chromosomes are DNA
2. Proof that genes lie on chromosomes was obtained by
   a. correlating aneuploids resulting from nondisjunction with inheritance patterns
   b. showing that some genes appear to be sex-linked
   c. showing that males have an X and a Y, while females have two Xs
   d. showing that in *Drosophila,* males are the heterogametic sex
3. An individual that produces both ova and sperm
   a. has the XY genotype
   b. reproduces asexually
   c. is said to be hermaphroditic
   d. both b and c above
4. What piece of evidence best indicates that two X chromosomes are required for a normal human female, even though one of the X chromosomes is inactivated into a Barr body?
   a. Turner females are usually infertile and have morphological abnormalities.
   b. XXY individuals show a Klinefelter syndrome.
   c. 47,XXX individuals are almost completely normal females.
   d. Lyonization occurs in 46,XX individuals.
5. The somatic cells of an individual having the karyotype 48,XXXY would have
   a. four Barr bodies
   b. three Barr bodies
   c. two Barr bodies
   d. no Barr bodies
6. Pedigrees showing X-linked recessive traits typically have
   a. only females being affected
   b. all daughters of an affected father being affected
   c. all sons of an unaffected mother being normal
   d. the trait appearing more frequently in males

7. In humans, what can be said about genes on a normal male's X chromosome?
   a. They are hemizygous.
   b. They are inactivated when the male's cells form Barr bodies.
   c. They are inherited as autosomal recessive traits.
   d. They are responsible for male sex type.
8. During spermatogenesis in a male fly, there is X chromosome nondisjunction during meiosis I. What types of sperm are produced?
   a. X- and Y-bearing sperm
   b. XX- and YY-bearing sperm
   c. XY- and no-sex-chromosome-bearing sperm
   d. XX-, XY-, X-, and Y-bearing sperm
9. Which of the following is true about organisms where sex type is determined by an X chromosome–autosome balance system?
   a. Sex type is determined by the number of X chromosomes. Organisms with two X chromosomes are female, while organisms with one X chromosome are male, regardless of the number of autosomes.
   b. Sex type is determined by the ratio of X chromosomes to autosomes. The Y chromosome plays no role in sex determination.
   c. A key gene called *SRY* on the Y chromosome determines sex type.
   d. The number of autosomes determines sex type. Organisms with one set of autosomes are male, while organisms with two sets of autosomes are female.
10. What unusual feature is associated with the *Xist* gene?
    a. It is active on both X chromosomes of a normal female.
    b. It is inactive on both X chromosomes of a normal female.
    c. It is active only on the nonlyonized X chromosome of a normal female.
    d. It is active only on the lyonized X chromosome of a normal female.
11. Which of the following statements is *not* true about sex determination in mammals?
    a. It is equivalent to testis determination.
    b. In males, it results from the action of testis-determining factor.
    c. Sex reversal occurs when the *SRY* gene is present in males.
    d. The *SRY* gene encodes the testis-determining factor.

Answers: 1b, 2a, 3c, 4a, 5c, 6d, 7a, 8c, 9b, 10d, 11c

Thought Questions

1. What is meant by crisscross inheritance, and what clinical significance does it have in pedigree analysis?
2. Develop a flowchart illustrating the decision-making path you would employ to ascertain whether a pedigree showed autosomal dominant, autosomal recessive, X-linked dominant, X-linked recessive, or Y-linked inheritance.
3. Define the term *lyonization,* give an example of it, and discuss its functional significance.
4. Why do you think most flowering plants are monoecious, whereas in most animals, the sexes are separate?
5. If all mammals have evolved from a common ancestor, how do you account for the variety of chromosome numbers they possess? By what mechanism do you think changes in chromosome number could have occurred? (Consider chromosome number variation resulting from nondisjunction and other accidental changes.)

6. Except for the loci involved with sex determination in mammalian males, there is little clear evidence for genetic loci on the Y chromosome. Why might this be the case, and what might this tell you about the Y chromosome in mammals?
7. What are some of the different mechanisms by which sex type can be determined? Do you think it is significant that so many different mechanisms can be employed?
8. Even though X inactivation occurs in XXY individuals, they do not have the same phenotype as XY males. Similarly, even though X inactivation occurs in XX individuals, they do not have the same phenotype as XO individuals. Why might this be the case?
9. An individual who phenotypically appears to have Klinefelter syndrome has a karyotype of XX,46. An individual who phenotypically appears to have Turner syndrome has a karyotype of XY,46. What common explanation might underlie each of these anomalies? (Hint: Review the text material on sex reversal.)

Thought Questions for Media

After reviewing the media at the *Essential iGenetics* website, try to answer these questions.

1. After which division (meiosis I or meiosis II) are haploid cells produced? At which point are haploid gametes produced?
2. If nondisjunction occurs in meiosis I, how many types of abnormal gametes are produced? Are any normal gametes produced?
3. If nondisjunction occurs in meiosis II, how many types of abnormal gametes are produced? Are any normal gametes produced?
4. The animation illustrates how nondisjunction in either meiosis I or meiosis II can lead to the production of monosomic and trisomic zygotes. How might nondisjunction be involved in generating a nullosomic (none of one type of chromosome) or a tetrasomic (four copies of one kind of chromosome) zygote?
5. In one meiosis in an *Aa Bb* individual, four gametes are produced. How many different genotypes are produced?
6. Based on the animation and considering your answer to question 1, how can four equal frequencies of gametes (*Ab*, *AB*, *aB*, *ab*) be obtained?
7. At what point during meiosis does the chromosome behavior that underlies independent assortment occur?

1. a. Why is it important to determine that none of the individuals that have married into Anna's or Jackson's family have a separate family history of deafness?
   b. How would your analysis of the pedigree of Jackson's family change if you learned that individuals II-3, II-5, and III-4 have a history of deafness?
   c. Since both congenital and progressive hearing loss are relatively uncommon traits, when a pedigree analysis is being performed, can you just assume that individuals marrying into the family have no family history of deafness? If not, how would you proceed?
2. What is the chance that if Anna and Jackson have two boys and two girls, they will all have normal hearing?
3. Anna may or may not be a carrier for deafness. Suppose Anna and Jackson indeed have two boys and two girls with normal hearing. Can you conclude anything about whether Anna is a carrier for deafness? If so, what? If not, why not?

SOLUTIONS TO TEXT PROBLEMS

3.1 Depict each of the following crosses, first using Mendelian and then using *Drosophila* notation (see Box 3.1, p. 59). Give the genotype and phenotype of the F_1 progeny that can be produced.

a. In humans, a mating between two individuals, each heterozygous for the recessive trait phenylketonuria, whose locus is on chromosome 12

b. In humans, a mating between a female heterozygous for both phenylketonuria and X-linked color blindness and a male with normal color vision who is heterozygous for phenylketonuria

c. In *Drosophila,* a mating between a female with white eyes, curled wings, and normal long bristles and a male that has normal red eyes, normal straight wings, and short, stubble bristles. In these individuals, curled wings result from a heterozygous condition at a gene whose locus is on chromosome 2, while the short, stubble bristles result from a heterozygous condition at a gene whose locus is on chromosome 3.

d. In *Drosophila,* a mating between a female from a true-breeding line that has eyes of normal size that are white, black bodies (a recessive trait on chromosome 2), and tiny bristles (a recessive trait called *spineless* on chromosome 3) and a male from a true-breeding line that has normal red eyes, normal grey bodies, normal long bristles but reduced eye size (a dominant trait called *eyeless* on chromosome 4)

Answer: The crosses and their progeny are tabulated below:

	Mendelian Notation	***Drosophila* Notation Long Version**	***Drosophila* Notation Short Version**
a.			
Allele symbols:	P = normal	p^+ = normal	+ = normal
	p = phenylketonuria	p = phenylketonuria	p = phenylketonuria
Cross:	$Pp \times Pp$	$p^+p \times p^+p$	×
F_1:	1 PP : 2 Pp : 1 pp	1 p^+p^+ : 2 p^+p : 1 pp	×
F_1 phenotypes:	¾ normal	¼ phenylketonuria	
b.			
Allele symbols:	C = normal	c^+ = normal	+ = normal
	c = color blind	c = color blind	c = color blind
Cross:	$Cc\ Pp \times CY\ Pp$	$c^+c\ p^+p \times c^+Y\ p^+p$	$+c\ +p \times +Y\ +p$
F_1:	1/16 $CC\ PP$	1/16 $c^+c^+\ p^+p^+$	1/16 ++ ++
	1/8 $CC\ Pp$	1/8 $c^+c^+\ p{+}p$	1/8 ++ +p
	1/16 $CC\ pp$	1/16 $c^+c^+\ pp$	1/16 ++ pp
	1/16 $Cc\ PP$	1/16 $c^+c\ p^+p^+$	1/16 +c ++
	1/8 $Cc\ Pp$	1/8 $c^+c\ p^+p$	1/8 +c +p
	1/16 $Cc\ pp$	1/16 $c^+c\ pp$	1/16 +c pp

(continued)

	Mendelian Notation	*Drosophila* Notation Long Version	*Drosophila* Notation Short Version
	$\frac{1}{16}$ CY *PP*	$\frac{1}{16}$ c^+Y p^+p^+	$\frac{1}{16}$ +Y ++
	$\frac{1}{8}$ CY *Pp*	$\frac{1}{8}$ c^+Y p^+p	$\frac{1}{8}$ +Y +*p*
	$\frac{1}{16}$ CY *pp*	$\frac{1}{16}$ c^+Y *pp*	$\frac{1}{16}$ +Y *pp*
	$\frac{1}{16}$ *c*Y *PP*	$\frac{1}{16}$ c^+Y p^+p^+	$\frac{1}{16}$ *c*Y ++
	$\frac{1}{8}$ *c*Y *Pp*	$\frac{1}{8}$ *c*Y p^+p	$\frac{1}{8}$ *c*Y +*p*
	$\frac{1}{16}$ *c*Y *pp*	$\frac{1}{16}$ *c*Y *pp*	$\frac{1}{16}$ *c*Y *pp*

F_1 phenotypes:
- $\frac{3}{8}$ normal females
- $\frac{1}{8}$ phenylketonuria females
- $\frac{3}{16}$ normal males
- $\frac{1}{16}$ phenylketonuria males
- $\frac{3}{16}$ color-blind males
- $\frac{1}{16}$ color-blind phenylketonuria males

c.

	Mendelian Notation	*Drosophila* Notation Long Version	*Drosophila* Notation Short Version
Allele symbols:	normal eyes = *W*	normal eyes = w^+	normal eyes = +
	white eyes = *w*	white eyes = *w*	white eyes = *w*
	normal wings = *c*	normal wings = C^+	normal wings = +
	curled wings = *C*	curled wings = *C*	curled wings = *C*
	normal bristles = *s*	normal bristles = S^+	normal bristles = +
	stubble bristles = *S*	stubble bristles = *S*	stubble bristles = *S*
Cross:	*ww Cc ss*	*ww* CC^+ S^+S^+	*ww C*+ ++
	×	×	×
	*W*Y *cc Ss*	w^+Y C^+C^+ SS^+	+Y ++ *S*+
F_1 genotypes:	$\frac{1}{8}$ *Ww Cc Ss*	$\frac{1}{8}$ w^+w CC^+ SS^+	$\frac{1}{8}$ +*w C*+ *S*+
	$\frac{1}{8}$ *Ww Cc ss*	$\frac{1}{8}$ w^+w CC^+ S^+S^+	$\frac{1}{8}$ +*w C*+ ++
	$\frac{1}{8}$ *Ww cc Ss*	$\frac{1}{8}$ w^+w *C*+*C*+ SS^+	$\frac{1}{8}$ +*w* ++ *S*+
	$\frac{1}{8}$ *Ww Cc ss*	$\frac{1}{8}$ w^+w C^+C^+ S^+S^+	$\frac{1}{8}$ +*w* ++ ++
	$\frac{1}{8}$ *w*Y *Cc Ss*	$\frac{1}{8}$ *w*Y CC^+ SS^+	$\frac{1}{8}$ *w*Y *C*+ *S*+
	$\frac{1}{8}$ *w*Y *Cc ss*	$\frac{1}{8}$ *w*Y CC^+ S^+S^+	$\frac{1}{8}$ *w*Y *C*+ ++
	$\frac{1}{8}$ *w*Y *cc Ss*	$\frac{1}{8}$ *w*Y C^+C^+ SS^+	$\frac{1}{8}$ *w*Y ++ *S*+
	$\frac{1}{8}$ *w*Y *cc ss*	$\frac{1}{8}$ *w*Y C^+C^+ S^+S^+	$\frac{1}{8}$ *w*Y ++ ++

F_1 phenotypes:
- $\frac{1}{8}$ stubble curled females
- $\frac{1}{8}$ curled females
- $\frac{1}{8}$ stubble females
- $\frac{1}{8}$ normal females
- $\frac{1}{8}$ white, stubble, curled males
- $\frac{1}{8}$ white curled males
- $\frac{1}{8}$ white stubble males
- $\frac{1}{8}$ white males

d.

	Mendelian Notation	*Drosophila* Notation Long Version	*Drosophila* Notation Short Version
Allele symbols:	white eye = *w*	white eye = *w*	white eye = *w*
	red eye = *W*	red eye = w^+	red eye = +
	black body = *b*	black body = *b*	black body = *b*
	grey body = *B*	grey body = b^+	grey body = +
	normal bristles = *S*	normal bristles = s^+	normal bristles = +

(continued)

	Mendelian Notation	*Drosophila* Notation Long Version	*Drosophila* Notation Short Version
	tiny bristles = s	tiny bristles = s	tiny bristles = s
	normal size eye = e	normal size eye = E^+	normal size eye = +
	reduced size eye = E	reduced size eye = E	reduced size eye = E
Cross:	$ww\ bb\ ss\ ee$	$ww\ bb\ ss\ E^+E^+$	$ww\ bb\ ss$ ++
	×	×	×
	$WY\ BB\ SS\ EE$	$w^+Y\ b^+b^+\ s^+s^+\ EE$	+Y ++ ++ EE
F_1 genotypes:	½ $Ww\ Bb\ Ss\ Ee$	½ $w^+w\ b^+b\ s^+s\ EE^+$	½ $+w\ +b\ +s\ +E$
	½ $wY\ Bb\ Ss\ Ee$	½ $wY\ b^+b\ s^+s\ EE^+$	½ $wY\ +b\ +s\ +E$
F_1 phenotypes:	½ eyeless females	½ white eyeless males	

3.2 In *Drosophila*, white eyes are a sex-linked character. The mutant allele for white eyes (w) is recessive to the wild-type allele for brick-red eye color (w^+). A white-eyed female is crossed with a red-eyed male. An F_1 female from this cross is mated with her father, and an F_1 male is mated with his mother. What will be the eye color of the offspring of these last two crosses?

Answer: The initial cross is $ww \times w^+$ Y, so that the F_1 females are ww^+ and the F_1 males are wY. The second set of crosses are therefore $w^+\ w \times w^+$ Y and wY × ww. The former will give all brick-red females (w^+–) and half white (wY) and half brick-red (w^+Y) males. The latter will give only white-eyed males and females (wY and ww).

3.3 One form of color blindness in humans is caused by a sex-linked recessive mutant gene (c). A woman with normal color vision (c^+) and whose father was color-blind marries a man of normal vision whose father was also color-blind. What proportion of their offspring will be color-blind? (Give your answer separately for males and females.)

Answer: Since fathers always give their X chromosome to their daughters, the woman must be heterozygous for the color blindness trait and is c^+c. As her husband received his X chromosome from his mother and has normal color vision, he is c^+Y. The cross is therefore $c^+c \times c^+$Y. All daughters will receive the paternal X bearing the c^+ allele and have normal color vision. As sons will receive the maternal X; half will be cY and be color-blind and half will be c^+Y and have normal color vision.

3.4 In humans, red-green color blindness is recessive and X-linked, whereas albinism is recessive and autosomal. What types of children can be produced as the result of marriages between two homozygous parents: a normal-visioned albino woman and a color-blind, normally pigmented man?

Answer: Let c and c^+ be the color-blind and normal vision alleles, respectively, and let a and a^+ be the albino and normal pigmentation alleles, respectively. Then the cross can be represented as $c^+c^+\ aa \times c$Y a^+a^+. As all the offspring will be a^+a, all will have normal pigmentation. The offspring will either be c^+c or c^+Y and have normal color vision. The daughters will, however, be carriers for the color blindness trait.

3.5 In *Drosophila*, vestigial (partially formed) wings (*vg*) are recessive to normal long wings (vg^+), and the gene for this trait is autosomal. The gene for the white-eye trait is on the X chromosome. Suppose a homozygous white-eyed, long-winged female fly is crossed with a homozygous red-eyed, vestigial-winged male.

a. What will be the appearance of the F_1 flies?
b. What will be the appearance of the F_2 flies?
c. What will be the appearance of the offspring of a cross of the F_1 back to each parent?

Answer:

a. The initial cross is $ww\ vg^+vg^+ \times w^+Y\ vgvg$. The F_1 consists of $wY\ vg^+vg$ (white, normal-winged) males and $ww^+\ vg^+vg$ (red, normal-winged) females.
b. The F_2 would be produced by crossing $wY\ vg^+vg$ males and $w^+w\ vg^+vg$ females. In both the male and the female progeny, ⅛ will be white and vestigial, ⅛ will be red and vestigial, ⅜ will be white and normal-winged, and ⅜ will be red and normal-winged.
c. If the F_1 males are crossed back to the female parent, the cross is $wY\ vg^+vg \times ww\ vg^+vg^+$. All the progeny would be white and normal-winged. If the F_1 females are crossed back to the male parent, the cross is $ww^+\ vg^+vg \times w^+Y\ vgvg$. Among the male progeny, there would be ¼ white, vestigial; ¼ red, vestigial; ¼ white, normal-winged; and ¼ red, normal-winged. Among the female progeny, half would be red and normal-winged and half would be red and vestigial.

3.6 In *Drosophila*, two red-eyed, long-winged flies are bred together and produce the offspring given in the following table:

	Females	Males
red-eyed, long-winged	¾	⅜
red-eyed, vestigial-winged	¼	⅛
white-eyed, long-winged	–	⅜
white-eyed, vestigial-winged	–	⅛

What are the genotypes of the parents?

Answer: From problem 3.5, we know that *w* is X-linked while *vg* is autosomal. This can also be determined by considering just one trait at a time and examining the frequency of progeny phenotypes. The ratio of long-winged to vestigial-winged progeny is 3:1 (¾ to ¼) in both sexes, while the ratio of red-eyed to white-eyed progeny is all to none in females and 1:1 in males. This is consistent with *vg* being autosomal and *w* being X-linked. The 3:1 ratio of long-winged to vestigial-winged progeny indicates that each parent is heterozygous at the *vg* locus. Since both parents have red eyes, both have (at least) one w^+ allele. Since half of the sons are white-eyed, the mother must be heterozygous. Therefore, the parents are $w^+w\ vg^+vg$ and $w^+Y\ vg^+vg$.

3.7 In chickens, a dominant sex-linked gene (*B*) produces barred feathers, and the recessive allele (*b*), when homozygous, produces nonbarred (solid color) feathers. Suppose a nonbarred cock is crossed with a barred hen.

a. What will be the appearance of the F_1 birds?

b. If an F_1 female is mated with her father, what will be the appearance of the offspring?

c. If an F_1 male is mated with his mother, what will be the appearance of the offspring?

Answer:

a. In poultry, sex type is determined by a ZZ (male) and ZW (female) system. The cross can be depicted as *bb* (nonbarred cock) × *BW* (barred hen). The F_1 progeny will be *bW* (nonbarred) hens and *Bb* (barred) cocks.

b. The cross can be represented as *bW* × *bb*. All the progeny will be nonbarred.

c. The cross can be represented as *Bb* × *BW*. The progeny will be ½ barred cocks (¼ *BB*, ¼ *Bb*), ¼ barred hens (*BW*), and ¼ nonbarred hens (*bW*).

3.8 A man suffering from defective tooth enamel, which results in brown-colored teeth, marries a normal woman. All their daughters have brown teeth, but the sons are normal. The sons marry normal women, and all their children are normal. The daughters marry normal men, and 50 percent of their children have brown teeth. Explain these facts.

Answer: Notice that the trait is transmitted from the father to his daughters, indicating crisscross inheritance. This is typical of an X-linked trait. Since the man marries a normal woman and all of their daughters have the trait, the trait must be dominant. The man's X chromosome bearing the defective tooth enamel allele is inherited by all of his daughters and none of his sons. All of his daughters therefore have defective tooth enamel and are heterozygous for the defective enamel allele. These daughters transmit the defective enamel allele 50 percent of the time, giving rise to 50 percent of their children being affected.

3.9 In humans, differences in the ability to taste phenylthiourea are due to a pair of autosomal alleles. Inability to taste is recessive to ability to taste. A child who is a nontaster is born to a couple who can both taste the substance. What is the probability that their next child will be a taster?

Answer: Since the inability to taste the substance is recessive, the nontaster child must be homozygous for the recessive allele, and each of his parents must have given the child a recessive allele. Since both parents can taste, they must also bear a dominant allele. Let *T* represent the dominant (taster) allele and *t* represent the recessive (nontaster) allele (Note: Mendelian notation is used here for convenience, but also because there is no value in assigning a normal (+) and abnormal allele.) Then the cross can be written as *Tt* × *Tt*. The chance that their next child will be a taster is the chance that the child will be *TT* or *Tt*, or ¾.

3.10 Cystic fibrosis is inherited as an autosomal recessive. Two parents without cystic fibrosis have two children with cystic fibrosis and three children without. The parents come to you for genetic counseling.

a. What is the numerical probability that their next child will have cystic fibrosis?

b. Their unaffected children are concerned about being heterozygous. What is the numerical probability that a given unaffected child in the family is heterozygous?

Answer:

a. Since the disease is autosomal recessive, and unaffected parents have affected offspring, both parents must be heterozygous. Let c^+ represent the normal allele and c represent the disease allele. Then the parental cross can be represented as $c^+c \times c^+c$, and there is a ¼ chance that any conception will produce a cc (affected) child.

b. If a child is not affected, the child is either c^+c^+ ($P = ⅓$) or c^+c ($P = ⅔$). Thus, there is a ⅔ chance that a nonaffected child is heterozygous.

3.11 Huntington disease is a human disease inherited as a Mendelian autosomal dominant. The disease results in choreic (uncontrolled) movements, progressive mental deterioration, and eventually death. The disease affects the carriers of the trait any time between 15 and 65 years of age. American folksinger Woody Guthrie died of Huntington disease, as did one of his parents. Marjorie Mazia, Woody's wife, had no history of this disease in her family. The Guthries had three children. What is the probability that a particular Guthrie child will die of Huntington disease?

Answer: Let H represent the disease allele, and let h represent the normal allele. Since we are told that one of Woody's parents died of the disease and are not told anything about the other parent, we may assume that only one parent had the disease allele. Thus, Woody must have been Hh. His children are the progeny of a cross that can be represented as $Hh \times hh$. Each will have a 50 percent chance of receiving the H allele.

3.12 Suppose gene A is on the X chromosome, and genes B, C, and D are on three different autosomes. Thus, $A–$ signifies the dominant phenotype in the male or female. An equivalent situation holds for $B–$, $C–$, and $D–$. The cross $AA\ BB\ CC\ DD$ ♀ × $aY\ bb\ cc\ dd$ ♂ is made.

a. What is the probability of obtaining an $A–$ individual in the F_1?

b. What is the probability of obtaining an a male in the F_1?

c. What is the probability of obtaining an $A–\ B–\ C–\ D–$ female in the F_1?

d. How many different F_1 genotypes will there be?

e. What proportion of F_2 individuals will be heterozygous for the four genes?

f. Determine the probabilities of obtaining each of the following types in the F_2: (1) $A–\ bb\ CC\ dd$ (female); (2) $aY\ BB\ Cc\ Dd$ (male), (3) $AY\ bb\ CC\ dd$ (male); (4) $aa\ bb\ Cc\ Dd$ (female).

Answer:

a. Since only a single trait is being followed, consider just part of the cross: $AA \times aY$. The progeny will either be AY or Aa, and all are $A–$. Thus, the chance of obtaining an $A–$ individual in the F_1 is 1.

b. As shown in (a), there is no chance ($P = 0$) of obtaining an aY individual in the F_1.

c. The F_1 progeny will be $A–\ Bb\ Cc\ Dd$. Half will be female, so $P = ½$.

d. Two, $Aa\ Bb\ Cc\ Dd$ (females) and $AY\ Bb\ Cc\ Dd$ males

e. For the X chromosome trait, the F_1 cross is $AY \times Aa$. Half of the female offspring (¼ of the total) will be heterozygous Aa individuals. For each of the autosomal traits, ½ of the offspring will be heterozygous (e.g., $Bb \times Bb$ gives ½ Bb individuals). Thus, the chance that an F_2 individual will be heterozygous at all four traits is $\frac{1}{4} \times \frac{1}{2} \times \frac{1}{2} \times \frac{1}{2} = \frac{1}{32}$.

f. Before determining the probabilities, consider that at any autosomal gene, there is a $\frac{1}{4}$ chance of obtaining either type of homozygote (e.g., *BB, bb*) and a $\frac{1}{2}$ chance of obtaining a heterozygote. At the *A* gene, the cross is *AY* × *Aa*, so there is a $\frac{1}{4}$ chance of obtaining an *AY* male, a $\frac{1}{4}$ chance of obtaining an *aY* male, a $\frac{1}{4}$ chance of obtaining an *Aa* female and a $\frac{1}{4}$ chance of obtaining an *AA* female (there will be a $\frac{1}{2}$ chance of obtaining an *A–* female). Then the chance of obtaining (1) an *A– bb CC dd* (female) is $P = (\frac{1}{2} \times \frac{1}{4} \times \frac{1}{4} \times \frac{1}{4}) = \frac{1}{128}$; (2) an *aY BB Cc Dd* (male) is $P = (\frac{1}{4} \times \frac{1}{4} \times \frac{1}{2} \times \frac{1}{2}) = \frac{1}{64}$; (3) an *AY bb CC dd* (male) is $P = (\frac{1}{4} \times \frac{1}{4} \times \frac{1}{4} \times \frac{1}{4}) = \frac{1}{256}$; and (4) an *aa bb Cc Dd* (female) is $P = (0 \times \frac{1}{4} \times \frac{1}{2} \times \frac{1}{2}) = 0$.

3.13 A Turner syndrome individual would be expected to have how many Barr bodies in the majority of cells?

Answer: Turner syndrome individuals have only one X, as they are XO. Consequently, no X is inactivated.

3.14 An XXY Klinefelter syndrome individual would be expected to have how many Barr bodies in the majority of cells?

Answer: All but one X chromosome is inactivated. An XXY individual, having two X chromosomes, has one inactivated.

3.15 In human genetics, the pedigree is used for analysis of inheritance patterns. Females are represented by a circle, and males by a square. The following figure presents three 2-generation family pedigrees for a trait in humans. Normal individuals are represented by unshaded symbols, and people with the trait by shaded symbols. For each pedigree (A, B, and C), state, by answering yes or no in the appropriate blank space, whether transmission of the trait can be accounted for on the basis of each of the listed simple modes of inheritance:

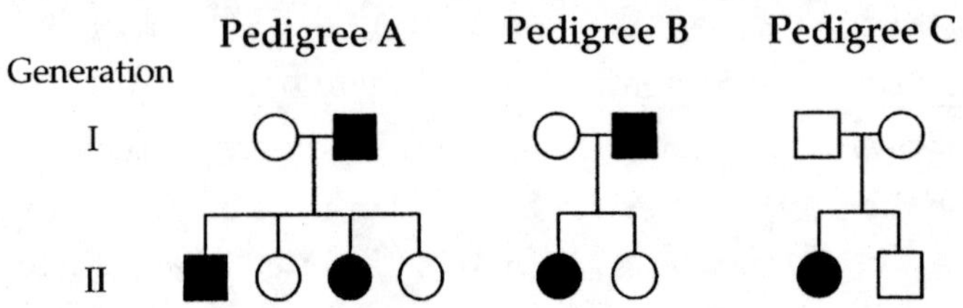

	Pedigree A	Pedigree B	Pedigree C
Autosomal recessive			
Autosomal dominant			
X-linked recessive			
X-linked dominant			

Answer: This problem raises the issue that the precise mode of inheritance of a trait often cannot be determined when a pedigree is small and the frequency of the trait in a population is unknown. For example, pedigree A could easily fit an autosomal dominant trait (*AA* and *Aa* = affected): The affected father would be heterozygous for the trait (*Aa*), the mother would be unaffected (*aa*), and half of their offspring would be affected. However, if the trait were autosomal recessive (i.e., *aa* = affected individuals) and the mother were heterozygous (*Aa*) and the father homozygous (*aa*), half of the offspring would still be affected. One could also fit the pedigree to an X-linked recessive trait: The mother would be heterozygous *Aa,* the father hemizygous *a*Y, and half of the progeny would be affected (either *aa* or *a*Y). An X-linked dominant trait would not fit the pedigree, as it would require all the daughters of the affected father to be affected (as they all receive their father's X) and not allow a son to be affected (as he does not receive his father's X). Pedigrees B and C can be solved by similar analytical reasoning.

	Pedigree A	**Pedigree B**	**Pedigree C**
Autosomal recessive	Yes	Yes	Yes
Autosomal dominant	Yes	Yes	No
X-linked recessive	Yes	Yes	No
X-linked dominant	No	No	No

3.16 Shaded symbols in the following pedigree represent a trait. Which of the progeny eliminate X-linked recessiveness as a mode of inheritance for the trait?

a. I-1 and I-2

b. II-4

c. II-5

d. II-2 and II-4

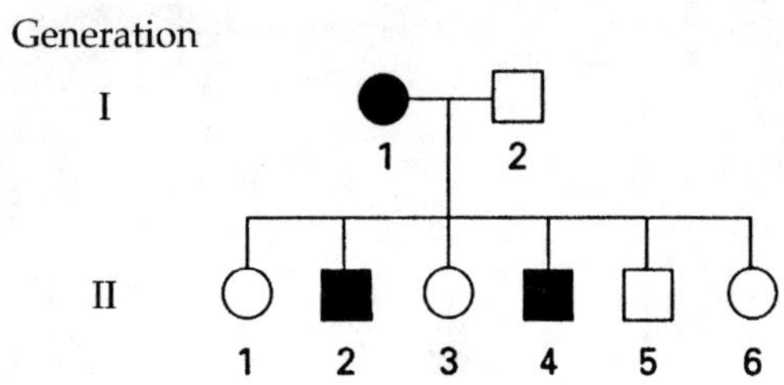

Answer: If the trait were X-linked recessive, then II-5 should be affected, as he would receive his mother's X chromosome. Therefore, the correct answer is (c), II-5.

3.17 When constructing human pedigrees, geneticists often refer to particular individuals by a number. The generations are labeled by Roman numerals and the individuals in each generation by Arabic numerals. For example, in the pedigree in the following figure, the female with the asterisk is I-2. Use this means to designate specific individuals in the pedigree. Determine the probable inheritance mode for the trait shown in the affected individuals (the shaded symbols) by answering the following questions. Assume that the condition is caused by a single gene.

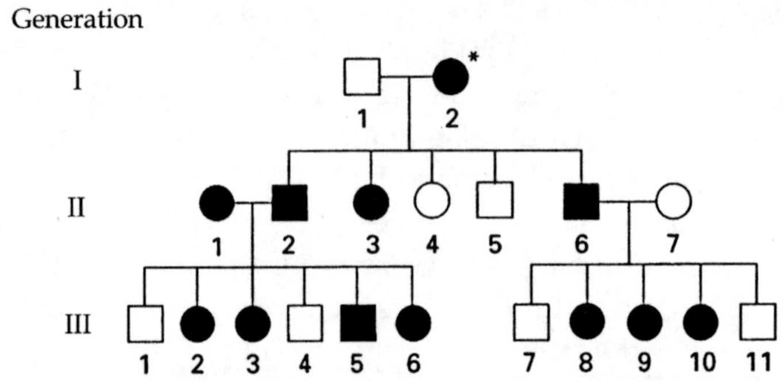

a. Y-linked inheritance can be excluded at a glance. What two other mechanisms of inheritance can be definitely excluded? Why can these be excluded?
b. Of the remaining mechanisms of inheritance, which is the most likely? Why?

Answer:

a. Y-linked inheritance can be excluded because females are affected. X-linked recessive inheritance can also be excluded, as an affected mother (I-2) has a normal son (II-5). Autosomal recessive inheritance can also be excluded, as in such a case, two affected parents, such as II-1 and II-2, could only have affected offspring, which they do not.
b. The two remaining mechanisms of inheritance are X-linked dominant and autosomal dominant. Genotypes can be written to satisfy both mechanisms of inheritance. Of these two, X-linked dominant inheritance may be more likely, as II-6 and II-7 have only affected daughters, indicating crisscross inheritance. If the trait were autosomal dominant, one would expect half of the daughters and half of the sons to be affected.

3.18 A three-generation pedigree for a particular human trait is shown in the following figure.

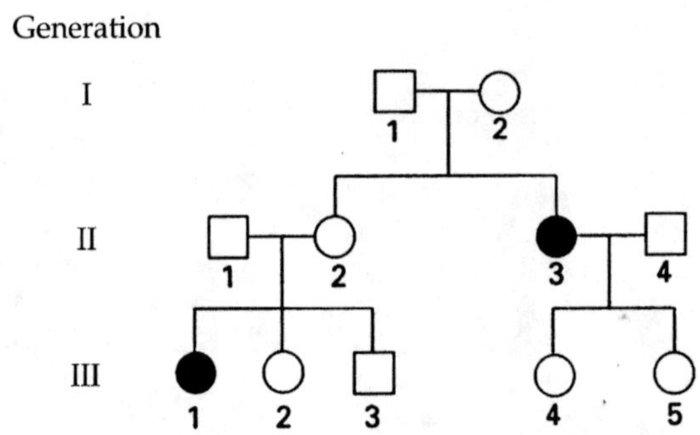

a. What is the mechanism of inheritance for the trait?
b. Which persons in the pedigree are known to be heterozygous for the trait?
c. What is the probability that III-2 is a carrier (heterozygous)?
d. If III-3 and III-4 marry, what is the probability that their first child will have the trait?

Answer:

a. The trait must be recessive, as unaffected parents (I-1 and I-2, II-1 and II-2) have affected offspring. It also must be autosomal, as affected female daughters are born to unaffected fathers.

b. Given the answer to (a), individuals I-1 and I-2, as well as individuals II-1 and II-2, must be heterozygous. Also, individuals III-4 and III-5 must be heterozygous, as they have an affected, homozygous mother.

c. One can denote the parents of III-2 as *Aa* × *Aa*. We know individual III-2 is *A*–, so she is either *AA* (P = ⅓) or *Aa* (P = ⅔).

d. By the same reasoning as described in (c), individual III-3 has a ⅔ chance of being *Aa*. The chance that he will contribute an *a* allele to his offspring is thus ⅔ × ½ = ⅓. The chance that individual III-4 will contribute an *a* allele to her offspring is ½, as she is heterozygous. Thus, the chance that progeny of III-3 and III-4 will be affected is ⅓ × ½ = ⅙.

3.19 For each of the more complex pedigrees shown in Figure 3.B, determine the probable mechanism of inheritance: autosomal recessive, autosomal dominant, X-linked recessive, X-linked dominant, or Y-linked.

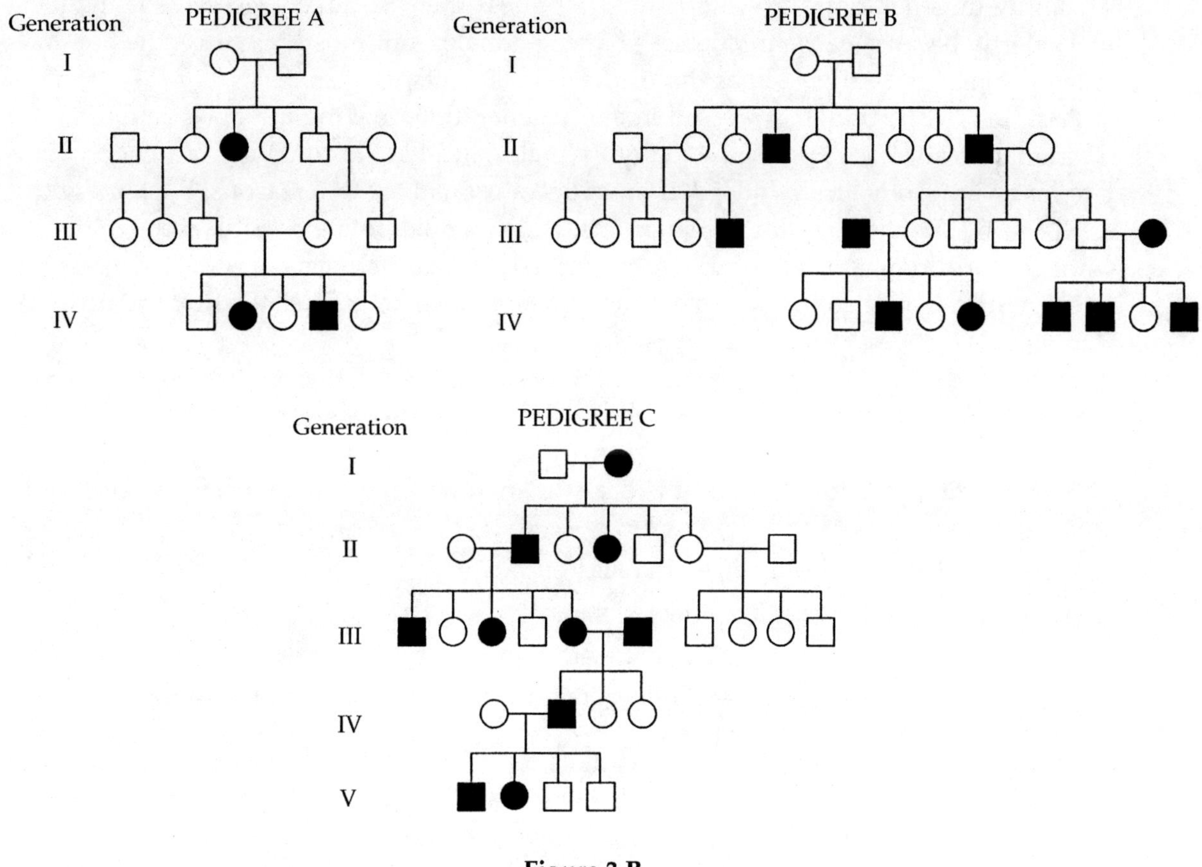

Figure 3.B

Answer: In complex pedigrees, an appropriate strategy is to ask initially whether the trait is Y-linked, then if it is dominant or recessive, and finally whether it is autosomal or X-linked. Proceeding logically through these choices helps limit the options. Before concluding, it is important to verify that the phenotype of every member of the pedigree fits the inferred inheritance pattern. Often, there may be only a single individual that can be used to distinguish one type of inheritance pattern from another.

Pedigree A: Clearly, the trait is not Y-linked, as males are not the sole transmitters of the trait. The most striking aspect of this pedigree is that the trait skips generations (i.e., unaffected parents have affected offspring). This indicates that the trait must be recessive. Now, one is left to determine if it is autosomal or X-linked. It does not fit an X-linked pedigree, as two affected females have normal (albeit heterozygous) fathers. For X-linked recessive pedigrees, both parents must contribute an abnormal allele, so that affected daughters must have affected fathers. The trait is therefore autosomal recessive. (Check by assigning appropriate genotypes to each pedigree member.)

Pedigree B: This pedigree also shows that the trait is not Y-linked and must be recessive, as twice there are unaffected parents who have affected offspring. The preponderance of affected males suggests that it might be X-linked. To determine this for certain, try to assign genotypes to see if such an inheritance mode fits. It does. Could the trait be autosomal recessive? This would be a possibility if the trait were relatively common, as one would have to suppose that the individuals who married into the family in generation II were heterozygotes. This may be the case, as several homozygotes do marry into the family in generation III. What is the best guess then? X-linked recessive is the most likely, partly because of the frequency of affected males, but mostly because the homozygous female in generation III has three affected male offspring.

Pedigree C: The trait is expressed in every generation, making it a good candidate for a dominant trait. It is expressed equally by males and females, making it a good candidate for an autosomal trait. Indeed, it must be autosomal for two reasons: First, individuals III-5 and III-6 have unaffected daughters, which would not be possible if the trait were X-linked and dominant; and second, individual IV-2, a male, has an affected male offspring, which would not be possible if the trait were X-linked. Thus, the trait is autosomal dominant.

In problems 3.20–3.22, select the correct answer.

3.20 If a genetic disease is inherited on the basis of an autosomal dominant gene, one would expect to find which of the following?

a. Affected fathers have only affected children.
b. Affected mothers never have affected sons.
c. If both parents are affected, all of their offspring have the disease.
d. If a child has the disease, one of his or her grandparents also had the disease.

Answer: Answer (a) is untrue because if the affected father is heterozygous, only half of his offspring should be affected. Answer (b) is untrue because if the mother is heterozygous, half of her offspring, regardless of sex type, should be affected. Answer (c) is untrue because if the parents are each heterozygous, ¼ of their offspring should be homozygous recessive and normal. Answer (d) is the most likely. Note, however, that if the mutation is new in either the child or his/her parents, his/her grandparent could have been unaffected.

3.21 If a genetic disease is inherited as an autosomal recessive, one would expect to find which of the following?

a. Two affected individuals never have an unaffected child.
b. Two affected individuals have affected male offspring but no affected female children.

c. If a child has the disease, one of his or her grandparents will have had it.

d. In a marriage between an affected individual and an unaffected one, all the children are unaffected.

Answer: Answer (a) is true because two affected individuals will always have affected children ($aa \times aa$ can give only aa offspring). Answer (b) is untrue, as an autosomal trait is inherited independent of sex type. Answer (c) need not be true, as the trait could be masked by normal dominant alleles through many generations before two heterozygotes marry and produce affected, homozygous offspring. Answer (d) could also be true. If the trait is rare, then it is likely that an unaffected individual marrying into the pedigree is homozygous for a normal allele. Since the trait is recessive, and the children receive the dominant, normal allele from the unaffected parent, the children will be normal. Answer (d) would not be true if the unaffected individual was heterozygous. In this case, half of the children would be affected.

3.22 Which of the following statements is *not* true for a disease that is inherited as a rare X-linked dominant trait?

a. All daughters of an affected male will inherit the disease.

b. Sons will inherit the disease only if their mother has the disease.

c. Both affected males and affected females will pass the trait to half the children.

d. Daughters will inherit the disease only if their father has the disease.

Answer: The only untrue statement is (d). Since daughters receive an X chromosome from each of their parents, they can inherit an X-linked dominant disease from either their mother or father.

3.23 Women who were known to be carriers of the X-linked recessive hemophilia gene were studied to determine the amount of time required for the blood-clotting reaction. It was found that the time required for clotting was extremely variable from individual to individual. The values obtained ranged from normal clotting time at one extreme, to clinical hemophilia at the other extreme. What is the most probable explanation for these findings?

Answer: Since hemophilia is an X-linked trait, the most likely explanation is that random inactivation of X chromosomes (lyonization) produces individuals with different proportions of cells with a functioning allele. Normal clotting times would be expected in females with a functional h^+ allele (i.e., females whose h X chromosome was very frequently inactivated). Clinical hemophilia would be expected in females without a functional h^+ allele (i.e., females whose h^+ X chromosome was very frequently inactivated). Intermediate clotting times would be expected to be proportional to the amount of h^+ function, which is related to the frequency of inactivation of the h^+ X chromosome.

4

EXTENSIONS OF MENDELIAN GENETIC ANALYSIS

CHAPTER OUTLINE

MULTIPLE ALLELES

ABO Blood Groups

Drosophila Eye Color

MODIFICATIONS OF DOMINANCE RELATIONSHIPS

Incomplete Dominance

Codominance

Molecular Explanations of Incomplete Dominance and Codominance

GENE INTERACTIONS AND MODIFIED MENDELIAN RATIOS

Gene Interactions That Produce New Phenotypes

Epistasis

ESSENTIAL GENES AND LETHAL ALLELES

GENE EXPRESSION AND THE ENVIRONMENT

Penetrance and Expressivity

Effects of the Environment

Nature Versus Nurture

REVIEW OF KEY TERMS, SYMBOLS, AND CONCEPTS

In your own words, write a brief, precise definition of each term in the groups on the following page. Check your definitions using the text. Then develop a concept map using the terms in each list.

THINKING ANALYTICALLY

While the reasoning methods used in this chapter are not very different from those used in the earlier chapters on mendelian genetics, there is much more information to consider. Phenotypes are now affected by multiple alleles, gene interactions, and environmental influences. To understand how all of these affect or contribute to a phenotype, *analyze one at a time.* First focus on understanding what each factor contributes to a phenotype. Then consider how several jointly affect a phenotype.

1	2	3
mutant	dominance	penetrance
multiple alleles	epistasis, epistatic	expressivity
allelic series	hypostasis, hypostatic	population
diploid	dihybrid cross	individual
wild-type allele	dominant epistasis	phenocopy
complete dominance	recessive epistasis	phenocopying agent
complete recessiveness	duplicate dominant epistasis	hereditary trait
incomplete dominance	duplicate recessive epistasis	nature vs. nurture
partial dominance	duplicate gene interaction	norm of reaction
codominance	complementary gene action	sex-limited trait
essential gene	9:3:3:1	sex-influenced trait
lethal allele	9:6:1	sex-linked gene
recessive lethal allele	9:3:4	autosomal gene
dominant lethal allele	9:7	
ABO blood group	12:3:1	
MN blood group	13:3	
blood-typing	9:6:1	
cellular antigen	15:1	
antibody		
agglutination		
universal donor, recipient		

As geneticists discover additional factors that affect a phenotype, the phenotype is often described in different, more revealing ways. For example, when a disease is first observed, it may be described only in terms of its inheritance pattern—say, as a dominantly inherited trait. As more is learned about the disease, its phenotype will be reassessed in light of the new knowledge. Consequently, as more is understood, different labels will be applied to the disease phenotype. These labels will indicate the kind and degree of interaction that occurs within a set of alleles at one gene (e.g., the types of dominance relationships), between several alleles at more than one gene (e.g., the types of epistasis), and between genetic and environmental influences (e.g., the level of penetrance and/or expressivity). Although complex, these labels are very useful: They quickly convey a coherent, informative description of the phenotype.

The major message of this chapter is that phenotypes can be affected by many factors. These factors include the set of alleles an organism inherits as well as the environment(s) in which it develops and lives.

QUESTIONS FOR PRACTICE

Multiple Choice Questions

1. In clover leaves, chevron pattern (a light-colored triangular leaf pattern) is controlled by seven different alleles at a single gene. From this information alone, what can be said about this trait?

a. Alleles at this gene show incomplete dominance.
b. There is a multiple allelic series with seven alleles that controls chevron pattern.
c. This gene shows epistasis.
d. One allele at this gene must be completely dominant.

2. The phenotype associated with a wild-type allele of a particular gene is
   a. the phenotype that is always found in nature
   b. a reference phenotype used by geneticists for comparison
   c. a special mutant phenotype
   d. the most common phenotype seen
3. A phenocopy is different from a particular heritable mutation in that
   a. a phenocopy cannot have the same phenotype as the mutation
   b. the biochemical anomaly in a phenocopy and a mutation must be different
   c. the phenocopy is always dominant
   d. a phenocopy is not a heritable trait
4. Neurofibromatosis is a dominantly inherited disease that can show mild, moderate, or severe symptoms. Every individual that inherits the dominant allele shows at least mild symptoms. This means that the disease allele shows
   a. variable penetrance and complete expressivity
   b. variable expressivity and complete penetrance
   c. variable penetrance and variable expressivity
   d. complete penetrance and complete expressivity
5. Two alleles at the *white* locus in *Drosophila melanogaster* are $w^{apricot}$, which has orange-colored eyes, and w^{coral}, which has pink-colored eyes. This is an example of
   a. codominance
   b. a multiple allelic series
   c. penetrance
   d. epistasis
6. One difference between epistasis and dominance is that
   a. epistasis occurs between two different genes while dominance occurs between alleles at one gene
   b. only epistasis is influenced by environmental interactions
   c. dominant traits are completely penetrant, while epistatic interactions may not be
   d. dominant traits may show variable penetrance, while epistatic interactions may not

For questions 7, 8, 9, and 10, choose the kind of inheritance that best explains the data. Explain your choices.

a. recessive lethal allele
b. dominant lethal allele
c. incomplete dominance; two allelic pairs
d. duplicate recessive epistasis; two allelic pairs
e. dominant epistasis; two allelic pairs

7. When crossed, two pure-breeding varieties of white kernel corn yield progeny with purple kernels. When these purple progeny are selfed, a count of the kernels in the F_2 yields an average of 270 purple kernels: 210 white kernels per ear.
8. When pure-breeding wheat with red kernels is crossed to the pure-breeding white variety, the progeny have pink kernels. One hundred F_2 plants exhibit the following phenotypes: 6 red, 25 dark pink, 38 pink, 25 light pink, 6 white.

9. Yellow mice crossed to pure-breeding agouti mice produce a 50:50 ratio of yellow to agouti offspring. Crosses between yellow mice never produce 100 percent yellow progeny, but rather yield 36 yellow and 15 agouti offspring.
10. At age 52, a man begins to show the symptoms of Huntington disease. Like his mother before him, he eventually dies from the disease. His wife is normal and there is no history of the disease in his wife's family or his father's family. Two of his four children also die from the disease, albeit not until late in their lives.

Answers: 1b, 2b, 3d, 4b, 5b, 6a, 7d, 8c, 9a, 10b

Thought Questions

1. Distinguish between dominance and epistasis and between epistasis and multiple alleles.
2. How would you distinguish whether a complex mutant phenotype was caused by one gene or alleles at several genes? (Hint: Review independent assortment.)
3. Tay-Sachs disease is fatal due to a lack of the enzyme hexoaminidase A. Phenotypically normal heterozygotes have about half as much enzyme activity as homozygotes for the normal allele. In what sense do alleles at the gene affected in Tay-Sachs disease show incomplete dominance, codominance, and recessive lethality?
4. A sex-linked gene controls coat color in cats. One allele determines black coat color while another allele determines orange coat color. A separate set of genes determines where color is produced (i.e., where white patches are). Use this information to explain why most calico (black, orange, and white-patched) cats are female. How do rare calico males arise? Do you expect them to be fertile?
5. In a very large natural population, could you ever presume to know all the allelic variants in a multiple allelic series? (Hint: Could new variants appear at any time?)
6. Devise a situation in which a multiple allelic series shows elements of codominance as well as dominant lethality.
7. Could a lethal allele, either dominant or recessive, be incompletely penetrant? (Hint: Consider small variations in the timing of events during development.)
8. As you will discover in Chapter 15, SNPs (single nucleotide polymorphisms) are very common, about one per several hundred base pairs of DNA. What does this tell you about the number of alleles that exist at one gene in a population? Do you expect the allelic differences to always show clear patterns of dominance or recessiveness relative to a "wild-type" allele?
9. Some traits are influenced by many genes as well as by the environment. How might you use gene chips to identify genes influencing such a trait. How would you then analyze the epistatic relationships between the genes influencing such a trait?
10. What different explanations are there for a trait to be incompletely penetrant? What approaches can be devised to test these explanations?

Thought Questions for Media

After reviewing the media at the *Essential iGenetics* website, try to answer these questions.

1. Based on viewing the animation, give an example where neither of two alleles at a gene is dominant or recessive. Using the alleles of your example, write the genotypes of a cross whose progeny show identical genotypic and phenotypic ratios. What type of dominance relationship is shown by these two alleles?

2. Correct the following false statement: "In incomplete dominance, a heterozygous individual exhibits the phenotype of both homozygotes."
3. Based on viewing the animation, give an example of a situation where two dominant alleles are fully expressed at the same time. Using the alleles of your example, write the genotypes of a cross whose progeny show identical genotypic and phenotypic ratios. What type of dominance relationship is shown by these two alleles?
4. Suppose you are able to show that a cross of two heterozygotes gives progeny in 1:2:1 genotypic and phenotypic ratios. What additional information do you need to distinguish between incomplete dominance and codominance?

1. n the modern courtroom, blood type tests can be used to exclude paternity, but it has become more difficult to use them convincingly to demonstrate paternity.
   a. Should the blood type tests have excluded Charlie Chaplin from paternity of Carol Ann? Why or why not?
   b. Suppose a defendant who is the alleged father takes a blood type test that does not exclude him from paternity consideration. What arguments might the lawyer of the defendant make in a modern courtroom to support the exclusion of the test results as "compelling" evidence? (Assume the tests are done correctly.)
   c. As mentioned in the iActivity, an individual of type A blood can be either homozygous or heterozygous. Suppose a blood-typing test allowed these two genetic conditions to be distinguished. Based on what you inferred about the ABO blood types of Joan and Carol Ann Berry in the iActivity, and supposing that you could identify their exact blood types, can you now predict the exact blood type of the father?
2. Why is it helpful to use both the M-N and the ABO blood group tests, instead of just one alone? Would a third blood type test be even more useful, or are two enough?

SOLUTIONS TO TEXT PROBLEMS

4.1 In rabbits, C = agouti coat color, c^{ch} = chinchilla, c^{h} = Himalayan, and c = albino. The four alleles constitute a multiple allelic series. The agouti C is dominant to the three other alleles, c is recessive to all three other alleles, and chinchilla is dominant to Himalayan. Determine the phenotypes of progeny from the following crosses:

a. $C/C \times c/c$
b. $C/c^{ch} \times C/c$
c. $C/c \times C/c$
d. $C/c^{h} \times c^{h}/c$
e. $C/c^{h} \times c/c$

Answer:

a. All agouti
b. ¾ agouti, ¼ chinchilla
c. ¾ agouti, ¼ albino
d. ½ agouti, ½ Himalayan
e. ½ agouti, ½ Himalayan

4.2 If a given population of diploid organisms contains only three alleles of a particular gene (say *w*, *w1*, and *w2*), how many different diploid genotypes are possible in the populations? List all possible genotypes of diploids (consider *only* these three alleles).

Answer: Six genotypes are possible: *w/w*, *w/w1*, *w/w2*, *w1/w1*, *w1/w2*, *w2/w2*.

4.3 The genetic basis of the ABO blood types seems most likely to be

a. multiple alleles
b. polyexpressive hemizygotes
c. allelically excluded alternates
d. three independently assorting genes

Answer: a

4.4 In humans, the three alleles I^A, I^B, and i constitute a multiple allelic series that determines the ABO blood group system, as described in this chapter. For the following problems, state whether the child mentioned can actually be produced from the marriage. Explain your answer.

a. an O child from the marriage of two A individuals
b. an O child from the marriage of an A to a B
c. an AB child from the marriage of an A to an O
d. an O child from the marriage of an AB to an A
e. an A child from the marriage of an AB to a B

Answer:

a. Yes, if both parents were I^Ai
b. Yes, if the A parent were I^Ai and the B parent were I^Bi
c. No, as neither parent has an I^B allele
d. No, as an O child must obtain an i allele from each parent, and one parent is I^AI^B
e. Yes, providing that the B parent is I^Bi (and the child is I^Ai)

4.5 A man is blood type O,M. A woman is blood type A,M, and her child is type A,MN. The man cannot be the father of this child because

a. O men cannot have type A children
b. O men cannot have MN children
c. an O man and an A woman cannot have an A child
d. an M man and an M woman cannot have an MN child

Answer: d. If both parents are M, an MN child is impossible because neither has an N allele. An A child is possible if the mother contributes I^A and the father contributes i.

4.6 A woman of blood group AB marries a man of blood group A whose father was group O. What is the probability that

a. their two children will both be group A?
b. one child will be group B and the other group O?
c. the first child will be a son of group AB and the second child a son of group B?

Answer: First determine the genotypes of the parents. The mother is type AB, so she must have the genotype I^AI^B. The father is type A, so he must have at least one I^A allele. The grandfather was type O, so he only had *i* alleles. Therefore, the cross must be $I^AI^B \times I^Ai$. The four equally likely genotypes (I^AI^A, I^Ai, I^BI^A, I^Bi) give rise to three phenotypes: A ($P = ½$), AB ($P = ¼$), or B ($P = ¼$).

a. The chance that two children will both be group A is ½ × ½ = ¼.
b. There is no chance of producing an O child, so $P = 0$.
c. The chance of a male AB child is ½ × ¼ = ⅛. The chance of a male B child is ½ × ¼ = ⅛. Therefore, the chance of both events happening is ⅛ × ⅛ = 1/64. As the first event is independent of the second, this is the chance of the events happening in the order specified.

4.7 If a mother and her child belong to blood group O, what blood group could the father *not* belong to?

Answer: Since the child is group O, its genotype is *ii*. Therefore, the father must have been able to contribute an *i* allele. The father could not have been type AB, whose genotype is I^AI^B.

4.8 A man of what blood group could *not* be the father of a child of blood type AB?

Answer: A man who has group O blood is genotypically *ii* and could not contribute either the I^A or I^B to an AB individual.

4.9 In snapdragons, red flower color (C^R) is incompletely dominant to white (C^W); the C^R/C^W heterozygotes are pink. A red-flowered snapdragon is crossed with a white-flowered one. Determine the flower color of

a. the F_1
b. the F_2
c. the progeny of a cross of the F_1 to the red parent
d. the progeny of a cross of the F_1 to the white parent

Answer:

a. The cross is $C^RC^R \times C^WC^W$; the F_1 are all C^RC^W and are pink.
b. The F_2 will be 1 C^RC^R : 2 C^RC^W : 1 C^WC^W and will be 1 red : 2 pink : 1 white.
c. $C^RC^W \times C^RC^R$ produces ½ C^RC^R (red) and ½ C^RC^W (pink) progeny.
d. $C^RC^W \times C^WC^W$ produces ½ C^RC^W (pink) and ½ C^WC^W (white) progeny.

4.10 In shorthorn cattle, the heterozygous condition of the alleles for red coat color (C^R) and white coat color (C^W) is roan coat color. If two roan cattle are mated, what proportion of the progeny will resemble their parents in coat color?

Answer: A cross between two roan cattle can be denoted as $C^RC^W \times C^RC^W$. The progeny will be ¼ C^RC^R, ½ C^RC^W, and ¼ C^WC^W. Thus, half will be roan progeny.

4.11 What progeny will a roan shorthorn have if bred to

a. a red?
b. a roan?
c. a white?

Answer: In shorthorn cattle, the color roan indicates a heterozygote (C^RC^W) because of incomplete dominance of each allele. Hence, the crosses can be represented as (a) $C^RC^W \times C^RC^R$, (b) $C^RC^W \times C^RC^W$, and (c) $C^RC^W \times C^WC^W$. (a) will give ½ red and ½ roan, (b) will give ¼ red, ½ roan, and ¼ white, and (c) will give ½ roan and ½ white.

4.12 In peaches, fuzzy skin (*F*) is completely dominant to smooth (nectarine) skin (*f*), and the heterozygous condition of oval glands at the base of the leaves (G^O) and no glands (G^N) gives round glands. A homozygous fuzzy, no-gland peach variety is bred to a smooth, oval-gland variety.

a. What will be the appearance of the F_1 peaches?
b. What will be the appearance of the F_2 peaches?
c. What will be the appearance of the offspring of a cross of the F_1 back to the smooth, oval-gland parent?

Answer:

a. The parental cross can be written as $FF\ G^NG^N \times ff\ G^OG^O$. The F_1 will be all $Ff\ G^NG^O$ and be fuzzy with round leaf glands.
b. As the alleles at the *G* gene show incomplete dominance, there will be a modified 9:3:3:1 ratio in the F_2. The progeny will be 3/16 fuzzy, oval-glanded ($F\text{–}\ G^OG^O$), 6/16 fuzzy, round-glanded ($F\text{–}\ G^OG^N$), 3/16 fuzzy, no-glanded ($F\text{–}\ G^NG^N$), 1/16 smooth, oval-glanded ($ff\ G^OG^O$), 2/16 smooth, round-glanded ($ff\ G^OG^N$), and 1/16 smooth, no-glanded ($ff\ G^NG^N$).
c. The cross can be written as $Ff\ G^NG^O \times ff\ G^OG^O$. The progeny will be ¼ fuzzy, oval-glanded ($Ff\ G^OG^O$), ¼ fuzzy, round-glanded ($Ff\ G^NG^O$), ¼ smooth, oval-glanded ($ff\ G^OG^O$), and ¼ smooth, round-glanded ($ff\ G^NG^O$).

4.13 In guinea pigs, short hair (*L*) is dominant to long hair (*l*), and the heterozygous condition of yellow coat (C^Y) and white coat (C^W) gives cream coat. A short-haired cream guinea pig is bred to a long-haired white guinea pig, and a long-haired cream baby guinea pig is produced. When the baby grows up, it is bred back to the short-haired cream parent. What phenotypic classes and in what proportions are expected among the offspring?

Answer: The initial cross can be diagrammed as $L\text{–}\ C^YC^W \times ll\ C^WC^W$. As a long-haired cream baby ($ll\ C^YC^W$) is produced, one can infer that the short-haired parent must have been *Ll*. The backcross can be diagrammed as $ll\ C^YC^W \times Ll\ C^YC^W$. There will be ⅛ short-haired yellow, ¼ short-haired cream, ⅛ short-haired white, ⅛ long-haired yellow, ¼ long-haired cream, and ⅛ long-haired white progeny.

4.14 The shape of radishes may be long (S^L/S^L), oval (S^L/S^S), or round (S^S/S^S), and the color of radishes may be red (C^R/C^R), purple (C^R/C^W), or white (C^W/C^W). If a long red radish plant is crossed with a round white plant, what will be the appearance of the F_1 and the F_2 plants?

Answer: The cross can be diagrammed as $S^L/S^L\ C^R/C^R \times S^S/S^S\ C^W/C^W$. The F_1 will be oval and purple ($S^L/S^S\ C^R/C^W$). The F_2 consists of 1/16 long red ($S^L/S^L\ C^R/C^R$), ⅛ oval red ($S^L/S^S\ C^R/C^R$), 1/16 round red ($S^S/S^S\ C^R/C^R$), ⅛ long purple ($S^L/S^L\ C^R/C^W$), ¼ oval purple ($S^L/S^S\ C^R/C^W$), ⅛ round purple ($S^S/S^S\ C^R/C^W$), 1/16 long white ($S^L/S^L\ C^W/C^W$), ⅛ oval white ($S^L/S^S\ C^W/C^W$), and 1/16 round white ($S^S/S^S\ C^W/C^W$).

4.15 In poultry, the dominant alleles for rose comb (R) and pea comb (P), if present together, give walnut comb. The recessive alleles of each gene, when present together in a homozygous state, give single comb. What will be the comb characters of the offspring of the following crosses?

a. $R/R\ P/p \times r/r\ P/p$
b. $r/r\ P/P \times R/r\ P/p$
c. $R/r\ p/p \times r/r\ P/p$

Answer:

a. ¾ walnut ($R/r\ P/–$), ¼ rose ($R/r\ p/p$)
b. ½ walnut ($R/r\ P/–$), ½ pea ($r/r\ P/–$)
c. ¼ walnut ($R/r\ P/p$), ¼ rose ($R/r\ p/p$), ¼ pea ($r/r\ P/p$), ¼ single ($r/r\ p/p$)

4.16 For the following crosses involving the comb character in poultry, determine the genotypes of the two parents.

a. A walnut crossed with a single produces offspring that are ¼ walnut, ¼ rose, ¼ pea, and ¼ single.
b. A rose crossed with a walnut produces offspring that are ⅜ walnut, ⅜ rose, ⅛ pea, and ⅛ single.
c. A rose crossed with a pea produces five walnut and six rose offspring.
d. A walnut crossed with a walnut produces one rose, two walnut, and one single offspring.

Answer: To approach solving this problem, first consider the possible genotypes of the parents and their offspring from the phenotypes they present. Then consider the phenotypic ratios in the offspring and determine the precise genotype.

a. $R/r\ P/p$ (walnut) × $r/r\ p/p$ (single)
b. $R/r\ p/p$ (rose) × $R/r\ P/p$ (walnut)
c. $R/–\ p/p$ (rose) × $r/r\ P/p$ (pea)
d. $R/r\ P/p \times R/r\ P/p$ (note that since a single $r/r\ p/p$ offspring is produced, both parents must have an r and a p allele)

4.17 In poultry, feathered (F) shanks (part of the legs) are dominant to clean (f), and white plumage of white leghorns (I) is dominant to black (i). Comb phenotypes and genotypes are given in Figure 4.4.

a. A feathered-shanked, white, rose-combed bird crossed with a clean-shanked, white, walnut-combed bird produces these offspring: 2 feathered, white, rose; 4 clean, white, walnut; 3 feathered, black, pea; 1 clean, black, single; 1 feathered, white, single; 2 clean, white, rose. What are the genotypes of the parents?
b. A feathered-shanked, white, walnut-combed bird crossed with a clean-shanked, white, pea-combed bird produces a single offspring that is clean-shanked, black, and single-combed. In additional offspring from this cross, what proportion may be expected to resemble each parent?

Answer: Approach this problem by considering each set of traits separately.

a. A cross between a feathered and a clean bird produces 6 feathered : 7 clean offspring (approximately 1:1), indicating the parents are $F/f \times f/f$. A cross between two white birds produces 9 white and 4 black offspring (approximately 3:1), indicating the parents are *I/i* × *I/i*. Since a cross between a rose and a walnut bird produces single-combed (as well as walnut, rose, and pea) offspring, each parent must have an *r* and a *p* allele. Thus, the parents must have been *R/r p/p* × *R/r P/p*. The complete genotypes of the parents were *F/f I/i R/r p/p* × *f/f I/i R/r P/p*.

b. The single-combed, clean-shanked, black offspring displays the phenotypes associated with the recessive alleles at the *F/f, I/i, R/r,* and *P/p* genes. Since its genotype is *f/f i/i r/r p/p*, each parent was able to give it recessive *f, i, r,* and *p* alleles. Considering this and the parents' phenotypes, the parents' genotypes must be *F/f I/i R/r P/p* and *f/f I/i r/r P/p*. To resemble the feathered-shanked, white, walnut-combed parent, the offspring must be *F/– I/– R/– P/–*. The chance of this is $\frac{1}{2} \times \frac{3}{4} \times \frac{1}{2} \times \frac{3}{4} = \frac{9}{64}$. To resemble the clean-shanked, white, pea-combed parent, the offspring must be *f/f I/– r/r P/–*. The chance of this is $\frac{1}{2} \times \frac{3}{4} \times \frac{1}{2} \times \frac{3}{4} = \frac{9}{64}$.

4.18 F_2 plants segregate $\frac{9}{16}$ colored : $\frac{7}{16}$ colorless. If a colored plant from the F_2 generation is chosen at random and selfed, what is the probability that there will be *no* segregation of the two phenotypes among its progeny?

Answer: Since there is a modified 9:3:3:1 ratio in the F_2, the F_1 cross must have been of the form *Aa Bb* × *Aa Bb*. An F_2 colored plant must have the genotype *A–, B–*, where *A* and *B* are dominant alleles that must be present for colored pigment to be produced. For "no segregation of the two phenotypes" to be observed in its progeny, the colored plant must be true-breeding (i.e., *AA BB*). Thus, the question can be rephrased as: "What is the chance that if an F_2 colored plant is picked, it is true-breeding?" Of *all* the plants in the F_2, only $\frac{1}{16}$ are *AA BB*. Among the $\frac{9}{16}$ *A– B–* plants, $\frac{1}{9}$ [$= (\frac{1}{16})/(\frac{9}{16})$] are true-breeding plants, so $P = \frac{1}{9}$.

4.19 In peanuts, a plant may be either "bunch" or "runner." Two different strains of peanut, V4 and G2, in which "bunch" occurred were crossed, with the following results:

V4 bunch × V4 bunch

↓

all bunch

G2 bunch × G2 bunch

↓

all bunch

The two true-breeding strains of bunch were crossed in the following way:

V4 bunch × G2 bunch

↓

F_1 runner

$F_1 \times F_1$

↓

F_2 9 runner : 7 bunch

What is the genetic basis of the inheritance pattern of runner and bunch in the F_2 peanuts?

Answer: The 9:7 ratio in the F_2 is a modified 9:3:3:1 ratio, where the *A– B–* genotypes are "runner" and the *A– bb, aa B–,* and *aa bb* genotypes are "bunch." This is an example of needing dominant alleles at each of two genes to observe a phenotype. When only recessive alleles at either of the two genes are present, they block (are epistatic to) the "runner" phenotype. Thus, this is an example of duplicate recessive epistasis.

4.20 In rabbits, one enzyme (the product of a functional gene *A*) is needed to produce a substance required for hearing. Another enzyme (the product of a functional gene *B*) is needed to produce another substance required for hearing. The genes responsible for the two enzymes are not linked. Individuals homozygous for either one or both of the nonfunctional recessive alleles, *a* or *b*, are deaf.

a. If a large number of matings were made between two double heterozygotes, what phenotypic ratio would be expected in the progeny?
b. This phenotypic ratio is a result of what well-known phenomenon?
c. What phenotypic ratio would be expected if rabbits homozygous recessive for trait A and heterozygous for trait B were mated to rabbits heterozygous for both traits?

Answer:

a. Since individuals homozygous for either one or both recessive alleles are deaf, individuals that are *aa B–, A– bb,* or *aa bb* will be deaf. Only *A– B–* individuals are able to hear. Thus, one will get a 9 hearing : 7 deaf phenotypic ratio.
b. This is duplicate recessive epistasis. Homozygous recessive alleles at either of two genes block hearing and are epistatic to the dominant alleles at the other gene.
c. The cross being described can be written as: *aa Bb* × *Aa Bb*. Use a branch diagram to show that there would be ⅝ deaf progeny (⅛ *Aa bb* + ½ *aa* – –) and ⅜ hearing (*Aa B–*).

4.21 In Doodlewags (hypothetical creatures), the dominant allele *S* causes solid coat color; the recessive allele *s* results in white spots on a colored background. The black coat color allele *B* is dominant to the brown allele *b*, but these genes are expressed only in the genotype *a/a*. Individuals that are *A/–* are yellow regardless of *B* alleles. Six pups are produced in a mating between a solid yellow male and a solid brown female. Their phenotypes are 2 solid black, 1 spotted yellow, 1 spotted black, and 2 solid brown.

a. What are the genotypes of the male and female parents?
b. What is the probability that the next pup will be spotted brown?

Answer: Two characteristics controlled by three genes are described here. Spotted coat (*where* pigment is positioned) is controlled by the *S/s* alleles. The pigmentation color itself (brown, black, or yellow) is controlled by the *B/b* and *A/a* genes, where *A–* is epistatic to (i.e., blocks the expression of) alleles at the *B/b* gene. Consider each characteristic separately and remember that when dealing with small numbers of progeny, precise Mendelian ratios are not always obtained.

a. Two solid-colored Doodlewags produce 4 solid and 2 spotted pups. For any spotted Doodlewags to be produced, both parents must have been heterozygous (i.e., *Ss*). The yellow male parent can initially be assigned the genotype *A– – –*, and the brown female must be *aa bb*. Since both black and brown pups are obtained, the yellow parent must be *Aa Bb*. Thus, the parents are *Ss Aa Bb* × *Ss aa bb*.

b. A spotted brown pup has the genotype *ss aa bb*.

P (*ss aa bb* pup)

$= P(s \text{ from each parent}) \times P(a \text{ from each parent}) \times P(b \text{ from each parent})$

$= (\frac{1}{2} \times \frac{1}{2}) \times (\frac{1}{2} \times 1) \times (\frac{1}{2} \times 1)$

$= \frac{1}{16}$

4.22 The allele *l* in *Drosophila* is recessive and sex-linked and lethal when homozygous or hemizygous (the condition in the male). If a female of genotype *L/l* is crossed with a normal male, what is the probability that the first two surviving progeny will be males?

Answer: The cross can be denoted as *L/l* × *L/Y*, where Y is the Y chromosome. The progeny of this cross are 1 *L/Y* : 1 *l/Y* : 1 *L/L* : 1 *L/l*. Of these, *l/Y* progeny die. Thus, one observes a 1 *L/Y* male : 2 *L/–* female progeny ratio. The chance of observing a surviving male offspring is ⅓. The chance of observing two males as the first two surviving progeny is $\frac{1}{3} \times \frac{1}{3} = \frac{1}{9}$.

4.23 A locus in mice is involved with pigment production; when parents heterozygous at this locus are mated, ¾ of the progeny are colored and ¼ are albino. Another phenotype concerns coat color; when two yellow mice are mated, ⅔ of the progeny are yellow and ⅓ are agouti. The albino mice cannot express whatever alleles they may have at the independently assorting agouti locus.

a. When yellow mice are crossed with albinos, they produce F_1 mice consisting of ½ albino, ⅓ yellow, and ⅙ agouti. What are the probable genotypes of the parents?
b. If yellow F_1 mice are crossed among themselves, what phenotypic ratio would you expect among the progeny? What proportion of the yellow progeny produced here would be expected to breed true?

Answer: Two loci are involved here. The first controls pigment production. From the information that a cross of heterozygous parents gives a 3 colored : 1 albino ratio, we can infer that the homozygous recessive condition is that of no pigmentation at all. If we denote this locus as *C/c*, this also means that *C–* is required for pigmentation. Another way to state this is that *cc* blocks the production of pigment.

The second locus determines whether the coat color is yellow or agouti. The 2 yellow : 1 agouti progeny ratio seen in a cross between two yellow mice is a modified 3:1 ratio from a monohybrid cross. It indicates recessive lethality. Specifically, it indicates that the yellow allele is dominant for coat color, but that when homozygous, it is lethal. If we denote this locus as *Y/y*, the *YY* condition is lethal, *Yy* individuals are yellow, and *yy* individuals are agouti.

Now consider the relationship between the two loci. If *cc* blocks the production of pigment, *cc* is epistatic to *Yy* and *yy* (*cc Yy* and *cc yy* mice will be albino).

a. From the above analysis, the phenotype of a yellow mouse indicates that its genotype must be *C– Yy*, and the phenotype of an albino mouse indicates that its genotype must be *cc y–*. To determine what the unknown alleles are, consider the two traits separately by considering color/no color (the *C/c* locus) and type of color (the *Y/y* locus). There are half albino and half colored offspring, indicating that the cross must have been *Cc* × *cc*. There is a 2:1 ratio of yellow to agouti offspring, indicating that the cross was *Yy* × *Yy*. Therefore, the parental genotypes are *Cc Yy* × *cc Yy*.

b. The yellow F_1 mice have the genotype *Cc Yy*. The progeny that will be obtained from the cross *Cc Yy* × *Cc Yy* are ¼ albino, ½ yellow, and ¼ agouti (remember that the *YY* progeny are inviable). None of the yellow mice will be true-breeding, as they are all *Yy*.

4.24 In *Drosophila melanogaster*, a recessive autosomal allele, ebony (*e*), produces a black body color when homozygous, and an independently assorting autosomal allele, black (*b*), also produces a black body color when homozygous. Flies with genotypes $e/e\ b^+/-$, $e^+/-\ b/b$ and $e/e\ b/b$ are phenotypically identical with respect to body color. Flies with genotype $e^+/-\ b^+/-$ have a grey body color. True-breeding $e/e\ b^+/b^+$ ebony flies are crossed with true-breeding $e^+/e^+\ b/b$ black flies.

a. What will be the phenotype of the F_1 flies?
b. What phenotypes and what proportions would occur in the F_2 generation?
c. What phenotypic ratios would you expect to find in the progeny of these backcrosses?
   i. F_1 × true-breeding ebony
   ii. F_1 × true-breeding black

Answer:

a. The F_1 flies will be $b/b^+\ e/e^+$ and be wild-type in color (grey).
b. The F_2 will show a 9 $b^+/-\ e^+/-$ (grey) : 3 $b^+/-\ e/e$ (black) : 3 $b/b\ e^+/-$ (black) : 1 $b/b\ e/e$ (black) ratio (i.e., 9⁄16 grey and 7⁄16 black).
c. The F_1 × true-breeding ebony can be denoted $b/b^+\ e/e^+ \times b^+/b^+\ e/e$ and will give a 1 $b^+/-\ e/e$: 1 $b^+/-\ e/e^+$ progeny ratio, ½ black and ½ grey. The F_1 × true-breeding black can be denoted $b^+/b\ e/e^+ \times b/b\ e^+/e^+$ and will give a 1 $b^+/b\ e^+/-$: 1 $b/b\ e^+/-$ progeny ratio, ½ grey and ½ black.

4.25 In four-o'clock plants, two genes, *Y* and *R*, affect flower color. Neither is completely dominant, and the two interact on each other to produce seven different flower colors:

Y/Y R/R = crimson	*Y/y R/R* = magenta
Y/Y R/r = orange-red	*Y/y R/r* = magenta-rose
Y/Y r/r = yellow	*Y/y r/r* = pale yellow

y/y R/R, *y/y R/r*, and *y/y r/r* = white

a. In a cross of a crimson-flowered plant with a white one (*y/y r/r*), what will be the appearance of the F_1 plants, the F_2 plants, and the offspring of the F_1 backcrossed to the crimson parent?
b. What will be the flower colors in the offspring of a cross of orange-red × pale yellow?
c. What will be the flower colors in the offspring of a cross of a yellow with a *y/y R/r* white?

Answer:

a. *Y/Y R/R* × *y/y r/r* will give a magenta-rose *Y/y R/r* F_1 and an F_2 that is 1⁄16 crimson (*Y/Y R/R*), ⅛ orange-red (*Y/Y R/r*), 1⁄16 yellow (*Y/Y r/r*), ⅛ magenta (*Y/y R/R*), ¼ magenta-rose (*Y/y R/r*), ⅛ pale yellow (*Y/y r/r*), and ¼ white (*y/y* –/–). A backcross of the F_1 to the crimson parent will give ¼ crimson (*Y/Y R/R*), ¼ magenta-rose (*Y/y R/r*), ¼ magenta (*Y/y R/R*), and ¼ orange-red (*Y/Y R/r*).

b. The cross can be denoted as *Y/Y R/r* × *Y/y r/r*. The progeny will be ¼ orange-red (*Y/Y R/r*), ¼ magenta-rose (*Y/y R/r*), ¼ yellow (*Y/Y r/r*), and ¼ pale yellow (*Y/y r/r*).

c. The cross can be denoted as *Y/Y r/r* × *y/y R/r*, and the progeny will be ½ *Y/y R/r* magenta-rose and ½ *Y/y r/r* pale yellow.

4.26 Two four-o'clock plants were crossed and gave the following offspring: ⅛ crimson, ⅛ orange-red, ¼ magenta, ¼ magenta-rose, and ¼ white. Unfortunately, the person who made the crosses was color-blind and could not record the flower colors of the parents. From the results of the cross, deduce the genotypes and flower colors of the two parents.

Answer: First determine the genotypes of the progeny. They are ⅛ *Y/Y R/R*, ⅛ *Y/Y R/r*, ¼ *Y/y R/R*, ¼ *Y/y R/r*, and ¼ *y/y –/–*. Now consider each gene separately. There are ¼ *Y/Y*, ½ *Y/y*, and ¼ *y/y* progeny, a 1:2:1 ratio, so the parental cross must have been *Y/y* × *Y/y*. Among the ¾ of the progeny whose R/*r* genotypes can be determined, half are *R/R* and half are *R/r*, indicating that the parental cross *R/R* × *R/r*. Thus, the parental cross was *Y/y R/R* (magenta) × *Y/y R/r* (magenta-rose).

4.27 Genes *A*, *B*, and *C* are independently assorting and control production of a black pigment.

a. Assume that *A*, *B*, and *C* act in a pathway as follows:

$$\text{colorless} \quad \xrightarrow{A} \quad \xrightarrow{B} \quad \xrightarrow{C} \quad \text{black}$$

The alternative alleles that give abnormal functioning of these genes are designated *a*, *b*, and *c*, respectively. A black *A/A B/B C/C* is crossed with a colorless *a/b b/b c/c* to give a black F_1. The F_1 is selfed. What proportion of the F_2 is colorless? (Assume that the products of each step except the last are colorless, so only colorless and black phenotypes are observed.)

b. Assume that *C* produces an inhibitor that prevents the formation of black by destroying the ability of *B* to carry out its function, as follows:

$$\text{colorless} \quad \xrightarrow{A} \quad \xrightarrow{B} \quad \text{black}$$
$$\uparrow$$
$$\text{C (inhibitor)}$$

A colorless *A/A B/B C/C* individual is crossed with a colorless *a/a b/b c/c*, giving a colorless F_1. The F_1 is selfed to give an F_2. What is the ratio of colorless to black in the F_2? (Only colorless and black phenotypes are observed, as in part [a].)

Answer:

a. For an individual to be black, it must have normal function at each step of the pathway. This is provided by the alleles *A*, *B*, and *C*. Thus, *A/– B/– C/–* individuals will be black, while all others (those having *a/a* and/or *b/b* and/or *c/c*) will be colorless. The chance of obtaining a black *A/– B/– C/–* individual from a cross of *A/a B/b C/c* × *A/a B/b C/c* is $3/4 \times 3/4 \times 3/4 = 27/64$. The proportion of the F_2 that is colorless is $1 - 27/64 = 37/64$. There will be a 37 colorless : 27 black ratio.

b. With this pathway, an individual is black only if it has the first two steps of the pathway (those provided by *A* and *B*) and lacks the inhibitor provided by *C* (i.e., if it is *A/– B/– c/c*). The chance of obtaining this genotype from a cross of *A/a B/b C/c* × *A/a B/b C/c* is $3/4 \times 3/4 \times 1/4 = 9/64$. The proportion of the F_2 that is colorless is $1 - 9/64 = 55/64$. There will be a 55 colorless : 9 black ratio. Thus, the ratio of black to colorless in the F_2 can be used to distinguish between hypotheses concerning these two pathways.

4.28 In *Drosophila,* a mutant strain has plum-colored eyes. A cross between a plum-eyed male and a plum-eyed female gives ⅔ plum-eyed and ⅓ red-eyed (wild-type) progeny flies. A second mutant strain of *Drosophila,* called stubble, has short bristles instead of the normal long bristles. A cross between a stubble female and a stubble male gives ⅔ stubble and ⅔ normal-bristled flies in the offspring. Assuming that the plum gene assorts independently from the stubble gene, what will be the phenotypes and their relative proportions in the progeny of a cross between two plum-eyed, stubble-bristled flies? (Both genes are autosomal.)

Answer: For either of the stubble or plum mutants, a cross of two mutant individuals gives a 2 mutant : 1 wild-type progeny ratio. This 2:1 ratio is a modified 3:1 ratio from a monohybrid cross, where the 25 percent homozygous mutant individuals die. We can denote the cross (using *P/p* for plum, *S/s* for stubble) as *P/p S/s* × *P/p S/s*. Any *PP* and/or *SS* progeny will be inviable. The kinds of progeny expected can be diagrammed as follows:

⅔ *S/s*
- ⅔ *P/p* → 4/9 *S/s P/p* (stubble plum)
- ⅓ *p/p* → 2/9 *S/s p/p* (stubble red)

⅓ *s/s*
- ⅔ *P/p* → 2/9 *s/s P/p* (long plum)
- ⅓ *p/p* → 1/9 *s/s p/p* (long red)

4.29 In sheep, white fleece (*W*) is dominant over black (*w*), and horned (*H*) is dominant over hornless (*h*) in males but recessive in females. If a homozygous horned white ram is bred to a homozygous hornless black ewe, what will be the appearances of the F_1 and F_2 sheep?

Answer: One can restate the information given in the problem as follows: In males, *H/h* and *H/H* result in horned animals, while *h/h* results in hornless animals. Both the notation and phenotypes are that of a typical dominant trait. In females, the fact that the trait is sex-influenced makes the situation quite different. In females, only *H/H* animals are horned. *H/h* and *h/h* animals are hornless. The white trait shows dominance in both sexes, with *W/–* animals being white and *ww* animals being black.

This information allows one to diagram the cross as *H/H W/W* male × *h/h w/w* female. The F_1 genotype will be *H/h W/w*. Both sex types will have white fleece, and while males will be horned, females will be hornless. The F_2 genotypes will be 9 *H/– W/–* : 3 *H/– w/w* : 3 *h/h W/–* : 1 *h/h w/w*. Because *H/h* and *H/H* show the dominant horned trait in males, the males will be 9/16 horned white, 3/16 horned black, 3/16 hornless white, and 1/16 hornless black. Female phenotypes can be determined with the following branch diagram:

¾ H/h or h/h hornless
- ¾ $W/–$ ⟶ 9/16 hornless white
- ¼ w/w ⟶ 3/16 hornless black

¼ H/H
- ¾ $W/–$ ⟶ 3/16 horned white
- ¼ w/w ⟶ 1/16 horned black

4.30 A horned black ram bred to a hornless white ewe produces the following offspring: Of the males, ¼ are horned and white; ¼ are horned and black; ¼ are hornless and white; and ¼ are hornless and black. Of the females, ½ are hornless and black, and ½ are hornless and white. What are the genotypes of the parents?

Answer: Start by using the symbols of problem 4.29 to denote what can be inferred about the genotypes of the parents and the offspring. Remember that in males, the horned condition is dominant, while in females, it is recessive. Hence, $H/–$ (= H/H or H/h) denotes the horned condition in males, while $h/–$ (h/h or H/h) denotes the hornless condition in females. Then the parents are $H/–\ w/w$ (male) × $h/–\ W/–$ (female). The male progeny are ¼ $H/–\ W/–$, ¼ $H/–\ w/w$, ¼ $h/h\ W/–$, and ¼ $h/h\ w/w$. The female progeny are ½ $h/–\ w/w$ and ½ $h/–\ W/–$. Since homozygous recessive (male) progeny are recovered, we know that each parent must have had (at least) one recessive h and one recessive w allele. Since none of the female progeny are H/H, only one parent had an H allele. Thus, the parents were $H/h\ w/w$ (male) × $h/h\ W/w$ (female).

5

GENE MAPPING IN EUKARYOTES

CHAPTER OUTLINE

REVIEW OF KEY TERMS, SYMBOLS, AND CONCEPTS

In your own words, write a brief, precise definition of each term in the groups on the following page. Check your definitions using the text. Then develop a concept map using the terms in each list.

THINKING ANALYTICALLY

When working on a mapping problem, start by systematically organizing the data. This will allow you to extract critical information by inspection. Arrange the data by using symbols to represent the traits, writing out the crosses and their progeny, and tabulating the number of offspring in each progeny class. After organizing the data, inspect two genes at a time. This often allows you to determine if the two genes are linked, and if linked, whether they are sex-linked. Any trait appearing in progeny in just one sex type should be checked for sex linkage.

Mapping genes using meiotic recombination and testcrosses. It is essential to remember that two genes that show 50 percent recombination in a testcross may either be on different chromosomes or be on the same chromosome but very far apart. The map distances between two genes can exceed 50 map units, but their recombination frequency cannot exceed 50 percent. If two genes showing 50 percent recombination are each linked to a third, intermediate gene (i.e., they show less than 50 percent recombination with the third gene), they will be on the same chromosome but more than 50 map units apart.

1	2	3
linkage, partial linkage	genetic map	tetrad analysis
linked genes	meiotic, mitotic recombination	ordered, unordered tetrads
linkage group	centimorgan (cM), map unit	linear tetrads
syntenic	recombination frequency	ascus, asci
chromosome	coupling vs. repulsion	*Neurospora crassa*
parental genotypes, classes	two-point testcross	*Saccharomyces cerevisiae*
recombinant genotypes, classes	three-point testcross	*Chlamydomonas reinhardtii*
genetic recombination	two-, three-, four-strand, double crossovers	ascospore
idependent assortment	interference	vegetative cell
crossing-over	coefficient of coincidence	random-spore analysis
chiasma, chiasmata	mapping function	parental ditype (PD) tetrad
cytological markers		nonparental-ditype (NPD) tetrad
genetic, gene markers		tetratype (T) tetrad
translocation		first-division segregation
reciprocal exchange		second-division segregation
four-chromatid stage		gene-gene distance
ordered tetrads		
ascus, asci		

In a three-point testcross, a systematic, methodical approach is essential.

1. Assign gene symbols to the phenotypes that are seen and rewrite the crosses.
2. Reorganize the progeny genotypes according to the frequency of each class.
3. Assign parental, single-crossover, and double-crossover classes—parentals will be most frequent, double crossovers will be least frequent.
4. Infer the gene order by comparing the parental and double-crossover classes.
5. Rewrite the testcross; infer which single crossover gives rise to which class.
6. Calculate the recombination frequency and map distance for each gene interval.
7. Calculate the coefficient of coincidence and the interference for the region.

When using the chi-square test to check for linkage, remember that this test, like most statistical tests, only serves as a guide for invalidation of a hypothesis. It does not validate or prove a hypothesis. Carefully consider the hypothesis you are testing and what the values of χ^2 and P mean.

Mapping genes using tetrad analysis. Many students find the analysis of tetrads quite challenging. Step back for a minute, though, and ask yourself, "Can I recognize the *principles* behind these methods?" Tetrad analysis is based on the normal events of meiosis in organisms in which all four products of a meiosis remain together.

A solid understanding of chromosome behavior during meiosis will enable you to master tetrad analysis. Review how chromosomes behave during meiosis. Consider how T, PD, and NPD tetrads are produced. Approaching the material this way will provide you with a solid, *mechanistic* understanding of this process.

Once you understand how meiosis relates to this mapping technique, you will be able to apply them more readily. Analyzing the results of crosses in which tetrads are produced is in many regards much more straightforward than analyzing the results of crosses in which they are not produced. Try employing the following systematic approach:

1. Assign gene symbols to the traits and write out the cross, considering just two loci at a time.
2. Decide which are parental-type spores and which are recombinant-type spores.
3. If you have unordered asci, calculate the recombination frequency (RF). RF = (# recombinant spores/total # spores) × 100%
4. If you have ordered tetrads, identify PD, T, and NPD tetrads.
   a. If PD = NPD (or is close), the genes are not linked.
   b. If PD >> NPD, the genes are linked
5. RF = [(½ T + NPD)/total # asci] × 100%
6. Draw a map.

QUESTIONS FOR PRACTICE

Multiple Choice Questions

1. Evidence that genetic recombination is associated with physical exchange between chromosome homologues was found by
   a. observing the existence of chiasmata during meiosis
   b. observing the pairing of homologues
   c. using a combined genetic and cytological approach
   d. showing that genetic recombination was accompanied by an exchange of identifiable chromosomal segments
   e. both c and d
2. Alleles are considered linked if recombination is less than 50 percent in
   a. a $P \times P \rightarrow F_1$ cross
   b. an $F_1 \times F_1 \rightarrow F_2$ cross
   c. an $F_2 \times F_2 \rightarrow F_3$ cross
   d. an $F_1 \times$ homozygous recessive cross
   e. any F_2 progeny × homozygous recessive cross
3. If two loci *A*/*a* and *B*/*b* are 9 map units apart, and an *A b*/*a B* individual is testcrossed,
   a. there will be 9 percent *A B* individuals
   b. there will be 91 percent *A b* individuals
   c. there will be 4.5 percent *a b* individuals
   d. there will be 91 percent recombinant individuals
4. If the genes *a* and *b* show 50 percent recombination, genes *a* and *c* show 35 percent recombination, and genes *b* and *c* show 32 percent recombination,
   a. genes *a* and *b* are on the same chromosome and 50 map units apart
   b. genes *a* and *b* are on different chromosomes
   c. genes *a* and *b* are on the same chromosome and 67 map units apart
   d. one cannot tell if genes *a* and *b* are on the same or different chromosomes from the given data

5. A chi-square test is performed with a hypothesis that two genes are unlinked. A chi-square value of 11.35 is obtained, which in turn gives a value of $P = 0.01$ (i.e., 1 percent). Based on this information, one should
   a. accept the hypothesis as being possible and expect that the genes might be unlinked
   b. accept the hypothesis as being proven true and know that the genes are unlinked
   c. reject the hypothesis as being unlikely and expect that the genes might be linked
   d. reject the hypothesis as being impossible and know that the genes are linked
6. In a three-point cross, interference is a measure of
   a. how often one chiasma physically impedes the occurrence of a second crossover in a particular region
   b. the accuracy of the map distances between three genes
   c. whether the data are considered reliable
   d. whether the genes are linked
7. In a region in which interference is 0.85,
   a. there are fewer double crossovers than expected
   b. there are more double crossovers than expected
   c. one observes the expected number of double crossovers
   d. 0.85 percent of testcross progeny show double-crossover phenotypes
8. If two genes *a* and *b* are 8.3 map units apart, one expects to find
   a. 8.3 percent recombinant gametes from a double heterozygous parent
   b. 91.7 percent recombinant gametes from a doubly heterozygous parent
   c. a chiasma between the *a* and *b* loci in 8.3% of the meioses
   d. a chiasma between the *a* and *b* loci in 91.7% of the meioses
9. Map distances are most accurate when genes are closely linked because as genes become further apart,
   a. they always show independent assortment
   b. the chance of multiple crossovers increases
   c. they no longer recombine
   d. there is chiasma interference
10. Tetrad analysis is a technique for
    a. determining base sequences in DNA
    b. measuring the number of chromatids at meiosis
    c. separating *Neurospora* mycelia
    d. mapping genes in organisms retaining the four meiotic products in one structure
11. Ordered tetrads occur in the asci of
    a. *Neurospora*
    b. *Chlamydomonas*
    c. *Saccharomyces*
    d. *Drosophila*
12. A single crossover between two gene loci during meiosis in *Neurospora* spore formation produces an ascus with a genetic configuration known as
    a. a tetratype
    b. a parental ditype
    c. a nonparental ditype
    d. a double recombinant

13. The formula (½ T + NPD)/(total # of tetrads) × 100% is used to calculate
    a. the length of a chromosome
    b. the location of the centromere
    c. the number of crossovers
    d. the percentage of recombinants
14. In a tetrad in which a single crossover between two loci has occurred, what percentage of the spores will be recombinant types?
    a. 0%
    b. 25%
    c. 50%
    d. 100%

Answers: 1e, 2d, 3c, 4c, 5c, 6a, 7a, 8a, 9b, 10d, 11a, 12a, 13d, 14c

Thought Questions

1. Do you think that mapping according to the methods described in this chapter will still be useful if the entire DNA sequence of an organism is known? (Hint: Does genetic and/or cytological mapping narrow the field of possibilities?)
2. Why do you think that sex linkage was described before other examples of linkage in animals? (Hint: Is this an example of a correlation of genetic and cytological data?)
3. Can one make genetic maps of haploid organisms? (Hint: Review *Neurospora* material.)
4. Does the decision to accept or reject a hypothesis at the 0.05 level mean that one is more likely to discard a correct hypothesis or retain an incorrect one?
5. Do all chromosomes have the same length in terms of map distances?
6. Consider two pairs of genes: *A* and *B*, and *C* and *D*. The number of base pairs of DNA between *A* and *B* is the same as the number of base pairs of DNA between *C* and *D*. Can you infer from this information alone whether these pairs of genes have the same or different map distances between them? Why or why not?
7. In some organisms, correlations have been made between the physical distance between two closely linked loci (i.e., the number of base pairs of DNA) and the map distance between them. If one measures the recombination frequency between two genes in another chromosomal region, can this information be used to infer the physical distance between them as well? (Hint: Consider the basis of interference and the observation that different regions of a chromosome show different amounts of crossing-over.)
8. Compared to a chromosomal region that shows no interference, in a chromosomal region that shows high levels of interference, would you expect there to be the same, greater, or fewer megabases of DNA per map unit?

Thought Questions for Media

After reviewing the media at the *Essential iGenetics* website, try to answer these questions.

1. Would the results in the three-point testcross animation be different if the gene order used in the analysis was the one given initially? How?

1. When the initial data generated using testcrosses were analyzed, a hypothesis of "no linkage" was tested.

a. What was the rationale for testing this hypothesis instead of testing a hypothesis of "linkage"?

b. Why was this hypothesis not accepted?

c. Why, in this case, does this suggest that the three genes are linked?

2. When mapping genes using data from trihybrid crosses, it is essential to determine the gene order. This is done by comparing the parental-type and double-crossover-type progeny.

a. In the trihybrid testcross, how do you know which progeny classes result from double-crossover events?

b. How does knowing which progeny classes result from a double crossover help you determine the gene order?

3. When mapping the distance between a pair of genes in the trihybrid crosses, the number of recombinants in one of the intervals is equal to the sum of the number of single crossovers in that interval plus the number of double crossovers. Why does the number of double crossovers need to be added?

4. From the data in this problem alone, although you can determine the relative locations of the *cu* and *m* genes to the *o* gene at map position 55, you can't determine their exact map position. That is, you cannot determine whether the *cu* gene is at position 49 and the *m* gene is at position 71 or the *cu* gene is at position 61 and the *m* gene is at position 39. Suppose you knew the position of a fourth gene at map position 47. What is the easiest way to determine the correct map positions of *cu* and *m?*

SOLUTIONS TO TEXT PROBLEMS

5.1 A cross $a^+a^+\ b^+b^+ \times aa\ bb$ results in an F_1 of phenotype $a^+\ b^+$; the following numbers are obtained in the F_2 (phenotypes):

$a^+\ b^+$	110
$a^+\ b$	16
$a\ b^+$	19
$a\ b$	15
Total	160

Are genes at the *a* and *b* loci linked or independent? What F_2 numbers would otherwise be expected?

Answer: If two genes are unlinked in a dihybrid cross, one expects a 9:3:3:1 ratio in the F_2. For 160 progeny, one would expect 90:30:30:10. Testing the hypothesis of no linkage using a chi-square test, one has 2 = $(110 - 90)^2/90 + (16 - 30)^2/30 + (19 - 30)^2/30 + (15 - 10)^2/10 = 17.5$, df = 3, $P < 0.001$. This hypothesis is rejected as being unlikely. It is likely that the genes are linked.

5.2 In corn, a dihybrid for the recessives *a* and *b* is testcrossed. The distribution of the phenotypes is as follows:

A B	122
A b	118
a B	81
a b	79

Are the genes assorting independently? Test the hypothesis with a chi-square test. Explain tentatively any deviation from expectation, and tell how you would test your explanation.

Answer: Use the χ^2 test to test the hypothesis that the genes are unlinked. With this hypothesis, one would expect the cross *Aa Bb* × *aa bb* to produce four equally frequent classes of progeny (i.e., a 1:1:1:1 ratio). With 400 total progeny, one would expect 100 progeny in each phenotypic class. One has:

Phenotype	**Observed (= *o*)**	**Expected (= *e*)**	**$(o - e)^2/e$**
A B	122	100	4.84
A b	118	100	3.24
a B	81	100	3.61
a b	79	100	4.41
			$\chi^2 = 16.1$; df = 3

From the χ^2 test, $P < 0.01$. That is, there is less than a 1 percent likelihood of observing this much deviation from expected values by chance alone. This indicates that one can reject the hypothesis of "no linkage" as being unlikely. Note that the chi-square test neither proves that the genes are unlinked nor proves that they are linked.

Linkage might seem reasonable until it is realized that the minority classes are not reciprocal classes (both carry the *aa* phenotype). If the segregation at each locus is considered, however, the *B/– : b/b* ratio is about 1:1 (203:197), while the *A/– : a/a* ratio is not (240:160). The departure of the observed from expected values is thus due to a reduced number of *a/a* individuals. This departure should be confirmed in other crosses that test the segregation at the *A/a* locus. It would be particularly informative to check if the lack of *a/a* corn plants is associated with ungerminated seeds or seedlings that die before they are able to mature.

5.3 In *Drosophila,* the mutant black (*b*) has a black body, and the wild type has a grey body; the mutant vestigial (*vg*) has wings that are much shorter and crumpled compared with the long wings of the wild type. In the following cross, the true-breeding parents are given together with the counts of offspring of F_1 females × black and vestigial males:

P black and normal × grey and vestigial
F_1 females × black and vestigial males

Progeny:	grey, normal	283
	grey, vestigial	1,294
	black, normal	1,418
	black, vestigial	241

From these data, calculate the map distance between the black and vestigial genes.

Answer: The cross is $b\ vg^+/b\ vg^+ \times b^+\ vg/b^+\ vg$, which gives an F_1 that is $b\ vg^+/b^+\ vg$. An F_1 female is crossed with a homozygous recessive $b\ vg/b\ vg$. The female is used as the doubly heterozygous parent, as no crossing-over occurs in male *Drosophila* flies. The classes can be grouped in reciprocal pairs as follows:

Nonrecombinants (parentals):	grey, vestigial	1,294
	black, normal	1,418
Recombinants (nonparentals):	grey, normal	283
	black, vestigial	241
	Total progeny	3,236

Recombination frequency = [(283 + 241)/3,236] × 100% = 16.2%. There are 16.2 map units between the genes.

5.4 Use the following two-point recombination data to map the genes concerned. Show the order and the length of the shortest intervals.

Gene Loci	% Recombination
a,b	50
a,c	15
a,d	38
a,e	8
b,c	50
b,d	13
b,e	50
c,d	50
c,e	7
d,e	45

Answer: Because of the effects of double crossovers, larger recombination frequencies are less accurate measures of map distances than smaller recombination frequencies between close neighbors. As a consequence, the map distances that can be derived from these data are not strictly additive (e.g., $a - d = 38$ and $a - e = 8$, but $d - e = 45$). The distances between nearest neighbors can be used to infer the following map.

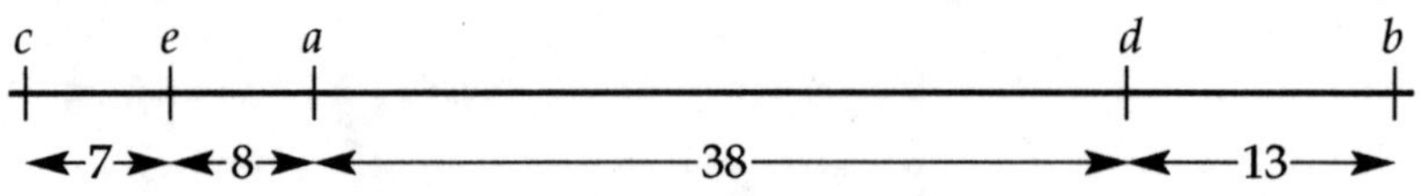

Remember that although recombination frequency cannot exceed 50 percent, map distances can exceed 50 map units. Thus, the map distance between *c* and *b* can be inferred to be 66 map units (7 + 8 + 38 + 13), although *c* and *b* show only 50 percent recombination.

5.5 A corn plant known to be heterozygous at three loci is testcrossed. The progeny phenotypes and frequencies are as follows:

+	+	+	455
a	*b*	*c*	470
+	*b*	*c*	35
a	+	+	33
+	+	*c*	37
a	*b*	+	35
+	*b*	+	460
a	+	*c*	475
	Total		2,000

Give the gene arrangement, linkage relations, and map distances.

Answer: First, note that since the offspring result from a testcross, the phenotypes of the offspring reflect the genotypes of the gametes of the heterozygous parent. Consequently, these phenotypes can be used to determine the gene arrangement in the heterozygous parent.

Second, notice that there are eight genotypic classes that appear in two frequencies. Four of the genotypic classes have about 460 individuals, while the other four have about 35 individuals. This would be expected if two of the genes were linked and the third was not. In this case, the chromosome bearing the two linked genes could undergo crossing-over and give rise to two frequencies of gametes. Because crossing-over between two closely linked genes is infrequent, gametes with noncrossover chromosomes are more frequent. The third gene assorts independently. The independent assortment of the third gene results in four, and not just two, genotypes present in each of two frequencies.

With this information, one can proceed in either of two ways. First, one can tabulate the data considering only two pairs of genes at one time. This allows one to view this cross as a series of two-point crosses.

a	*b*	*c*	# progeny
a	*b*		470 + 35 = 505
+	*b*		460 + 35 = 495
a	+		475 + 33 = 508
+	+		455 + 37 = 492
a		*c*	470 + 475 = 945
+		*c*	35 + 37 = 72
a		+	33 + 35 = 68
+		+	455 + 460 = 915
	b	*c*	470 + 35 = 505
	+	*c*	475 + 37 = 512
	b	+	460 + 35 = 495
	+	+	455 + 33 = 488

Notice that there is a 1:1:1:1 ratio in the offspring when one considers just the *a* and *b* genes or just the *b* and *c* genes. This is what one would expect in a dihybrid cross with two independently assorting genes. Thus, *a* assorts independently of *b*, and *b* assorts independently of *c*. This is not the case for *a* and *c*, however. Here, there are two frequencies of data indicative of two linked genes showing recombination. The recombination frequency between *a* and *c* is RF = (72 + 68)/(72 + 68 + 945 + 915) × 100% = 7%, and there are 7 map units between *a* and *c*.

Alternatively, one can use the reasoning presented above to infer that the four progeny classes with about 460 individuals contain parental types of chromosomes for two linked genes, as well as an independently assorting chromosome with the third gene. By inspection of the most frequent phenotypes, one sees that only *a c* and + + chromosomes are found, while the *b*/+ gene assorts independently from these noncrossover chromosomes. Thus, the *a* and *c* loci are linked, and the *b* gene is unlinked. The least frequent progeny classes contain *a* + and + *c* recombinant chromosomes, so that the percent recombination between *a* and *c* is RF = (35 + 37 + 33 + 35)/2,000 × 100% = 7%. Just as was seen earlier, the map distance between *a* and *c* is 7 map units.

One can diagram these results as

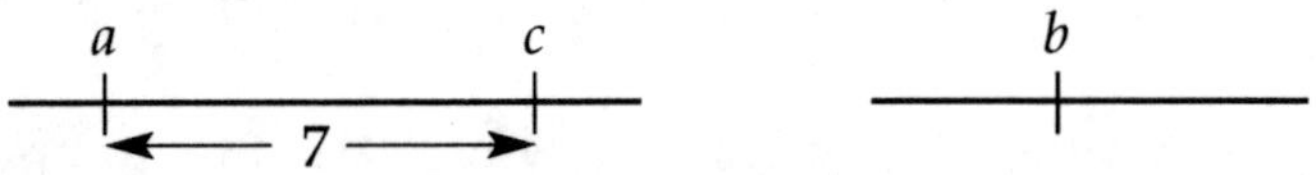

5.6 Genes *a* and *b* are linked, with 10 percent recombination. What would be the phenotypes, and the probability of each, among progeny of the following cross?

$$\frac{a\ b^+}{a^+\ b} \times \frac{a\ b}{a\ b}$$

Answer: The $a\ b^+/a^+\ b$ parent will give 90 percent parental-type ($a\ b^+$ or $a^+\ b$) gametes and 10 percent recombinant-type ($a\ b$ or $a^+\ b^+$) gametes. The genotypes of these gametes will determine the offspring phenotypes, as the $a\ b/a\ b$ parent will give only *a b* gametes. Therefore, there will be 45% $a\ b^+$, 45% $a^+\ b$, 5% $a\ b$, and 5% $a^+\ b^+$ offspring.

5.7 Genes *a* and *b* are sex-linked and are located 7 mu apart in the X chromosome of *Drosophila*. A female of genotype $a^+\ b/a\ b^+$ is mated with a wild-type ($a^+\ b^+/Y$).

a. What is the probability that one of her sons will be either $a^+\ b^+$ or $a\ b^+$ in phenotype?

b. What is the probability that one of her daughters will be $a^+\ b^+$ in phenotype?

Answer: Since the genes are 7 mu apart, the female will have 93 percent parental-type gametes (46.5% $a^+\ b$ and 46.5% $a\ b^+$) and 7 percent recombinant-type gametes (3.5% *a b* and 3.5% $a^+\ b^+$). The wild-type male will give either an X chromosome bearing a^+b^+ or a Y chromosome. As the genes are X-linked, the phenotype of the sons will reflect the chromosome they receive from the mother. The phenotype of the daughters will be $a^+\ b^+$, as they receive the $a^+\ b^+$ X chromosome from their father.

a. $P = 3.5\%\ (a^+\ b^+) + 46.5\%\ (a\ b^+) = 50\%$

b. $P = 100\%$

5.8 Genes *a* and *b* are on one chromosome, 20 mu apart; *c* and *d* are on another chromosome, 10 mu apart. Genes *e* and *f* are on yet another chromosome and are 30 mu apart. Cross a homozygous *A B C D E F* individual with an *a b c d e f* one, and cross the F_1 back to an *a b c d e f* individual. What are the chances of getting individuals of the following phenotypes in the progeny?

a. *A B C D E F*

b. *A B C d e f*

c. *A b c D E f*

d. *a B C d e f*

e. *a b c D e F*

Answer: The frequency of recombination between two genes can be inferred from the map distance between them. For example, the 20 mu between *a* and *b* indicate that in an *AB/ab* individual, gametes will have 80 percent parental-type chromosomes (40 percent each of *AB, ab*) and 20 percent recombinant-type chromosomes (10 percent each of *Ab, aB*). The cross under consideration is *AB/ab CD/cd EF/ef* × *a b/ab cd/cd ef/ef.*

a. The chance of obtaining the parental-type *AB, CD,* and *EF* chromosomes is $P = 0.40 \times 0.45 \times 0.35 = 0.063$, or 6.3%.

b. The chance of obtaining *AB, Cd,* and *ef* chromosomes is $P = 0.40 \times 0.05 \times 0.35 = 0.007$, or 0.7%.

c. The chance of obtaining *Ab, cD,* and *Ef* chromosomes is $P = 0.10 \times 0.05 \times 0.15 = 0.00075$, or 0.075%.

d. The chance of obtaining *aB, Cd,* and *ef* chromosomes is $P = 0.10 \times 0.05 \times 0.35 = 0.00175$, or 0.175%.

e. The chance of obtaining *ab, cD,* and *eF* chromosomes is $P = 0.40 \times 0.05 \times 0.15 = 0.003$, or 0.3%.

5.9 Genes *d* and *p* occupy loci 5 map units apart in the same autosomal linkage group. Gene *h* is a separate autosomal linkage group and therefore segregates independently of the other two. What types of offspring are expected, and what is the probability of each, when individuals of the following genotypes are testcrossed?

a. $\dfrac{D\ P}{d\ p}\ \dfrac{h}{h}$ b. $\dfrac{d\ P}{D\ p}\ \dfrac{H}{h}$

Answer:

a. 95% of the progeny will have *D P* or *d p* parental-type chromosomes, and 5% of the progeny will have *D p* or *d P* recombinant-type chromosomes. All of the progeny will be *h.* There will be 47.5% *D P h*, 47.5% *d p h*, 2.5% *D p h*, 2.5% *d P h.*

b. 23.75% *d P H*, 23.75% *d P h*, 23.75% *D p H*, 23.75% *D p h*, 1.25% *D P H*, 1.25% *D P h*, 1.25% *d p H*, 1.25% *d p h*

5.10 A hairy-winged (*h*) *Drosophila* female is mated with a yellow-bodied (*y*), white-eyed (*w*) male. The F_1 are all wild type. The F_1 progeny are then crossed, and the F_2 that emerge are as follows:

Females:	wild type	757
	hairy	243
Males:	wild type	390
	hairy	130
	yellow	4
	white	3
	hairy, yellow	1
	hairy, white	2
	yellow, white	360
	hairy, yellow, white	110

Give genotypes of the parents and the F_1, and note the linkage relations and distances where possible.

Answer: Careful inspection of the data shows that the F_2 males, but not the F_2 females, show the yellow and white traits. In contrast, the hairy trait is shown in an approximately 3:1 ratio in both females and males (females = 757 normal : 243 hairy; males = 757 normal : 243 hairy). This indicates that the yellow and white genes are sex-linked, while hairy is not.

The normal phenotype of the F_1 heterozygote indicates that the nonhairy, nonyellow, and nonwhite traits are dominant. Therefore, let h and h^+ represent the hairy and nonhairy traits, respectively, y and y^+ represent the yellow and normal body color traits, respectively, and w and w^+ represent the white- and red-eyed traits, respectively. Then, given the phenotypes of the parents and F_1 and the sex linkage of y and w, the crosses can be written as

P: $y^+ \ w^+/y^+ \ w^+ \ h/h$ females $\times$ $y \ w/\mathrm{Y} \ h^+/h^+$ males

F_1: $y^+ \ w^+/y \ w \ h^+/h$ females $\times$ $y^+ \ w^+/\mathrm{Y} \ h^+/h$ males

Now determine the map distance between y and w. This can be inferred from the frequency of recombinants among the F_2 males.

Phenotype of F_2 Males	Genotype	X-Chromosome Recombinant?	Number
wild type	$y^+ \ w^+/\mathrm{Y} \ h^+/–$	no	390
hairy	$y^+ \ w^+/\mathrm{Y} \ h/h$	no	130
yellow	$y \ w^+/\mathrm{Y} \ h^+/–$	yes	4
white	$y^+ \ w/\mathrm{Y} \ h^+/–$	yes	3
hairy, yellow	$y \ w^+/\mathrm{Y} \ h/h$	yes	1
hairy, white	$y^+ \ w/\mathrm{Y} \ h/h$	yes	2
yellow, white	$y \ w/\mathrm{Y} \ h^+/–$	no	360
hairy, yellow, white	$y \ w/\mathrm{Y} \ h/h$	no	110

The recombination frequency between *y* and *w* is $[(4 + 3 + 1 + 2)/1{,}000] \times 100\% = 1\%$, so there is 1 map unit between these genes. Note that since recombination can only be scored in the F_2 males, the number of recombinants is divided by the number of *male* progeny, and not the total number of progeny.

5.11 For each of the following tabulations of testcross progeny phenotypes and numbers, state which locus is in the middle, and reconstruct the genotype of the tested triple heterozygotes.

					b.					c.			
A	*B*	*C*	191		*C*	*D*	*E*	9		*F*	*G*	*H*	110
a	*b*	*c*	180		*c*	*d*	*e*	11		*F*	*g*	*h*	114
A	*b*	*c*	5		*C*	*d*	*e*	35		*F*	*g*	*h*	37
a	*B*	*C*	5		*c*	*D*	*E*	27		*f*	*G*	*H*	33
A	*B*	*c*	21		*C*	*D*	*e*	78		*F*	*G*	*h*	202
a	*b*	*C*	31		*c*	*d*	*E*	81		*f*	*g*	*H*	185
A	*b*	*C*	104		*C*	*d*	*E*	275		*F*	*g*	*H*	4
a	*B*	*c*	109		*c*	*D*	*e*	256		*f*	*G*	*h*	0

Answer: For each of the data sets, there are eight classes of progeny appearing in four frequencies, as is expected for linkage among three genes. Since recombination between closely linked genes is infrequent, the least frequent classes result from crossovers between each of the genes in the triple heterozygote (i.e., double crossovers). The most frequent classes are parental types, where no crossover occurred between the linked genes.

To help you determine which gene is in the middle in this and in subsequent problems, consider the example shown below, which shows a double crossover between two nonsister chromatids. Relative to the parental-type chromosomes, the recombinant chromosomes have the middle *A/a* alleles switched:

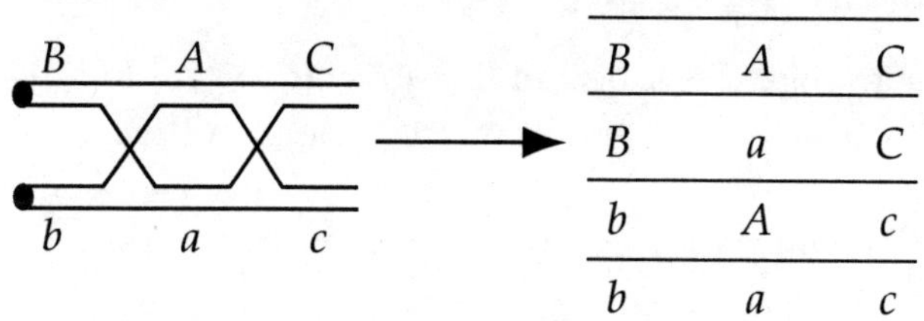

In three-point crosses, comparison of one of the double-crossover classes (the least frequent classes) with one of the parental classes (the most frequent classes) can allow you to infer which gene is in the middle. Depending on how they are chosen, one of two scenarios will be found. In one scenario, alleles at two genes will be identical, while alleles at the third gene will be different. In another scenario, alleles at two genes will be different, while alleles at the third gene will be identical. In each case, the third gene is the gene in the middle. The middle gene is unlike the other two because its alleles have been "switched."

a. *a* is in the middle. The genotype of the triple heterozygote is *BAC/bac*.
b. *d* is in the middle. The genotype is *CdE/cDe*.
c. *f* is in the middle. The genotype is *GFh/gfH*.

5.12 The following numbers were obtained for testcross progeny in *Drosophila* (phenotypes):

+	*m*	+	218
w	+	*f*	236
+	+	*f*	168
w	*m*	+	178
+	*m*	*f*	95
w	+	+	101
+	+	+	3
w	*m*	*f*	1
Total			1,000

Construct a genetic map.

Answer: Notice that among the eight phenotypic classes, there are four frequencies of progeny. This would be expected if the three genes were linked and a triply heterozygous parent were testcrossed. Since crossovers between closely linked genes occur less often than no crossovers, the most frequent classes of progeny will be those having nonrecombinant, or parental-type, chromosomes. Thus, the + *m* + and *w* + *f* chromosomes are parental-type chromosomes and indicate that the triply heterozygous parent was obtained as follows:

P: *w* + *f* /*w* + *f* × + *m* +/+ *m* +

F_1: *w* + *f* /+ *m* + × *w m f* /*w m f*

Since recombination between closely linked genes is infrequent, the least frequent classes result from crossovers between each of the genes in the triple heterozygote—that is, double crossovers (+ + +, *w m f*). The remaining classes will have resulted from single crossovers.

We use the strategy described in the answer to problem 5.11 to determine which gene is in the middle. This indicates that the *m* gene is in the middle:

One parental-type chromosome:	*w*	+	*f*
One double-crossover chromosome:	*w*	*m*	*f*
Comparison of alleles:	same	different	same

As *m* is unlike the other two, it is in the middle. While the genes are written in the correct order in this problem, this may not always be the case. (Always determine the correct order before proceeding.)

Now, diagram the crossovers in the triple heterozygote to infer the progeny classes that are associated with each crossover type:

Single crossover between *w, m:*

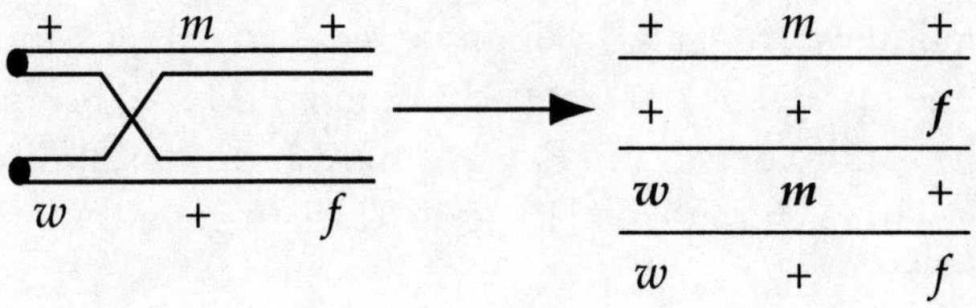

Single crossover between *m, f:*

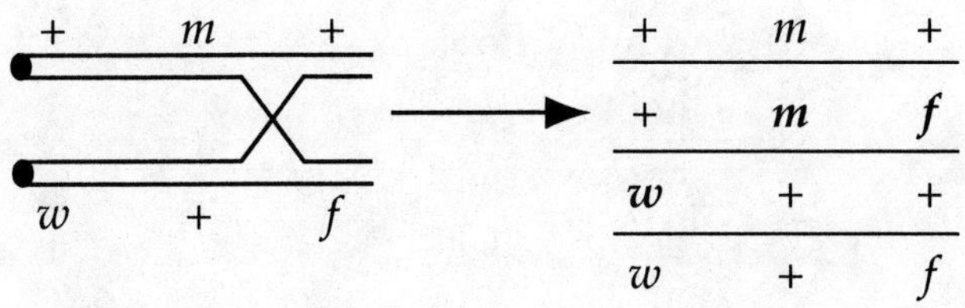

Double crossover:

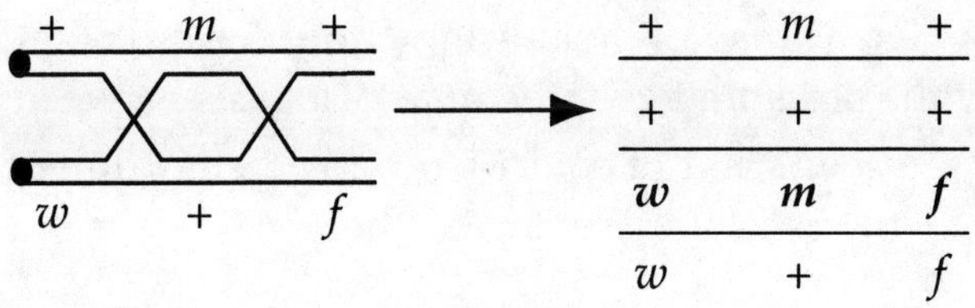

With this information, one can reevaluate the progeny and determine how much recombination occurred in each gene interval:

+	*m*	+	218	parental
w	+	*f*	236	parental
+	+	*f*	168	single crossover between *w, m*
w	*m*	+	178	single crossover between *w, m*
+	*m*	*f*	95	single crossover between *m, f*
w	+	+	101	single crossover between *m, f*
+	+	+	3	crossover between *w, m* and between *m, f*
w	*m*	*f*	1	crossover between *w, m* and between *m, f*
	Total		1,000	

Number of recombinants in *w, m* interval: 168 + 178 + 3 + 1 = 350
Recombination frequency between *w, m* = (350/1,000) × 100% = 35%

Number of recombinants in *m, f* interval: 95 + 101 + 3 + 1 = 200
Recombination frequency between *m, f* = (200/1,000) × 100% = 20%

This information can be represented in the following map:

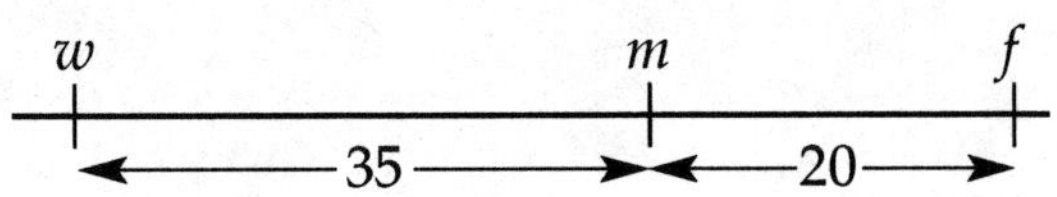

5.13 Three of the many recessive mutations in *Drosophila melanogaster* that affect body color, wing shape, or bristle morphology are black (*b*) body versus grey in the wild type; dumpy (*dp*), obliquely truncated wings versus long wings in the wild type; and hooked (*hk*) bristles at the tip versus not hooked in the wild type. From a cross of a dumpy female with a black, hooked male, all the F_1 were wild-type for all three characters. The testcross of an F_1 female with a dumpy, black, hooked male gave the following results:

wild type	169
black	19
black, hooked	301
dumpy, hooked	21
hooked	8
hooked, dumpy, black	172
dumpy, black	6
dumpy	304
Total	1,000

a. Construct a genetic map of the linkage group (or groups) these genes occupy. If applicable, show the order and give the map distances between the genes.
b. Determine the coefficient of coincidence for the portion of the chromosome involved in the cross. How much interference is there?

Answer:

a. First, use the symbols for the genes to write out the crosses that were done. It does not matter, at this point, what order you assign to the genes. Just be consistent in the order you choose so that you don't get confused.

P: + + *dp*/+ + *dp* (female) × *b hk* +/*b hk* + (male)
F_1 testcross: + + *dp*/*b hk* + (female) × *b hk dp*/*b hk dp* (male)

Now rewrite the data to reflect parental-type, single-crossover (SCO), and double-crossover (DCO) classes (based on frequency of each class).

Phenotype	**Gamete Genotype**	**Number**	**Class**
dumpy	+ + *dp*	304	parental
black, hooked	*b hk* +	301	parental
hooked, dumpy, black	*b hk dp*	172	SCO
wild type	+ + +	169	SCO
dumpy, hooked	+ *hk dp*	21	SCO
black	*b* + +	19	SCO
hooked	+ *hk* +	8	DCO
dumpy, black	*b* + *dp*	6	DCO

Now determine the gene order by comparing one of the double crossovers to one of the parental types (see solution to problem 5.12):

parental-type genotype:	*b*	*hk*	*dp*
DCO genotype:	+	*hk*	+
	same	different	different

The *b* gene is unlike the other two, so it is in the middle. The correct gene order is *dp* – *b* – *hk*. Continue to use this gene order. The F_1 female that was testcrossed was *dp* + +/+ *b hk*. Consequently, the *dp b hk* and + + + progeny are single crossovers between *dp* and *b*, while the *dp* + *hk* and + *b* + progeny are single crossovers between *b* and *hk*. Recombination frequencies (RF) can be calculated as

$$\text{RF}(dp - b) = [(172 + 169 + 6 + 8)/1{,}000] \times 100\% = 35.5\%$$
$$\text{RF}(b - hk) = [(21 + 19 + 6 + 8)/1{,}000] \times 100\% = 5.4\%$$

(Remember to include the double crossovers in this calculation, as each double crossover has a crossover in the interval being considered.) The map distance between *dp* and *b* is 35.5 mu, and the map distance between *b* and *hk* is 5.4 mu. The map distance between *dp* and *hk* is 35.5 + 5.4 = 40.9 mu.

b. The coefficient of coincidence (c.o.c.) is

$$\begin{aligned} \text{c.o.c.} &= \text{frequency observed doubles/frequency expected doubles} \\ &= (14/1{,}001)/(0.355 \times 0.054) \\ &= 0.014/0.01917 \\ &= 0.73 \end{aligned}$$

$$\begin{aligned} \text{Interference} &= 1 - \text{c.o.c.} \\ &= 1 - 0.73 \\ &= 0.27 \end{aligned}$$

5.14 Two normal-looking *Drosophila* flies are crossed and yield the following phenotypes among the progeny:

Females:				2,000
Males:	+	+	+	3
	a	*b*	*c*	1
	+	*b*	*c*	839
	a	+	+	825
	a	*b*	+	86
	+	+	*c*	90
	a	+	*c*	81
	+	*b*	+	75
			Total	4,000

Give parental genotypes, gene arrangement in the female parent, map distances, and the coefficient of coincidence.

Answer: The differential appearance of the traits in males and females indicates that the traits are sex-linked (i.e., they are X-linked). The female progeny are normal, since they received their father's X chromosome, which was + + +/Y. As they are phenotypically normal, they are not helpful in a mapping analysis. The male progeny received an X chromosome from the mother and therefore can be used to analyze the results of recombination in the mother.

The most frequent progeny classes in the males are + *b c* and *a* + +. These are the parental-type chromosomes. The least frequent classes (the double-crossover classes) are + + + and *a b c*. Comparison of these two classes (see the solutions to problems 5.11 and 5.12) indicates that gene *a* is in the middle, and the correct order is *b* – *a* – *c* (or *c* – *a* – *b*). The parental genotypes (with the correct gene order) were *c* + *b* / + *a* + (female) × +++/Y (male). Now analyze the data to determine which progeny classes result from crossing-over in each gene interval and calculate the recombination frequencies in each interval. (SCO = single crossover; DCO = double crossover.)

Gamete Genotype	**Number**	**Class**
c + *b*	839	parental
+ *a* +	825	parental
+ *a* *b*	86	SCO (*a* – *b*)
c + +	90	SCO (*a* – *b*)
c *a* +	81	SCO (*c* – *a*)
+ + *b*	75	SCO (*c* – *a*)
c *a* *b*	1	DCO (*a* – *b*, *c* – *a*)
+ + +	3	DCO (*a* – *b*, *c* – *a*)
Total	2,000	

Note that, since recombination cannot be scored in the female progeny, and only the 2,000 male progeny are being considered, the total that is used as the divisor in the calculation of recombination frequency (RF) is 2,000:

RF (*a* – *b*) = [(86 + 90 + 1 + 3)/2,000] × 100% = 9.0%

RF (*c* – *a*) = [(81 + 75 + 1 + 3)/2,000] × 100% = 8.0%

With this information, draw the following map:

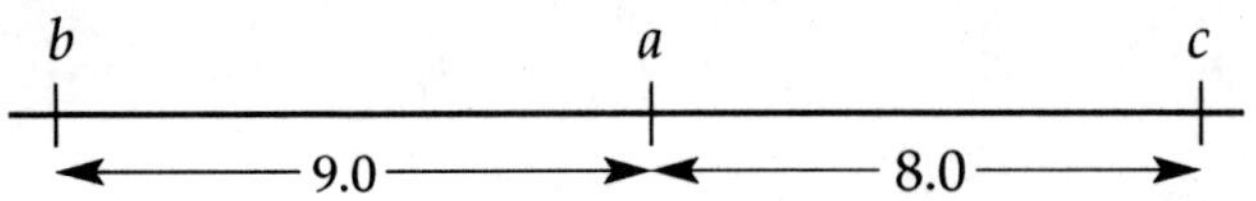

To calculate the coefficient of coincidence (c.o.c.), compare the frequency of actual double crossovers to that expected based on the crossover frequency observed in each interval:

$$
\begin{aligned}
\text{c.o.c.} &= \text{observed DCO frequency/expected DCO frequency} \\
&= [(3 + 1)/2{,}000]/(0.09 \times 0.08) \\
&= 0.002/0.0072 \\
&= 0.28
\end{aligned}
$$

5.15 The following questions make use of this genetic map:

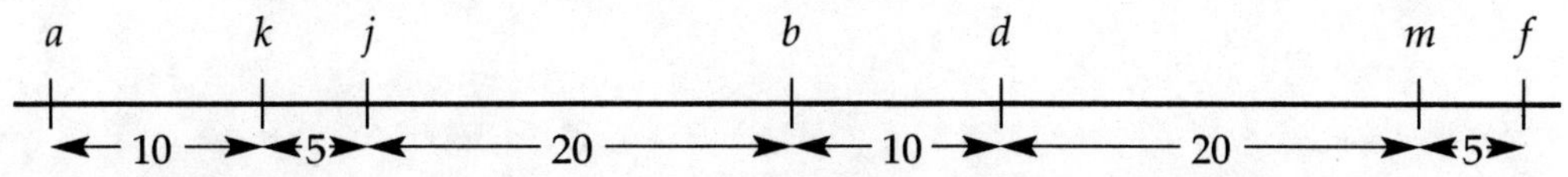

a. Calculate the frequency of *j b* gametes from a *J B* / *j b* genotype.
b. Calculate the frequency of *A M* gametes from an *a M* / *A m* genotype.
c. Calculate the frequency of *J B D* gametes from a *j B d* / *J b D* genotype.
d. Calculate the frequency of *J B d* gametes from a *j B d* / *J b D* genotype.
e. Calculate the frequency of *j b d* / *j b d* genotypes in a *j B d* / *J b D* × *j B d* / *J b D* mating.
f. Calculate the frequency of *A k F* gametes from an *A K F* / *a k f* genotype.

Answer:

a. There are 20 map units between *j* and *b*, so there will be 20 percent recombinant-type and 80 percent parental-type gametes. In a *J B/j b* parent, *j b* gametes are half of the parental-type gametes, or 40 percent of the total gametes.
b. There are 65 map units between *A* and *M*, so these loci will show independent assortment, as the frequency of recombinant-type gametes cannot exceed 50 percent. In an *a M/ A m* parent, the *A M* gametes are half of the recombinant-type gametes. There will be 25 percent *A M* gametes.
c. In a *j B d* / *J b D* individual, *J B D* gametes are produced by a double crossover: a crossover in the interval *j* – *b* and a crossover in the interval *b* – *d*. As there are 20 map units between *j* and *b* and 10 map units between *b* and *d*, the frequency of double crossovers is expected to be 0.20 × 0.10 × 100% = 2%. *J B D* gametes are half of the double crossovers produced, so will be seen 1 percent of the time.
d. In a *j B d/ J b D* individual, *J B d* gametes are one-half of the gametes produced by a single crossover in the interval *j* – *b*. Since there are 20 map units between these genes, 10 percent of the gametes will be *J B d*.
e. By the reasoning in (c), *j b d* gametes will be seen 1 percent of the time in a *j B d/J b D* individual. To obtain a *j b d/j b d* genotype, one must obtain such gametes from both parents. The frequency of this is (0.01 × 0.01) × 100% = 0.01%.
f. Based on the map distances, one expects 10 percent recombination between *a* and *k* and 50 percent recombination between *k* and *f*. (Again, note that even though *k* and *f* are more than 50 map units apart, one observes only 50 percent recombination between these two genes.) In an *A K F/a k f* individual, an *A k F* gamete results from a double crossover: one crossover between *a* and *k* (10 percent chance) and one crossover between *k* and *f* (50 percent chance). The chance of both crossovers occurring simultaneously is 0.10 × 0.50 = 0.05, or 5%. Since *A k F* gametes are half of the double crossovers produced, they will be 2.5 percent of the total.

5.16 A female *Drosophila* carries the recessive mutations *a* and *b* in repulsion on the X chromosome (she is heterozygous for both). She is also heterozygous for an X-linked recessive lethal allele, *l*. When she is mated to a true-breeding, normal male, she yields the following progeny:

Females:	1,000	+	+
Males:	405	*a*	+
	44	+	*b*
	48	+	+
	2	*a*	*b*

Draw a chromosome map of the three genes in the proper order and with map distances as nearly as you can calculate them.

Answer: First, consider what happens when a female that is heterozygous for an X-linked lethal is crossed to a normal male. One can diagram such a cross as

P: *l*/+ female × +/Y male

F_1: females: *l*/+ and +/+ (phenotypically normal)
males: +/Y (normal) and *l*/Y (dead, not recovered)

In such a cross, one-half of the male progeny are not recovered due to the presence of the lethal allele. Only progeny bearing the + allele are recovered. Thus, the 499 males that are recovered in this cross are half of the expected male progeny.

Second, consider that the lethal-bearing chromosome can be recovered in the female progeny. Since the lethal allele is a recessive mutation, it is "rescued" by the normal + allele contributed to the female progeny by the father's X chromosome. However, as the father's X chromosome bears the + alleles of each gene, the females are phenotypically normal and are not helpful in a recombination analysis.

Now, analyze the phenotypes of the male progeny that are recovered to infer map distances and the gene order. Since one-half of the male progeny are not recovered, each of the four classes seen represents one of the two reciprocal classes of progeny recovered in a three-point cross. Include the third (*l*/+) locus and assign each genotype to a progeny class based on its frequency:

Male progeney:	405	*a*	+	+	parental type
	44	+	*b*	+	single crossover
	48	+	+	+	single crossover
	2	*a*	*b*	+	double crossover

Comparison of the parental and double-crossover classes indicates that *b* is in the middle, and the correct order is *a* – *b* – *l* (see solution to problem 5.12 for determining the middle locus). Since one of the parental-type chromosomes was *a* + +, the other must have been + *b l*, the reciprocal. The heterozygous female was therefore *a* + +/+ *b l*.

The 44 + *b* + progeny are obtained from gametes arising from single crossovers between *b* and *l*. They are half of the total single crossovers in that interval. The other half were not recovered, as they bore the *l* allele. Similarly, the 48 + + + progeny are half of the single crossovers in the interval *a* – *b*. This information can be used to calculate recombination frequencies (RF) and draw a map.

$$RF\ (a - b) = [(48 + 2)/499] \times 100\% = 10\%$$
$$RF\ (b - l) = [(44 + 2)/499] \times 100\% = 9.2\%$$

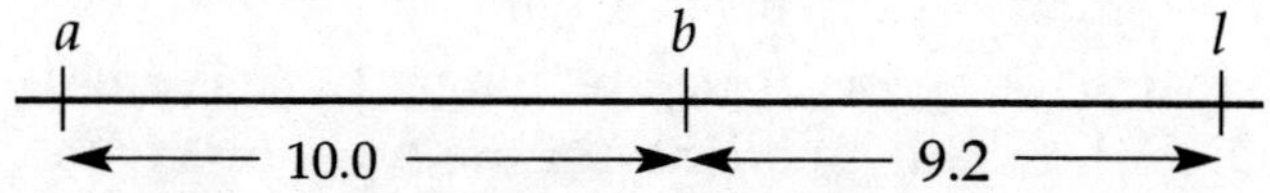

5.17 A farmer who raises rabbits wants to break into the Easter market. He has stocks of two true-breeding lines. One is hollow and long-eared but not chocolate, and the second is solid, short-eared, and chocolate. Hollow (*h*), long ears (*le*), and chocolate (*ch*) are all recessive, autosomal, and linked as in the following map:

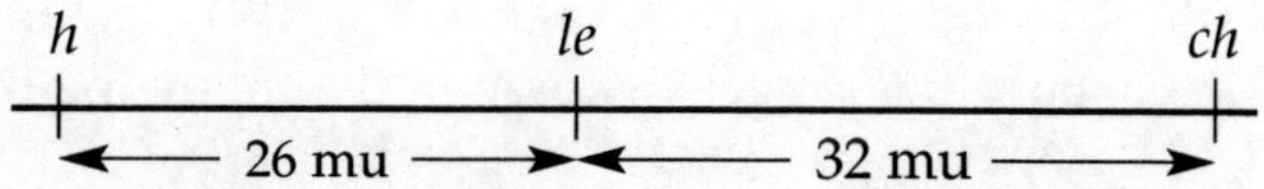

The farmer can generate a trihybrid by crossing his two lines, and at great expense he is able to obtain the services of a male homozygous recessive at all three loci to cross with his F_1 females.

The farmer has buyers for both solid and hollow bunnies; however, all must be chocolate and long-eared. Assuming that interference is zero, if he needs 25 percent of the progeny of the desired phenotypes to be profitable, should he continue with his breeding? Calculate the percent of the total progeny that will be the desired phenotypes.

Answer: The crosses under consideration are

P: *h le Ch* /*h le Ch* × *H Le ch* /*H Le ch*

F_1: *h le Ch* /*H Le ch* × *h le ch* /*h le ch*

To be profitable, the farmer needs 25 percent of the progeny to be either *h le ch/h le ch* or *H le ch/h le ch*. Hence, he is concerned with the probability of obtaining a *h le ch* or a *H le ch* gamete from an *h le Ch/H Le ch* individual. Gametes that are *h le ch* arise from a single crossover between *le* and *ch*. Since there are 32 mu between *le* and *ch*, there will be 16 percent *h le ch* individuals (one-half of the recombinants between *le* and *ch*). Gametes that are *H le ch* arise from a crossover in each interval. Since there is no interference, their frequency will be half of the expected double-crossover frequency, or (0.5 × 0.32 × 0.26) × 100% = 4.16%. The expected frequency of *h le ch* and *H le ch* gametes is 4.16% + 16% = 20.16%. The farmer should stop breeding and cut his losses.

5.18 In *Drosophila,* many different mutations have been isolated that affect a normally brick-red eye color caused by the deposition of brown and bright-red pigments. Two X-linked recessive mutations are *w* (white eyes, map position 1.5) and *cho* (chocolate-brown eyes, map position 13.0), with *w* epistatic to *cho.*

a. A white-eyed female is crossed with a chocolate-eyed male, and the normal, red-eyed F_1 females are crossed with either wild-type or white-eyed males. Determine the frequency of the progeny types produced in each cross.

b. The recessive mutation *st* causes scarlet (bright-red) eyes, and maps to the third chromosome at position 44. Mutant flies with only *st* and *cho* alleles have white eyes, and *w* is epistatic to *st.* Suppose a true-breeding *w* male is crossed with a true-breeding *cho, st* female. Determine the frequency of the progeny types you would expect if the F_1 females are crossed with true-breeding scarlet-eyed males.

Answer:

a. The cross is *w* +/*w* + (white female) × + *cho*/Y (chocolate male), and the F_1 females are *w* +/+ *cho* (red). Since the *w* and *cho* genes are 11.5 map units apart, the F_1 females will have 11.5 percent recombinant-type gametes (5.75% each + + and *w cho*) and 88.5 percent parental-type gametes (44.25% each *w* + and + *cho*). When these gametes are combined with those of wild-type (50% each + + and Y) or white-eyed (50% each *w* + and Y) males, one obtains the progeny shown in the following branch diagrams.

Cross: *w* + / + *cho* × + + /Y

Gametes of *w* +/+ *cho*	Gametes of + +/Y	Progeny
0.4425 *w* +	0.5 + +	→ 0.22125 *w* +/+ + (red female)
	0.5 Y	→ 0.22125 *w* +/Y (white female)
0.4425 + *cho*	0.5 + +	→ 0.22125 + *cho*/+ + (red female)
	0.5 Y	→ 0.22125 + *cho*/Y (chocolate male)
0.0575 + +	0.5 + +	→ 0.02875 + +/+ + (red female)
	0.5 Y	→ 0.02875 + +/Y (red male)
0.0575 *w cho*	0.5 + +	→ 0.02875 *w cho*/+ + (red female)
	0.5 Y	→ 0.02875 *w cho*/Y (white male)

Results: 50% red females; 2.875% red males; 25% white males; 22.125% chocolate males

Gametes of *w* +/+ *cho*	Gametes of *w* +Y		Progeny
0.4425 *w* +	0.5 + +	→	0.22125 *w* +/*w* + (white female)
	0.5 Y	→	0.22125 *w* +/Y (white male)
0.4425 + *cho*	0.5 *w* +	→	0.22125 + *cho*/*w* + (red female)
	0.5 Y	→	0.22125 + *cho*/Y (chocolate male)
0.0575 + +	0.5 + +	→	0.02875 + +/*w* + (red female)
	0.5 Y	→	0.02875 + +/Y (red male)
0.0575 *w cho*	0.5 *w* +	→	0.02875 *w cho*/*w* + (white female)
	0.5 Y	→	0.02875 *w cho*/Y (white male)

Results:	25% red females	2.875% red males
	25% white females	25% white males
		22.125% chocolate males

b. The initial cross is *w* + /Y; +/+ × + *cho*/ + *cho*; *st*/*st* with F_1 females being *w* +/+ *cho*; +/*st*. These trihybrid females will have eight types of gametes: 88.5 percent will be parental types (22.125% each of (*w* +; +), (*w* +; *st*), (+ *cho*; +), (+ *cho*; *st*)), and 11.5 percent will be recombinants between *w* and *cho* (2.875% each of (+ +; +), (+ +; *st*), (*w cho*; +), (*w cho*; *st*)). The results obtained when the F_1 female is crossed to a true-breeding scarlet male (with (+ +; *st*) and (Y; *st*) gametes) are shown in the following branch diagrams.

Cross: *w* + / + *cho*; *st*/ + × + ;+ /Y; *st*/*st*

Gametes of *w* +/+ *cho*; *st*/+	Gametes of + +/Y/*st*		Progeny
0.2215 *w* +	0.5 + +; *st*	→	0.110625 *w* +/+ +; +/*st* (red female)
	0.5 Y; *st*	→	0.110625 *w* +/Y; +/*st* (white male)
0.2215 + *cho*	0.5 *w* +; *st*	→	0.110625 *w* +/+ +; *st*/*st* (scarlet female)
	0.5 Y; *st*	→	0.110625 *w* +/Y; *st*/*st* (white male)
0.2215 + +	0.5 + +; *st*	→	0.110625 + *cho*/+ +; +/*st* (red female)
	0.5 Y; *st*	→	0.110625 + *cho*/Y; +/*st* (chocolate male)

(continued)

Gametes of		
w +/+ *cho; st*/+	+ +/Y/*st*	Progeny
0.2215 *w cho*	0.5 *w* +; *st*	→ 0.110625 + *cho*/+ +; *st*/*st* (scarlet female)
	0.5 Y; *st*	→ 0.110625 + *cho*/Y; *st*/*st* (white male)
0.02875 *w* +	0.5 + +; *st*	→ 0.014375 + +/+ +; +/*st* (red female)
	0.5 Y; *st*	→ 0.014375 + +/Y; +/*st* (red male)
0.02875 + *cho*	0.5 *w* +; *st*	→ 0.014375 + +/*w* +; *st*/*st* (scarlet female)
	0.5 Y; *st*	→ 0.014375 + +/Y; *st*/*st* (scarlet male)
00.02875 + +	0.5 + +; *st*	→ 0.014375 *w cho*/+ +; +/*st* (red female)
	0.5 Y; *st*	→ 0.014375 *w cho*/Y; +/*st* (white male)
0.02875 *w cho*	0.5 *w* +; *st*	→ 0.014375 *w cho*/+ +; *st*/*st* (scarlet female)
	0.5 Y; *st*	→ 0.014375 *w cho*/Y; *st*/*st* (white male)

Results:	25% red females	1.4375% red males
	25% scarlet females	1.4375% scarlet males
		36.0625% white males
		11.0625% chocolate males

5.19 In *Saccharomyces, Neurospora,* and *Chlamydomonas,* what meiotic events give rise to PD, NPD, and T tetrads?

Answer: It is useful here to visualize the chromosomes in a tetrad at the time of crossing-over, and refer to text Figure 5.14, p. 114. PD, or parental ditypes, can arise from tetrads in which no crossovers have taken place between the markers being observed, as well as from tetrads in which a double crossover has taken place between the *same* two nonsister chromatids, called a two-strand double crossover. NPD, or nonparental ditypes, also arise from tetrads with double crossovers between the two markers. With NPD tetrads, each of the two crossovers occurs on a separate pair of nonsister chromatids, called a four-strand double. T, or tetratype, tetrads, occur from either a single crossover between any pair of nonsister chromatids or a double crossover involving three chromatids, called a three-strand double.

5.20 Double exchanges between two loci can be of several types, called two-strand, three-strand, and four-strand doubles.

a. Four recombination gametes would be produced from a tetrad in which the first of two exchanges is depicted in the following figure. Draw in the second exchange.

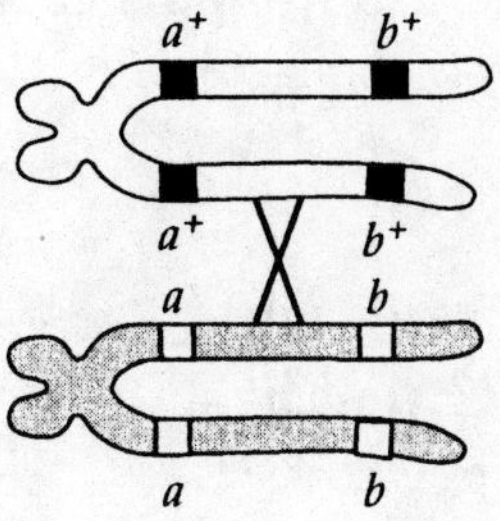

b. In the following figure, draw in the second exchange so that four nonrecombination gametes would result.

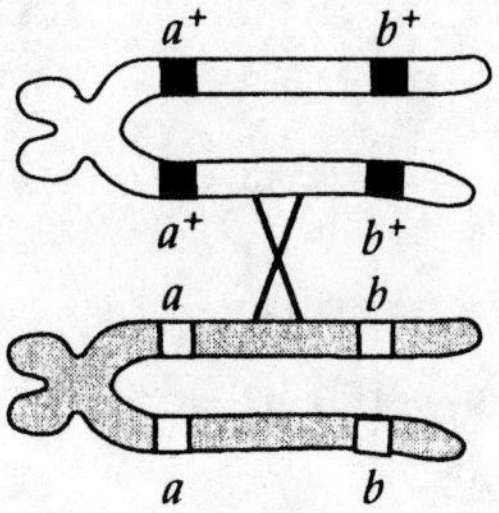

Answer: A solution to parts (a) and (b) can follow from trying to diagram a number of different crossovers and their resolution (by following a chromosome "through" the crossover from left to right). It is also possible to reason through these as follows: In part (a), four recombinant gametes are needed. The crossover shown occurs between two nonsister chromatids and gives two of these. A crossover between the remaining two nonsister chromatids would give two more, making a total of four. In part (b), four nonrecombinant gametes are needed. The crossover shown gives two recombinant gametes, and so a second crossover must be drawn between these same two nonsister chromatids to change them back to nonrecombinant-type chromosomes. Shown below is a diagram of a four-strand double crossover, the solution to (a), and a two-strand double crossover, the solution to (b), with resolution for the crossover events. Note that the diagram shown in (b) is equivalent to a two-strand double crossover between any two nonsister chromatids.

(a)

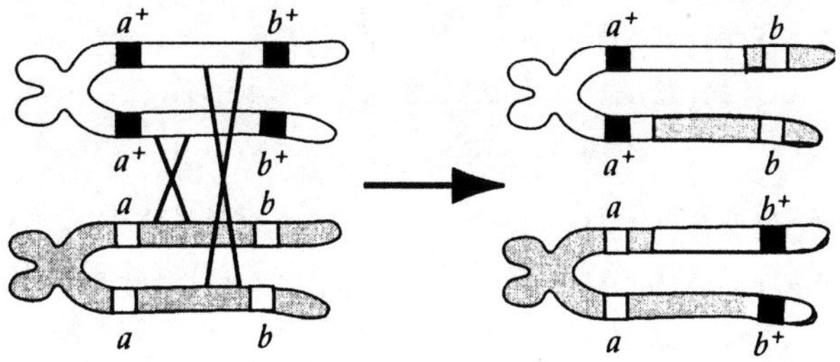

(b)

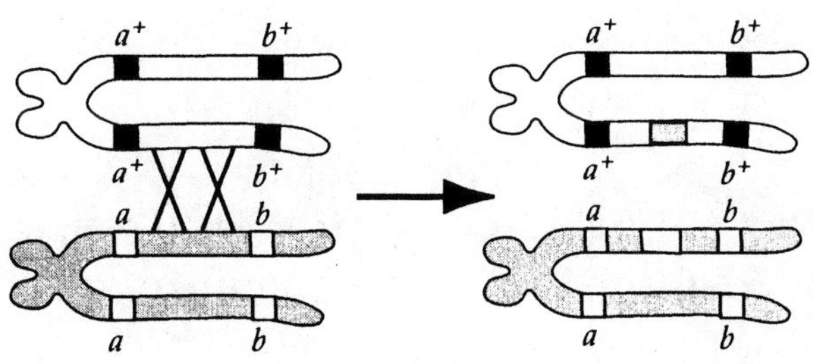

5.21 The following asci were obtained from the cross *leu* + × + *rib* in yeast. Draw the linkage map and determine the map distance.

110		45		6		39	
leu	+	*leu*	*rib*	+	+	*leu*	+
+	*rib*	*leu*	+	*leu*	*rib*	+	*rib*
leu	+	+	+	*leu*	*rib*	+	+
+	*rib*	+	*rib*	+	+	*leu*	*rib*

Answer: Categorizing these unordered asci shows that there are 110 PD, 6 NPD, and 45 + 39 = 84 T tetrads. Since PD >> NPD, the loci are linked. The percent recombination between them is given by

$$\begin{aligned} RF &= [\tfrac{1}{2}\,T + NPD)/(\text{total \# asci})] \times 100\% \\ &= [[\tfrac{1}{2} \times 84) + 6]/(110 + 6 + 45 + 39)] \times 100\% \\ &= (48/200) \times 100\% \\ &= 24\% \end{aligned}$$

The map distance between *leu* and *rib* is 24 mu.

5.22 The genes *a*, *b*, and *c* are linked in *Neurospora crassa*. The following asci were obtained from the cross *a b* + × + + *c*.

45	5	146	1	10	20	15	58
a b +	*a b* +	*a b* +	*a b* +	*a b* +	*a b* +	*a b* +	*a b* +
+ *b c*	*a* + +	*a b* +	+ + +	*a* + *c*	+ + *c*	*a b c*	+ *b* +
a + +	+ *b c*	+ + *c*	*a b c*	+ *b* +	*a b* +	+ + +	*a* + *c*
+ + *c*	+ + *c*	+ + *c*	+ + *c*	+ + *c*	+ + *c*	+ + *c*	+ + *c*

Determine the correct gene order. Calculate all gene-gene distances.

Answer: To solve this problem, consider only one pair of loci at a time, determine the gene-gene and gene-centromere distances for that pair, and then relate the findings for both pairs of loci. For the *a* and *b* loci, one can assign the type of tetrad as shown below:

45		5		146		1		10		20		15	
T		T		PD		PD		T		PD		PD	
a	*b*	*a*	*b*	*a*	*b*	*a*	*b*	*a*	*b*	*a*	*b*	*a*	*b*
+	*b*	*a*	+	*a*	*b*	+	+	*a*	+	+	+	*a*	*b*
a	+	+	*b*	+	+	*a*	*b*	+	*b*	*a*	*b*	+	+
+	+	+	+	+	+	+	+	+	+	+	+	+	+

For the cross *a b* × + +, then, there are 146 + 1 + 20 + 15 = 182 PD, 45 + 5 + 10 + 58 = 118 T, and 0 NPD tetrads. The recombination frequency between them is given by

$$\begin{aligned} RF &= [(\tfrac{1}{2}\,T + NPD)/\text{total \# asci}] \times 100\% \\ &= [(\tfrac{1}{2} \times 118)/300] \times 100\% \\ &= 19.7\% \end{aligned}$$

For the cross *b* + × + *c*, we have

45	5	146	1	10	20	15	58
T	T	PD	T	PD	PD	T	PD
b +	*b* +	*b* +	*b* +	*b* +	*b* +	*b* +	*b* +
b c	+ +	*b* +	+ +	+ *c*	+ *c*	*b c*	*b* +
+ +	*b c*	+ *c*	*b c*	*b* +	*b* +	+ +	+ *c*
+ *c*	+ *c*	+ *c*	+ *c*	+ *c*	+ *c*	+ *c*	+ *c*

There are 146 + 10 + 20 + 58 = 234 PD, 45 + 5 + 1 + 15 = 66 T, and 0 NPD tetrads. The recombination frequency between them is given by

$$\begin{aligned} RF &= [(\tfrac{1}{2} \times 66)/300] \times 100\% \\ &= 11\% \end{aligned}$$

For the cross *a* + × + *c*, we have

45	5	146	1	10	20	15	58
PD	PD	PD	T	T	PD	T	T
a +	*a* +	*a* +	*a* +	*a* +	*a* +	*a* +	*a* +
+ *c*	*a* +	*a* +	+ +	*a c*	+ *c*	*a c*	+ +
a +	+ *c*	+ *c*	*a c*	+ +	*a* +	+ +	*a c*
+ *c*	+ *c*	+ *c*	+ *c*	+ *c*	+ *c*	+ *c*	+ *c*

There are 45 + 5 + 146 + 20 = 216 PD, 1 + 10 + 15 + 58 = 84 T, and 0 NPD tetrads. The recombination frequency between them is given by

$$\begin{aligned} RF &= [(\tfrac{1}{2} \times 84)/300] \times 100\% \\ &= 14\% \end{aligned}$$

Using the calculations above, *a* – *b* is 19.7 map units, *a* – *c* is 14 map units, and *b* – *c* is 11 map units. Thus, the gene order is *b* – *c* – *a*, and one has the following map (taking into consideration that, due to increasing numbers of multiple crossovers, more accurate RFs are obtained over smaller intervals):

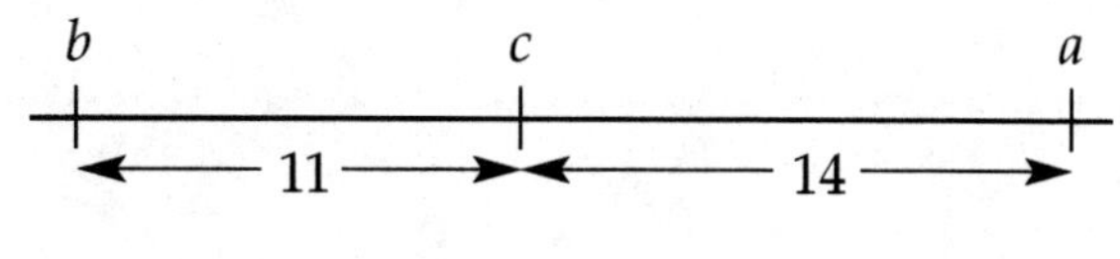

6

GENE MAPPING IN BACTERIA AND BACTERIOPHAGES

CHAPTER OUTLINE

REVIEW OF KEY TERMS, SYMBOLS, AND CONCEPTS

In your own words, write a brief, precise definition of each term in the groups on the following page. Check your definitions using the text. Then develop a concept map using the terms in each list.

THINKING ANALYTICALLY

While this chapter focuses on the genetics of bacteria and bacteriophages, many of the core concepts introduced here are used throughout the study of genetics. Of particular importance are the genetic principles that underlie selection, recombination, and complementation. These have had a major impact on how we think about gene structure and function: Geneticists working in many

1	2	3
transfer of genetic material	transfer of genetic material	bacteriophage cross
conjugation	transduction, cotransduction	turbid, clear plaques
plating	phage vectors	permissive host
colony	transductants	nonpermissive host
titer	phage lysate	bacterial lawn
minimal medium	virulent vs. temperate phage	host range property
complete medium	lysogenic pathway	intergenic mapping
auxotroph	prophage	"beads-on-a-string" view
prototroph	lysogeny	intragenic mapping
replica plating	lytic cycle	fine-structure mapping
transconjugants, exconjugant	generalized transduction	homoallelic mutations
sex factor	transducing phage	heteroallelic mutations
F^-, F', *Hfr*, F^+ strains	selected, unselected marker	point, deletion, revertant mutations
plasmid, episome	transformation	hot spot
F-duction (sexduction)	natural, engineered transformation	*cis-trans* test
merodiploid	heteroduplex DNA	complementation test
interrupted mating	cotransformation	cistron
donor vs. recipient		gene as a unit of function
F-pili (sex-pili)		
F factor origin		

different research areas are concerned with how to *select* for or against specific phenotypes; analysis of recombination within a gene (*intragenic* recombination) formed a foundation for our current understanding of gene structure; the concept of a gene as a unit of function was developed by using complementation tests to measure whether two mutants affect the same function. It is therefore essential to thoroughly understand these genetic principles as they are developed using bacterial and phage systems in this chapter.

Having stressed the importance of the concepts presented in this chapter, it is also important to be forewarned that much of the material used to develop these concepts is complicated and will not necessarily be intuitive. For example, there are many refinements of bacteria and phage types and special characteristics that are associated with one or another aspect of bacterial or phage growth and mating. Carefully read the chapter sections that describe these characteristics until you thoroughly understand them. The names are quite descriptive, and once you use them a few times in solving problems, you will become adept at using them. Approach using the terms with more than rote memorization. Use them in context until you fix upon what is practically a visual image of them.

As with the preceding chapters, carefully read the statements and conditions of a particular problem so that you gain a clear understanding of what it asks and what data it gives you to work with. Then organize and reorganize the data, and sometimes reorganize again, using scratch paper and pencil until you can see how to use the data to solve the problem. If you still get stuck, try representing the processes that are used in the problem with diagrams, and ask yourself how the data fit into the biological process you've drawn out.

The problems in this chapter are especially well organized to help you build up your understanding of the material. Work through each of them sequentially to fully develop your understanding and analytical skills.

QUESTIONS FOR PRACTICE

Multiple Choice Questions

1. A strain, such as a strain of *E. coli*, that requires nutritional or other kinds of supplements for growth and/or survival is known as
   a. a prototroph
   b. a heterotroph
   c. a pleiotroph
   d. an auxotroph
2. A conjugating bacterial cell that typically transfers only part of its *F* factor and some chromosomal genes is
   a. F^+
   b. F'
   c. *Hfr*
   d. F^-
3. The process by which cells take up genetic material from the extracellular environment and incorporate it into their genetic complement is called
   a. transduction
   b. transformation
   c. translocation
   d. DNA transfusion
4. In bacteria, conjugation involves
   a. the union of two bacterial genomes
   b. the fusion of two cells of opposite mating types
   c. a virus-mediated exchange of DNA
   d. the transfer of DNA from one cell into another
5. A self-replicating genetic element found in the cytoplasm of bacteria is a
   a. sex-pilus
   b. plasmid
   c. viral capsid
   d. cotransductant
6. Which of the following is true about cells having *F* factors?
   a. In both F^+ and *Hfr* cells, fertility genes are transferred first.
   b. In both F^+ and *Hfr* cells, genes are transferred through sex-pili.
   c. In *Hfr* cells, bacterial genes closest to the origin are transferred first.
   d. In F' cells, only *F* factor genes are transferred.
7. Phage that can grow using either a lytic or lysogenic pathway is called
   a. virulent
   b. temperate
   c. intemperate
   d. transductant

8. Recombination that occurs between two alleles at one gene is
   a. unheard of
   b. intergenic recombination
   c. intragenic recombination
   d. more frequent than recombination between two closely linked genes
9. A complementation test measures
   a. whether two mutants undergo intergenic recombination
   b. whether two mutants undergo intragenic recombination
   c. whether two mutants are linked
   d. whether two mutants affect the same function
10. *E. coli* K12(λ) cells will only support the growth of T4 phage that are r^+. If two different mutant *rII* phage coinfect an *E. coli* K12(λ) cell and lysis occurs,
   a. intragenic recombination must have occurred
   b. complementation must have occurred
   c. the two *rII* phages must have mutations in different cistrons
   d. both a and b are true
   e. both b and c are true
   f. a, b, and c are true
11. Which of the statements accurately describe an episome and a plasmid?
   a. An episome is a plasmid that can integrate into the host chromosome.
   b. A plasmid is a circular extrachromosomal element.
   c. An *F* factor is a plasmid that is also an episome.
   d. All episomes are *F* factors, but not all *F* factors are plasmids.
   e. a, b, and c, but not d, are accurate
   f. a, b, c, and d are accurate

Answers: 1d, 2c, 3b, 4d, 5b, 6c, 7b, 8c, 9d, 10e, 11d

Thought Questions

1. Explain why two genes that are close together on a chromosome show a high frequency of cotransduction and cotransformation.
2. Compare the life cycles of T4, P1 and λ. Which is (are) temperate? Which is (are) virulent? Which can undergo lysogeny?
3. What is the significance of bacterial transformation and conjugation in relation to the rapidity with which certain infectious organisms adapt to changing environmental conditions or factors, such as antibiotics? (Hint: Sexual reproduction promotes genetic variability. How do parasexual systems compare?)
4. Explain how intragenic recombination is different from complementation.
5. Consider how we may select for prototrophic and drug-resistant bacterial strains. Can you think of clever (or even not so clever) ways to attempt to select for (a) a pesticide-resistant strain of beetles, (b) bacteria that are able to degrade a toxic chemical, (c) very sweet sweet-corn, and (d) a strain of cats that doesn't respond to catnip?
6. When two true-breeding white-flowered sweet peas are crossed, a purple F_1 is produced. When the F_1 is selfed, the F_2 is 9 purple : 7 white. As described in Chapter 4, the 9:7 ratio is

characteristic of duplicate recessive epistasis. Why is duplicate recessive epistasis also called complementary gene action?

7. Suppose a linear piece of DNA with genes (in order) a^+ b^+ c^+ enters a bacterial cell that is a^- b^- c^-. How many crossovers are required to convert the bacterial cell to (a) a^+ b^+ c^+, (b) a^+ b^+ c^-, (c) a^+ b^- c^+, (d) a^- b^+ c^-? How would your answers change if the DNA was circular? In each case, which would be the most frequent and the least frequent event?
8. P1 and λ phage have been modified to serve as cloning vectors by removing selected genes (thereby making room for DNA inserts) and introducing specific mutations to remove some functions. Carefully consider their life cycles, and speculate what features of their life cycles it might be useful to retain in a cloning vector and what features it might be necessary to remove?

Thought Questions for Media

After reviewing the media at the *Essential iGenetics* website, try to answer these questions.

1. The *Mapping Bacterial Genes by Conjugation* animation shows that the *leu*$^+$ and *thr*$^+$ genes are transferred about 8 minutes after conjugation begins but that we can't tell which one is transferred first. What does this tell you about the limits of resolution of mapping by using the interrupted mating technique?
2. What are the units used to relate the distance between two bacterial genes mapped by interrupted mating? Do you expect these units to be proportional to the number of DNA base pairs between the two loci?
3. Could you perform an interrupted mating experiment between an F^+ strain and an F^- strain? Why or why not?
4. Why is it necessary to use diploids to perform a complementation test?
5. Suppose you had two true-breeding mutant lines of deaf rabbits. How would you determine if the cause of deafness in each line was the same?
6. To assess if two genes complement each other, is it important to obtain recombinants between them?

1. Suppose that in the initial cross between the prototrophic strain to the F^- strain that was *leu*$^-$, *arg*$^-$, *his*$^-$, *and trp*$^-$, the media used for selection of *leu*$^+$ recombinants was prepared incorrectly. On plate A, streptomycin was mistakenly left out of the medium, and only supplemental arginine, histidine, and tryptophan were added. On plate B, streptomycin was added, but leucine was mistakenly added along with the other supplements (arginine, histidine, and tryptophan). Compared to all of the other plates used to select for recombinants at the other loci, you observe a much larger number of colonies on plates A and B.
   a. Why are there so many more colonies on plates A and B, compared to the others?
   b. Suppose you picked several dozen individual colonies from each of plates A and B. Would you expect the genotypes of all of the colonies from both plates to be the same? Would you expect the genotypes of all of the colonies from just one of the plates to be the same?
   c. Could any of the colonies on plates A and B be recombinants?
   d. What would you expect their genotype(s) to be?
   e. How could you experimentally address your answers to (c) and (d)?

2. As presented in the iActivity, multiple *Hfr* strains are used to map the relative location of genes distributed throughout the *E. coli* chromosome.
   a. Does each *Hfr* strain have one or more than one *F* factor inserted in its *E. coli* chromosome?
   b. If, in an *Hfr* strain, a bacterial gene is far from the *F* factor site of insertion, will it be transferred frequently to an F^- strain?
   c. Why are multiple *Hfr* strains needed to map genes distributed throughout the *E. coli* chromosome?
   d. When maps are being constructed, why doesn't it matter if the *F* factors in the different *Hfr* strains are inserted in opposite directions (relative to each other) in the *E. coli* chromosome?

SOLUTIONS TO TEXT PROBLEMS

6.1 In $F^+ \times F^-$ crosses, the F^- recipient is converted to a donor with very high frequency. However, it is rare for a recipient to become a donor in $Hfr \times F^-$ crosses. Explain why.

Answer: The frequency with which a recipient is converted to a donor reflects the frequency with which a complete *F* factor is transferred. In $F^+ \times F^-$ crosses, only the *F* factor is transferred, and this occurs relatively quickly. In $Hfr \times F^-$ crosses, transfer starts at the origin within the *F* element and then must proceed through the bacterial chromosome before reaching the *F* factor. For the entire *F* factor to be transferred, the whole chromosome would have to be transferred. This would take about 100 minutes, and usually, the conjugal unions break apart before then.

6.2 With the technique of interrupted mating, four *Hfr* strains were tested for the sequence in which they transmitted a number of different genes to an F^- strain. Each *Hfr* strain was found to transmit its genes in a unique order, as shown in the accompanying table (only the first six genes transmitted were scored for each strain).

Order of Transmission	*Hfr* Strain 1	2	3	4
First	*O*	*R*	*E*	*O*
	F	*H*	*M*	*G*
	B	*M*	*H*	*X*
	A	*E*	*R*	*C*
	E	*A*	*C*	*R*
Last	*M*	*B*	*X*	*H*

What is the gene sequence in the original strain from which these *Hfr* strains derive? Indicate on a diagram the origin and polarity of each of the four *Hfr* strains.

Answer: The diagram below shows the gene sequence in the original F^+ strain and the locations of the four different *F* factor insertions. (Note that only one insertion exists in a particular *Hfr* strain.)

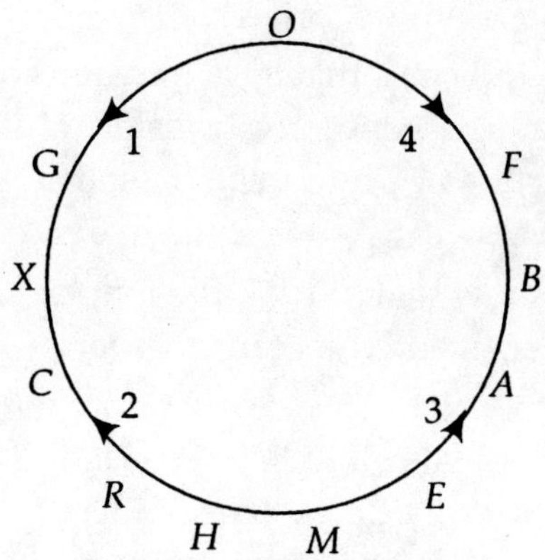

6.3 At time zero, an *Hfr* strain (strain 1) was mixed with an F^- strain, and at various times after mixing, samples were removed and agitated to separate conjugating cells. The cross may be written as

Hfr 1:	a^+	b^+	c^+	d^+	e^+	f^+	g^+	h^+	str^S
F^-:	a	b	c	d	e	f	g	h	str^R

(No order is implied in listing the markers.) The samples were then plated onto selective media to measure the frequency of h^+ str^R recombinants that had received certain genes from the *Hfr* cell. The graph of the number of recombinants against time is shown in the accompanying figure.

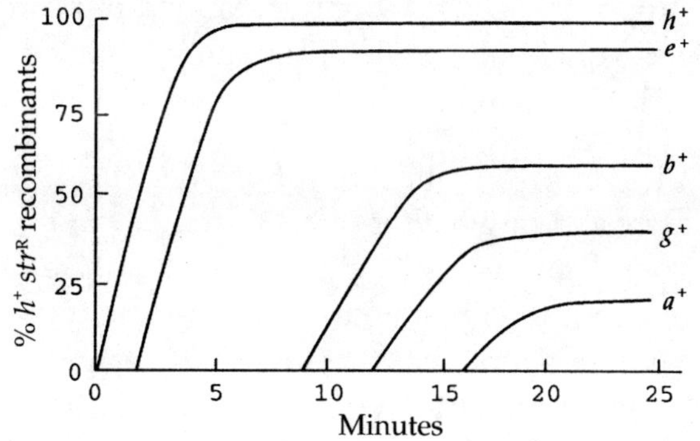

a. Indicate whether each of the following statements is true or false.
   i. All F^- cells that received a^+ from the *Hfr* in the chromosome transfer process must also have received b^+
   ii. The order of gene transfer from *Hfr* to F^- was a^+ (first), then g^+, then b^+, then e^+, then h^+.
   iii. Most e^+ str^R recombinants are likely to be *Hfr* cells.
   iv. None of the b^+ str^R recombinants plated at 15 minutes are also a^+.
b. Draw a linear map of the *Hfr* chromosome indicating
   i. the point of nicking (the origin) and direction of DNA transfer
   ii. the order of the genes a^+, b^+, e^+, g^+, and h^+
   iii. the shortest distance between consecutive genes on the chromosomes

Answer:

a. i. True. The graph indicates that a^+ is transferred last, starting about 16 minutes after conjugal pairing. The genes are transferred linearly in the order h^+ (starting at 0 minutes), e^+ (starting at 2 minutes), b^+ (starting at 9 minutes), g^+ (starting at 12 minutes), and then a^+.

ii. False. The data show that a^+ took the longest time to be transferred and so was the last gene to be received. The correct order of gene transfer was h^+ (first), e^+, b^+, g^+, and a^+ (last).

iii. False. For the recipient cell to be converted to F^+ or *Hfr*, it must receive a complete copy of the *F* factor. Since only part of the *F* factor is transferred at the beginning of conjugation, and e^+ is transferred quite quickly after conjugal pairing, most e^+ str^R recombinants will have only a small part of the donor chromosome. The remaining part of the *F* factor is transferred only after the entire bacterial chromosome is transferred. Because of turbulence, the conjugal pairing is almost always disrupted before transfer of an entire chromosome is completed. Hence, the chance that complete transfer will occur is quite remote.

iv. True. The graph shows that a^+ is not transferred until after about 16 minutes has elapsed.

b.

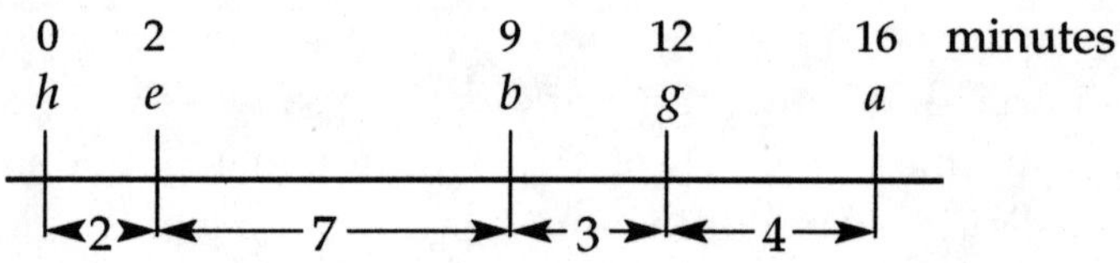

6.4 If an *E. coli* auxotroph *A* could grow only on a medium containing thymine, and an auxotroph *B* could grow only on a medium containing leucine, how would you test whether DNA from *A* could transform *B*?

Answer: Strain *A* is *thy⁻ leu⁺*, while strain *B* is *thy⁺ leu⁻*. To test if DNA from *A* can transform *B*, one is interested in whether one can transform the *leu⁻* allele of *B* to *leu⁺*. To do this, add DNA from strain *A* to a leucine-fortified culture of *B*. Incubate long enough for transformation to occur, then plate out these potentially transformed *B* cells on minimal medium or on medium supplemented only with thymine. As shown in the table below, one can select for transformants by plating on such media, as growth on such media requires *leu⁺*.

	Medium Type			
Strain	**Minimal**	**Plus Leucine**	**Plus Thymine**	**Plus Thymine and Leucine**
leu^+ thy^+	+	+	+	+
leu^+ thy^-	–	–	+	+
leu^- thy^+	–	+	–	+
leu^- thy^-	–	–	–	+

6.5 Wild-type phage T4 grows on both *E. coli B* and *E. coli K12*(λ), producing turbid plaques. The *rII* mutants of T4 grow on *E. coli B*, producing clear plaques, but do not grow on

E. coli K12(λ). This host range property permits the detection of a very low number of r^+ phages among a large number of *rII* phages. With this sensitive system, it is possible to determine the genetic distance between two mutations within the same gene, in this case the *rII* locus. Suppose *E. coli B* is mixedly infected with *rIIx* and *rIIy*, two separate mutants in the *rII* locus. Suitable dilutions of progeny phages are plated on *E. coli B* and *E. coli K12*(λ). A 0.1-mL sample of a thousandfold dilution plated on *E. coli B* produces 672 plaques. A 0.2-mL sample of undiluted phage plated on *E. coli K12*(λ) produces 470 turbid plaques. What is the genetic distance between the two *rII* mutations?

Answer: The plaques produced on *E. coli K12*(λ) are r^+, while those on *E. coli B* may be either r^+ or r^-. Thus, the total number of progeny can be inferred from the number of plaques formed on *E. coli B*. Since *E. coli B* is coinfected with *rIIx* and *rIIy*, the only way to obtain an r^+ progeny phage is to have a crossover within the *rII* locus. The progeny resulting from a crossover would be ½ r^+ and ½ *rIIxy* recombinants. The number of recombinant phages is twice the number of r^+ phages, which can be assayed for by growth on *E. coli K12*(λ).

$$\begin{aligned}\text{\# recombinant progeny in 1 mL} &= 2 \times (\text{number of } r\text{+ phage/mL})\\ &= 2 \times (470/0.2)\\ &= 4{,}700/\text{mL}\end{aligned}$$

$$\begin{aligned}\text{total \# of progeny in 1 mL} &= (\text{dilution factor}) \times (\text{\# progeny phage/mL})\\ &= 1{,}000 \times (672/0.1)\\ &= 6.72 \times 10^6/\text{mL}\end{aligned}$$

$$\begin{aligned}\text{RF} &= [4{,}700/(6.72 \times 10^6)] \times 100\%\\ &= 0.07\%\end{aligned}$$

The map distance between *rIIx* and *rIIy* is 0.07 mu.

6.6 Construct a map from the following two-factor phage cross data (show map distance):

Cross	**% Recombination**
$r1 \times r2$	0.10
$r1 \times r3$	0.05
$r1 \times r4$	0.19
$r2 \times r3$	0.15
$r2 \times r4$	0.10
$r3 \times r4$	0.23

Answer: Recall that, as discussed in Chapter 5, the most accurate map distances are those obtained over the shortest intervals.

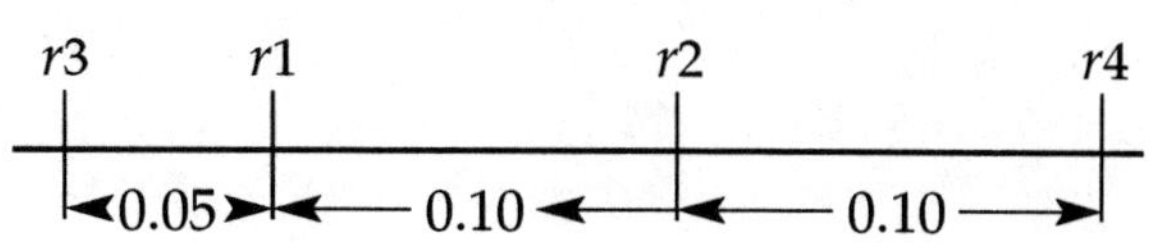

6.7 The following two-factor crosses were made to analyze the genetic linkage between four genes in phage λ: *c*, *mi*, *s*, and *co*.

Parents	Progeny
c + × + *mi*	1,213 *c* +, 1,205 + *mi*, 84 + +, 75 *c mi*
c + × + *s*	566 *c* +, 808 + *s*, 19 + +, 20 *c s*
co + × + *mi*	5,162 *co* +, 6,510 + *mi*, 311 + +, 341 *co mi*
mi + × + *s*	502 *mi* +, 647 + *s*, 65 + +, 56 *mi s*

Construct a genetic map of the four genes.

Answer: Analyze the data as you would a set of two-factor crosses:

Cross	Number of Progeny	# Recombinants	mu
c + × + *mi*	2,577	159	6.2
c + × + *s*	1,413	39	2.8
co + × + *mi*	12,324	652	5.3
mi + × + *s*	1,270	121	9.5

Two maps are compatible with these data:

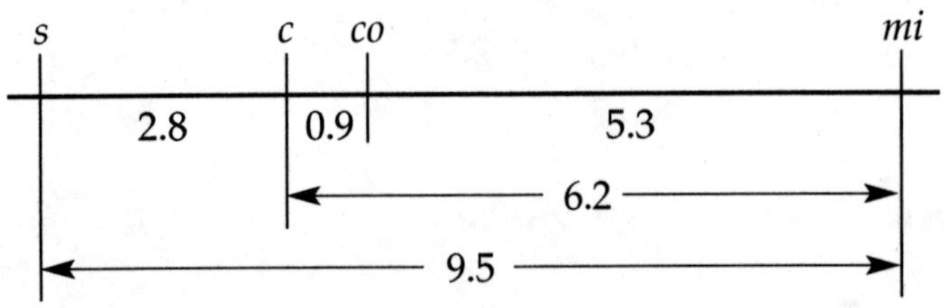

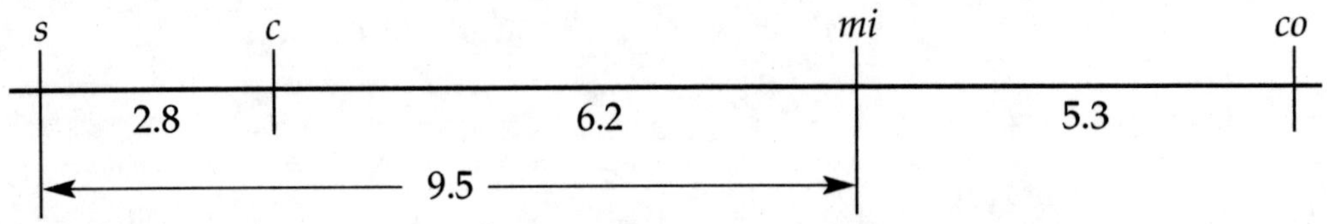

6.8 Three gene loci in T4 that affect plaque morphology in easily distinguishable ways are *r* (rapid lysis), *m* (minute), and *tu* (turbid). A culture of *E. coli* is mixedly infected with two types of phage: *r m tu* and $r^+ m^+ tu^+$. Progeny phage are collected, and the following genotype classes are found:

Class	Number
r^+ m^+ tu^+	3,729
r^+ m^+ tu	965
r^+ m tu^+	520
r m^+ tu^+	172
r^+ m tu	162
r m^+ tu	474
r m tu^+	853
r m tu	3,467
	10,342

Construct a map of the three genes. What is the coefficient of coincidence, and what does the value suggest?

Answer: The phage progeny numbers are handled just as they are in any three-point cross. First assign each progeny class to a parental, single- or double-crossover class, then determine the gene order by comparing double-crossover classes to parental classes, then determine the distances between adjacent genes and the coefficient of coincidence.

The parental-type chromosomes are *r m tu* and r^+ m^+ tu^+, and the gene order is *m-r-tu*.

$$\text{RF}(m - r) = [(162 + 520 + 474 + 172)/10{,}342] \times 100\% = 12.8\%$$

There are 12.8 mu between *m* and *r*.

$$\text{RF}(r - tu) = [(162 + 965 + 853 + 172)/10{,}342] \times 100\% = 20.8\%$$

There are 20.8 mu between *r* and *tu*.

A map of these loci is drawn below.

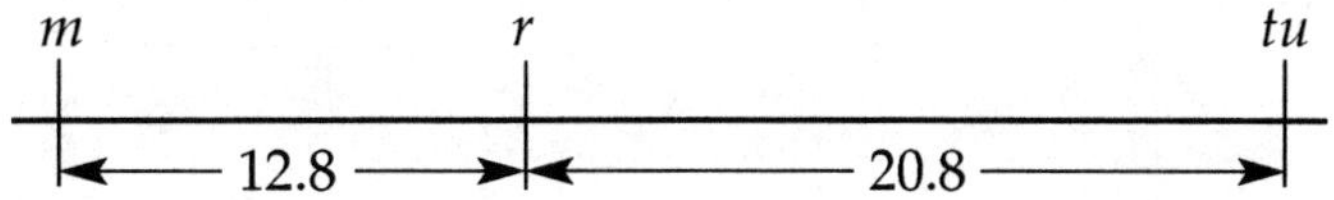

$$\text{c.o.c.} = (162 + 172)/(0.208 \times 0.128 \times 10{,}342) = 334/275 = 1.21$$

A c.o.c. value greater than 1 indicates (1) an absence of chiasma interference in the region between *m* and *tu* and (2) that the presence of one crossover in this region enhances the occurrence of a second crossover nearby. This phenomenon is called negative interference because the interference ($I = 1 - \text{c.o.c.}$) is a negative number.

6.9 The *rII* mutants of bacteriophage T4 grow in *E. coli B* but not in *E. coli K12*(λ). The *E. coli* strain *B* is doubly infected with two *rII* mutants. A 6×10^7 dilution of the lysate is plated on *E. coli B*. A 2×10^5 dilution is plated on *E. coli K12*(λ). Twelve plaques appeared on strain *K12*(λ) and 16 on strain *B*. Calculate the amount of recombination between these two mutants.

Answer: The plaques that appear on strain *B* include both r^+ and r^- phage and so can be used to determine the total number of phage progeny in the lysate. The plaques that appear on strain *K12*(λ) must be r^+, which result from recombination between the two mutant chromosomes. These can be used to determine the number of recombinant progeny. The recombinants consist of two types of phage: r^+ and r^- double mutants. Thus, the r^+ recombinants are half of the recombinants in the tested sample. See text pp. 138–140 particularly Figure 6.17.

$$
\begin{aligned}
\text{RF} &= \text{number of recombinants/total number of progeny} \\
&= (2 \times \text{number of recombinant } r^+ \text{ phage})/\text{total number of progeny} \\
&= [2 \times 12 \times (2 \times 10^5)]/[16 \times (6 \times 10^7)] \times 100\% = 0.5\%
\end{aligned}
$$

6.10 Wild-type (r^+) strains of T4 produce turbid plaques, whereas *rII* mutant strains produce larger, clearer plaques. Five *rII* mutations (*a–e*) in the *A* cistron of the *rII* region of T4 give the following percentages of wild-type recombinants in two-point crosses:

Cross	% of Wild-Type Recombinants
$a \times b$	0.2
$a \times c$	0.9
$a \times d$	0.4
$b \times c$	0.7
$e \times a$	0.3
$e \times d$	0.7
$e \times c$	1.2
$e \times b$	0.5
$b \times d$	0.2
$d \times c$	0.5

What is the order of the mutational sites, and what are the map distances between the sites?

Answer: Between any two *rII* mutants *rIIx and rIIy* there are two products of intragenic recombination: rII^{++} and rII^{xy}. Thus, the frequency of wild-type recombinants is half the total recombinant frequency. These data give the following map:

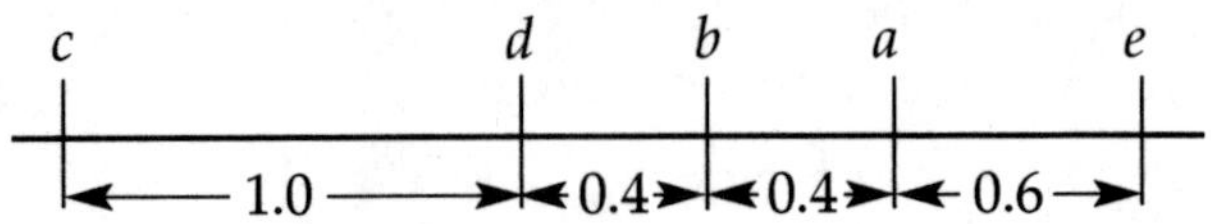

6.11 Some adenine-requiring mutants of yeast are pink because of the intracellular accumulation of a red pigment. Diploid strains were made by mating haploid mutant strains. The diploids exhibited the following phenotypes:

Cross	Diploid Phenotypes
1×2	pink, adenine-requiring
1×3	white, prototrophic
1×4	white, prototrophic
3×4	pink, adenine-requiring

How many genes are defined by the four different mutants? Explain.

Answer: When two haploid strains mate to form a diploid cell, each contributes its genome. Consequently, there are two sets of genes in a diploid cell, and complementation between two mutants can be observed. If two mutations are in the same unit of function (i.e., the same gene), then they will not complement when together in a diploid cell and the mutant phenotype will be exhibited. If two mutations are in different genes, they will complement each other in a diploid cell and the wild-type phenotype will result.

In the data presented here, mutations 1 and 3 and mutations 1 and 4 complement each other, indicating that mutations 1 and 3 are in different genes and mutations 1 and 4 are in different genes. On the other hand, mutations 1 and 2 and mutations 3 and 4 fail to complement each other. Thus, mutations 1 and 2 are in one gene, while mutations 3 and 4 are in a different gene. There are two genes.

6.12 In *Drosophila* mutants *A, B, C, D, E, F,* and *G,* all have the same phenotype: the absence of red pigment in the eyes. In pairwise combinations in complementation tests, the following results were produced, where + = complementation and – = no complementation.

	A	*B*	*C*	*D*	*E*	*F*	*G*
G	+	–	+	+	+	+	–
F	–	+	+	–	+	–	
E	+	+	–	+	–		
D	–	+	+	–			
C	+	+	–				
B	+	–					
A	–						

a. How many genes are present?
b. Which mutants have defects in the same gene?

Answer: Mutants that fail to complement each other are in the same gene. Mutants *G* and *B* fail to complement each other, mutants *C* and *E* fail to complement each other, and mutants *A, F,* and *D* fail to complement each other. Therefore, there are three genes, with mutants *G* and *B* in one, mutants *A, F,* and *D* in a second, and mutants *C* and *E* in a third.

6.13 A homozygous white-eyed Martian fly (w_1/w_1) is crossed with a homozygous white-eyed fly from a different stock (w_2/w_2). It is well known that wild-type Martian flies have red eyes. This cross produces all white-eyed progeny. State whether each of the following is true or false. Explain your answer.

a. w_1 and w_2 are allelic genes.

b. w_1 and w_2 are nonallelic.

c. w_1 and w_2 affect the same function.

d. The cross was a complementation test.

e. The cross was a *cis-trans* test.

f. w_1 and w_2 are allelic by terms of the functional test.

The F_1 white-eyed flies are allowed to interbreed, and when you classify the F_2, you find 20,000 white-eyed flies and 10 red-eyed progeny. Concerned about contamination, you repeat the experiment and get exactly the same results. How can you best account for the presence of the red-eyed progeny? As part of your explanation, give the genotypes of the F_1 and F_2 generation flies.

Answer:

a. True. In the cross $w_1/w_1 \times w_2/w_2$, the offspring are w_1w_2. That they are white-eyed indicates that w_1 and w_2 are mutations in the same function (i.e., the same gene). Thus, w_1 and w_2 are allelic genes.

b. False. If the genes were nonallelic, each mutation would be in a different function. The two mutations would each be in different steps of a pathway, with each mutant retaining the function of a different step. Since the progeny obtain a set of genes from each mutant parent, they would have the function of both steps and have red eyes. Since they have white eyes, this statement is false.

c. True.

d. True. A complementation test examines whether two mutations affect the same function.

e. True. The F_1 is a *trans*-heterozygote.

f. True. If two genes affect the same function, they are allelic.

The complementation test indicates that the two mutations affect the same function and are therefore alleles at the same gene. However, it does not indicate whether they are *identical* alleles or not. To evaluate if two allelic mutations lie in exactly the same position in a gene, one needs to assess whether recombination can occur between them. If two alleles are identical and lie in the same site, *intragenic* recombination between them cannot occur. If they lie in different sites, intragenic recombination can occur, giving rise to wild-type recombinant chromosomes. The appearance of wild-type F_2 progeny results from intragenic recombination, as illustrated below.

If w_1 and w_2 lie in different sites in the same gene, the cross and the F_1 progeny can be written as

$$\text{P: } \frac{w_1\ +}{w_1\ +} \times \frac{+\ w_2}{+\ w_2} \qquad \text{F}_1\text{: } \frac{w_1\ +}{+\ w_2}$$

A rare intragenic crossover between the two mutant sites can be diagrammed as follows:

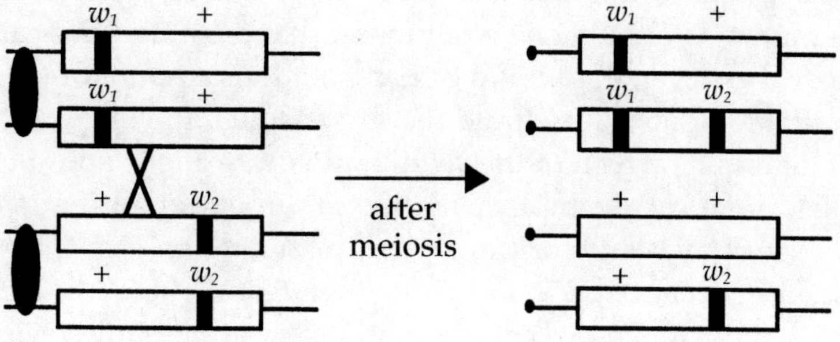

This crossover gives rise to gametes that are $w_1\ w_2$ and + +. The $w_1\ w_2$ gamete would not be detected ($w_1\ w_2/w_1$ + and $w_1\ w_2/+\ w_2$ flies are white-eyed, as are their noncrossover siblings), but a + + gamete would give rise to a red-eyed offspring that has a genotype of + +/w_1 + or + +/+ w_2. It is interesting to note that from the frequency of the white-eyed progeny, one can determine the recombination frequency (RF) between the two point mutants, just as in Benzer's *cis-trans* tests in the *rII* region of T4.

RF (w_1, w_2) = $[(2 \times 10)/20{,}010] \times 100\% = 0.1\%$. The w_1 and w_2 mutation sites are 0.1 mu apart.

6.14 A large number of mutations in *Drosophila* alter the normal deep-red eye color. As we discussed in Chapter 4 (p. 79), even alleles at one gene (*white, w*) can display a variety of phenotypes.

You are given a wild-type, deep-red strain and six independently isolated, true-breeding mutant strains that have varying shades of brown eyes, with the assurance that each mutant strain has only a single mutation. How would you determine

a. whether the mutation in each strain is dominant or recessive?
b. how many different genes are affected in the six mutant strains?
c. which mutants, if any, are allelic?
d. whether any of these mutants are alleles of genes already known to affect eye color?

Answer:

a. Cross each mutation to the wild-type, deep-red strain and observe the phenotypes of the progeny. A dominant mutation, by definition, is one that appears in a heterozygote. If the progeny show deep-red eyes, the mutation must be recessive. If the progeny show brownish eyes, the mutation is dominant.
b. Set up pairs of crosses between the mutants to perform complementation tests. Mutations that affect the same gene function will produce brown-eyed progeny when crossed and belong to the same complementation group. Mutations that affect different functions will produce deep-red-eyed progeny when crossed and belong to different complementation groups. Counting the number of different complementation groups will give the number of genes that are affected.
c. Allelic mutations are those that are members of the same complementation group, as determined in (b).
d. One could assess if a particular mutant is allelic to a known eye color gene by performing complementation tests between it and mutants at all known eye color genes.

This would involve crossing the mutant to mutants from a collection of strains with known eye color mutations and observing the progeny of each cross. If the progeny have a mutant eye color, one would infer that the mutations carried in the two strains are allelic. However, this would be a tremendous amount of work. There are many eye color mutations, and this would require a large number of crosses. It would be faster to first determine which of the six mutations are allelic and then choose a representative allele from each complementation group and identify its chromosomal location using a set of two- and/or three-point mapping crosses. Once this is done, examine published genetic maps of the *Drosophila* genome (e.g., at http://www.flybase.org) and ask if any known eye color mutations lie in the same region. Then obtain strains with these eye color mutations and perform complementation tests between these mutant strains and a representative mutant from each complementation group identified with the new eye color mutations.

7

NON-MENDELIAN GENETICS

CHAPTER OUTLINE

ORIGIN OF MITOCHONDRIA AND CHLOROPLASTS

RULES OF NON-MENDELIAN INHERITANCE

EXAMPLES OF NON-MENDELIAN INHERITANCE

Shoot Variegation in the Four O'Clock
The [*poky*] Mutant of *Neurospora*
Yeast *petite* Mutants
Non-Mendelian Inheritance in *Chlamydomonas*
Human Genetic Diseases and Mitochondrial DNA Defects
Infectious Heredity: Killer Yeast

CONTRASTS TO NON-MENDELIAN INHERITANCE

Maternal Effect
Genomic Imprinting

REVIEW OF KEY TERMS, SYMBOLS, AND CONCEPTS

In your own words, write a brief, precise definition of each term in the groups on the following page. Check your definitions using the text. Then develop a concept map using the terms in each list.

THINKING ANALYTICALLY

After developing skills to analyze nuclear inheritance patterns, non-Mendelian patterns of inheritance are quite a twist at first. However, since non-Mendelian inheritance patterns are usually due to the extranuclear inheritance of mitochondrial or chloroplast genomes, the patterns seen are well defined. In general, they follow from the cytoplasmic contributions of one (female) parent to the offspring.

Consideration of the variation in inheritance patterns and the variation in the structure and function of the mitochondrial and chloroplast genomes provides a valuable perspective on nuclear inheritance patterns and nuclear genome structure and function. After organizing the factual information given in this chapter, take the time to compare and contrast the organization, replication, and expression of mitochondrial and chloroplast genes with that of nuclear genes. The salient features of an important biological process can often be better understood by examining variations in it.

1	2	3
extranuclear inheritance, gene	mitochondria	chloroplast
non-Mendelian inheritance, gene	[*poky*] in *Neurospora*	shoot variegation in four o'clocks
uniparental inheritance	*petite, grande* in yeast	leukoplast
maternal inheritance	nuclear, neutral, suppressive *petite*	
biparental inheritance	protoperithecia	
cytohet	controlled cross	
maternal effect	heteroplasmy	
nuclear genotype	Leber's hereditary optic neuropathy	
maternal effect gene	Kearns-Sayre syndrome	
genomic (parental) imprinting	myoclonic epilepsy and ragged-red fiber disease	
infectious inheritance		
endosymbiont hypothesis		
cyanobacterium		
purple nonsulfur photosynthetic bacterium		
erythromycin resistance in *Chlamydomanas*		
killer yeast		
L, M viruses		
shell coiling direction in *Limnaea peregra*		
PWS, Angelman syndrome		

Unlike extranuclear inheritance patterns that are controlled by extranuclear genes, maternal effects and genomic imprinting are controlled by nuclear genes. Keep this distinction clear. In considering maternal effects, it will help to diagram the generations of a cross and illustrate how contributions from the parental generation are important for the development of the progeny. Carefully read the material in the chapter and do the problems to discover how to distinguish a maternal effect from an extranuclear, maternally inherited trait.

QUESTIONS FOR PRACTICE

Multiple Choice Questions

1. DNA is found in
   a. the nucleus of eukaryotes
   b. mitochondria
   c. chloroplasts
   d. all subcellular organelles
   e. all of the above except d

2. In general, extranuclear genes show
   a. maternal inheritance
   b. Mendelian inheritance
   c. maternal effects
   d. dominance
3. Which of the following is *not* a typical characteristic of extranuclear traits?
   a. They show uniparental inheritance.
   b. They cannot be mapped relative to nuclear genes.
   c. They show non-Mendelian segregation in crosses.
   d. They are affected by substitution of a nucleus with a different genotype.
4. Two brothers have a hereditary disease associated with a particular lesion in mitochondrial DNA. One brother is more severely affected than the other. Which of the following is *not* a plausible explanation for this observation?
   a. The brothers have different degrees of heteroplasmy.
   b. The brothers have different proportions of two mitochondrial types.
   c. The brothers do not have identical nuclear genomes.
   d. Different mitochondrial genes are affected in the two brothers.
5. Which of the following is *not* true about genomic imprinting?
   a. It is an example of extranuclear inheritance.
   b. Expression patterns of imprinted genes can vary, depending on their parental source.
   c. Expression patterns of imprinted genes are correlated with their methylation states.
   d. Genomic imprinting can affect the severity of disease symptoms.

Answers: 1e, 2a, 3d, 4d, 5a

Thought Questions

1. What are the tenets of the endosymbiont hypothesis? What arguments can you marshal to support or detract from this hypothesis?
2. Speculate on the evolutionary origin of plant chloroplasts, compared to the presumed origin of mitochondria.
3. Is there any evidence that there are paternal contributions to mtDNA inheritance? What, if any, significance might these contributions have?
4. Distinguish between biparental and uniparental inheritance. Which organisms have organelles that show biparental inheritance?
5. One often thinks of non-Mendelian inheritance being associated with genes in mitochondrial or chloroplast DNA. What other examples of non-Mendelian inheritance exist? What is the significance of the examples you cite?
6. Recall that environmental factors, as well as other genetic factors, were used to explain variable expressivity. What role might genomic imprinting play?
7. How do you account for the maternal effect in shell coiling in the snail *Limnaea peregra?* What features of this phenomenon indicate that it is controlled via a maternal effect and not via extranuclear inheritance?

Thought Questions for Media

After reviewing the media at the *Essential iGenetics* website, try to answer these questions.

1. A four o'clock plant shows variegation. That is, part of the plant has only green leaves, part has only white leaves, and part has mixed green and white leaves. Suppose you fertilized flowers in the green-only region of the plant with pollen from flowers in the white-only part of the plant. Based on the animation, what outcomes would you expect?
2. What is the difference between a leukoplast and a chloroplast?
3. In the maternal effect animation, dextral (right-handed) coiling is dominant to sinistral (left-handed) coiling, but the coiling pattern is determined by the maternal contribution, not the genotype of the zygote. Suppose you had a population of *Dd* dextral snails.
   a. What crosses would you do to obtain a *Dd* sinistral snail? (Remember that you can only tell the progeny's coiling phenotype and must infer the genotype by testcrossing or selfing!)
   b. What crosses would you do to obtain a *dd* sinistral snail?
   c. Could you ever obtain a *DD* sinistral snail?
4. How does maternal effect differ from maternal inheritance, since both involve cytoplasmic contributions from the mother?

1. The iActivity provided data indicating that individuals with MERFF show one or more symptoms that can include hearing loss, dementia, muscle jumps, seizures, unsteadiness, vision problems, and loss of cognitive skills.
   a. How would you characterize MERFF in terms of penetrance, expressivity, and pleiotrophy?
   b. Based on what you know about the mode of inheritance of MERFF, what hypotheses might you form to explain its symptomatic variability?
   c. How might you use methods similar to those presented in the iActivity to test your hypotheses?
2. The iActivity presented data that allowed you to distinguish between heteroplasmic maternal inheritance and autosomal dominant inheritance with variable expressivity.
   a. What methods were used in the iActivity to distinguish between these modes of inheritance?
   b. Would it ever be possible to distinguish between these modes of inheritance based on pedigree analysis alone?
   c. What challenges do mitochondrial inheritance present for the analyses of pedigrees?
3. Suppose a MERFF-like disorder was caused by a mutation in a nuclear gene whose product was imported into the mitochondrion. What types of inheritance might be associated with this disorder?

SOLUTIONS TO TEXT PROBLEMS

7.1 What features of non-Mendelian inheritance distinguish it from the inheritance of nuclear genes?

Answer: Non-Mendelian inheritance due to extranuclear genes can be distinguished from nuclear inheritance by four characteristics. First, reciprocal crosses of extranuclear genes give different results because uniparental inheritance is typically seen. These results are distinct from the results of reciprocal crosses of sex-linked nuclear genes and from the results of reciprocal crosses of autosomal nuclear genes (which give the same results regardless of sex type). Second, extranuclear genes cannot be mapped to chromosomes in the nucleus. New extranuclear mutations do not show linkage to any known nuclear linkage group. Third, ratios typical of Mendelian segregation are not found. Fourth, extranuclear inheritance is indifferent to nuclear substitution.

7.2 Distinguish between maternal effect and non-Mendelian inheritance.

Answer: Maternal effect is the determination of gene-controlled characters by the maternal genotype prior to the fertilization of the egg cell. A maternal effect is seen when nuclear genes of the mother function to specify some characteristic of the zygote. For example, a maternal effect could result from a maternally transcribed gene whose product was localized at one pole of the embryo and was responsible for the polarity of the developing embryo.

In contrast to nuclear genes that can function in the mother, the genes involved in non-Mendelian inheritance are extranuclear and are found in mitochondria and chloroplasts. If the mitochondria are inherited with transmission of the mother's cytoplasm, mitochondrial genes will show maternal inheritance, a type of extranuclear inheritance.

7.3 Reciprocal crosses between two types of the evening primrose, *Oenothera hookeri* and *Oenothera muricata*, produce the following effects on the plastids:

O. hookeri female × *O. muricata* male → yellow plastids

O. muricata female × *O. hookeri* male → green plastids

Explain the difference in these results, noting that the chromosome constitution is the same in both types.

Answer: Both offspring have the same hybrid nuclear constitution, but differ in the maternal cytoplasm they inherited. Thus, factors in the cytoplasm are likely to be the cause of the phenotypic difference that is seen. The two possible types of cytoplasmic factors are those that would cause a maternal effect and those that are inherited maternally. If the difference in phenotype is due to a maternal effect, the difference is due to the deposition in the oocyte of gene products from the mother's nuclear genome. For these results to be explained by a maternal effect, one would have to hypothesize (1) that a such a maternally contributed gene product affected plastid color and (2) that this product was different in *O. hookeri* and *O. muricata.* If the difference in phenotype is due to maternal inheritance, the plastids contributed by one female parent would have to be different from the plastids contributed by the other. Specifically, yellow plastids would have been contributed by the *O. hookeri* female and green plastids would have been contributed by the *O. muricata* female.

It is noteworthy that in most angiosperms (flowering plants), the plastids are inherited only from the maternal parent. Thus, maternal inheritance of plastids would explain the results presented in this problem. However, in the genus *Oenothera,* biparental inheritance (inheritance from both the male and the female parent) of plastids has been reported (see text). If biparental inheritance does not occur all of the time (i.e., maternal inheritance occurs some of the time), many progeny would be just as described in the results presented here. When plastids are contributed biparentally, and the plastids contributed by the male and female parent segregate to different daughter cells during mitosis, a green patch of tissue would be seen in the yellow progeny from the cross *O. hookeri* female × *O. muricata* male, and a yellow patch of tissue would be seen in the green progeny from the cross *O. muricata* female × *O. hookeri* male.

7.4 A series of crosses are performed with a recessive mutation in *Drosophila* called *tudor*. Homozygous *tudor* animals appear normal and can be generated from the cross of two heterozygotes, but a true-breeding *tudor* strain cannot be maintained. When homozygous *tudor* males are crossed to homozygous *tudor* females, both of which appear to be phenotypically normal, a normal-appearing F_1 is produced. However, when F_1 males are crossed to wild-type females, or when F_1 females are crossed to wild-type males, no progeny are produced. The same results are seen in the F_1 progeny of homozygous *tudor* females crossed to wild-type males. The F_1 progeny of homozygous *tudor* males crossed to wild-type females appear normal, and they are capable of issuing progeny when mated either with each other or with wild-type animals.

a. How would you classify the *tudor* mutation? Why?

b. What might cause the *tudor* phenotype?

Answer:

a. The *tudor* mutation is a maternal effect mutation. Homozygous *tudor* mothers give rise to sterile progeny, regardless of their mate.

b. The "grandchildless" phenotype results from the absence of some maternally packaged component in the egg needed for the development of the F_1's germ line.

7.5 A form of male sterility in corn is maternally inherited. Plants of a male-sterile line crossed with normal pollen give male-sterile plants. Some lines of corn carry a dominant, so-called restorer (*Rf*) gene that restores pollen fertility in male-sterile lines.

a. If a male-sterile plant is crossed with pollen from a plant homozygous for gene *Rf*, what will be the genotype and phenotype of the F_1?

b. If the F_1 plants of part (a) are used as females in a testcross with pollen from a normal plant (*rf/rf*), what would be the result? Give genotypes and phenotypes and designate the type of cytoplasm.

Answer:

a. Let the normal cytoplasm be denoted by [*N*] and the male-sterile cytoplasm be denoted by [*Ms*]. The cross can be written as [*Ms*] *rf/rf* female × [*N*] *Rf/Rf* male. The F_1 would be [*Ms*] *Rf/rf* and be male-fertile.

b. The cross is [*Ms*] *Rf/rf* female × [*N*] *rf/rf* male. Half of the progeny would be [*Ms*] *Rf/rf* and half would be [*Ms*] *rf/rf*. Thus, half of the progeny will be male-fertile and half will be male-sterile.

7.6 Distinguish between *nuclear* (segregational), *neutral*, and *suppressive petite* mutants of yeast.

Answer: Petites are a class of mutations that affect mitochondrial function. *Neutral petites* are able to grow on a medium that will support fermentation, but cannot grow on a medium that supports only aerobic respiration because they lack nearly all of their mitochondrial DNA. *Nuclear petites* result from a nuclear mutation that affects mitochondrial function (e.g., in a nuclear gene that encodes a subunit of a mitochondrial protein). These mutations, if crossed to a wild-type (*grande*) strain, will show a typical 2:2 Mendelian segregation pattern. *Neutral* and *suppressive petites* result from mutations in the mitochondrial genome. They show uniparental inheritance (but not maternal inheritance): When crossed to a normal cell, and (*grande*) diploids go through meiosis, a 0:4 ratio of *petite:grande* is seen. *Suppressive petites,* unlike *neutral petites,* do have an effect on the wild type. A diploid formed from a *suppressive petite* and a normal cell will have respiratory properties intermediate between the *petite* and normal. Mitosis in this diploid results in mostly *petites* (up to 99%) with a respiratory-deficient phenotype, and meiosis results in a 4:0 ratio of *petite:grande.* At a molecular level, the *suppressive petite* mutations result from partial deletions of the mitochondrial DNA.

7.7 In yeast, a haploid *nuclear* (segregational) *petite* is crossed with a *neutral petite.* Assuming that both strains have no other abnormal phenotypes, what proportion of the progeny ascospores are expected to be *petite* in phenotype if the diploid zygote undergoes meiosis?

Answer: Nuclear genes will show Mendelian segregation, so that +/*petite* should show 2:2 segregation. One should see ½ *petite* and ½ wild-type (*grande*) progeny.

7.8 When grown on a medium containing acriflavin, a yeast culture produces a large number of very small (*tiny*) cells that grow very slowly. How would you determine whether the slow-growth phenotype was the result of a cytoplasmic factor or a nuclear gene?

Answer: This problem is formally very similar to determining the mode of inheritance for yeast *petite* mutants. Acriflavin intercalates between base pairs and might be used to introduce *petite*-like mutations. If the *tiny* phenotypes are due to a nuclear gene, then meiosis in cells formed from a cross of *tiny* × wild type should result in a 2:2 segregation of *tiny*:wild-type ascospores. If, in contrast, an extranuclear gene is involved, one would expect a 0:4 or 4:0 ratio and see only wild-type or *tiny* progeny.

7.9 *Drosophila melanogaster* has a sex-linked, recessive, mutant gene called *maroon-like* (*ma-l*). Homozygous *ma-l* females or hemizygous *ma-l* males have light-colored eyes because of the absence of the active enzyme xanthine dehydrogenase, which is involved in the synthesis of eye pigments. When heterozygous *ma-l*$^+$/*ma-l* females are crossed with *ma-l* males, all the offspring are phenotypically wild-type. However, half the female offspring from this cross, when crossed back to *ma-l* males, give all *ma-l* progeny. The other half of the females, when crossed to *ma-l* males, give all phenotypically wild-type progeny. What is the explanation for these results?

Answer: Since *ma-l* is sex-linked, we know it is nuclearly inherited. One can therefore diagram the crosses and the results. In the first cross, *ma-l*$^+$/*ma-l* × *ma-l*/Y gives ¼ *ma-l*/*ma-l* females, ¼ *ma-l*$^+$/*ma-l* females, ¼ *ma-l*$^+$/Y males, and ¼ *ma-l*/Y males. To explain why the half of the total F_1 progeny that are *ma-l* hemizygotes or homozygotes do not have light-colored eyes, consider the possibility of a maternal effect. The parental female was *ma-l*$^+$/*ma-l* and so could have provided normal xanthine dehydrogenase (or stable, normal xanthine dehydrogenase mRNA or the biochemical product of the xanthine dehydrogenase reaction) to the embryo. The results of the crosses indicate that this maternally transmitted product is sufficient for the F_1 progeny to have a normal phenotype. When the F_1 females are backcrossed to the *ma-l* males, two different crosses are possible. One is *ma-l*$^+$/*ma-l* × *ma-l*/Y. Just like the initial cross, this will give all wild-type progeny. The other cross is *ma-l*/*ma-l* × *ma-l*/Y, which will give all *maroon-like* progeny (as the mother can no longer supply the *ma-l*$^+$ product to her progeny).

7.10 When females of a particular mutant strain of *Drosophila melanogaster* are crossed to wild-type males, all the viable progeny flies are females. Hypothetically, this result could be the consequence of either a sex-linked, male-specific lethal mutation or a maternally inherited factor that is lethal to males. What crosses would you perform to distinguish between these alternatives?

Answer: If the mutation is a sex-linked, male-lethal mutation, the first cross can be diagrammed as *L*/*L* × +/Y, giving *L*/+ (females) and *L*/Y (dead males). If the F_1 females are crossed to normal males, the cross is *L*/+ × +/Y. The progeny are ¼ *L*/+ (normal females), ¼ +/+ (normal females), ¼ +/Y (normal males), and ¼ *L*/Y (dead males), a 2:1 ratio of females to males. If the male lethality is associated with a maternally inherited factor lethal to males, the F_1 females should possess this factor (as they received cytoplasm from their mother). Consequently, just like their mother, the F_1 females should have no male offspring when mated to wild-type males.

7.11 Reciprocal crosses between two *Drosophila* species, *D. melanogaster* and *D. simulans*, produce the following results:

melanogaster ♀ × *simulans* ♂ → females only

simulans ♀ × *melanogaster* ♂ → males, with few or no females

Propose a possible explanation for these results.

Answer: When *melanogaster* females are crossed with *simulans* males, *melanogaster* cytoplasm is given to the offspring. Female progeny will have an X from the *melanogaster* as well as the *simulans* parent, while male progeny will have an X from the *melanogaster* parent but only a Y (with few structural genes) from the *simulans* parent. Female progeny survive selectively because of nuclear gene products encoded on the *simulans* X chromosome which are needed for hybrid survival in *melanogaster* cytoplasm. When a *simulans* female is crossed with a *melanogaster* male, *simulans* cytoplasm is given to the offspring. Female progeny will have a *melanogaster* as well as a *simulans* X, while male progeny will have a *simulans* X and a *melanogaster* Y. Since few or no females are recovered, it appears that the *melanogaster* X encodes products that (generally) cause lethality in the *simulans* cytoplasm.

7.12 Some *Drosophila* are very sensitive to carbon dioxide; administering it to them anesthetizes them. The sensitive flies have a cytoplasmic particle called *sigma* that has many properties of a virus. Resistant flies lack *sigma*. The sensitivity to carbon dioxide shows strictly maternal inheritance. What would be the outcome of the following two crosses: (a) sensitive female × resistant male and (b) sensitive male × resistant female?

Answer: In part (a), all of the progeny will inherit the *sigma* factor from the (sensitive) female parent. Consequently, all the progeny will be sensitive. In part (b), the resistant female parent lacks the *sigma* factor. All of the progeny will also lack the factor and so be resistant.

7.13 The pedigree in the following figure shows a family in which an inherited disease called Leber's hereditary optic atrophy is segregating. This condition causes blindness in adulthood. Studies have recently shown that the mutant gene causing Leber's hereditary optic atrophy is located in the mitochondrial genome.

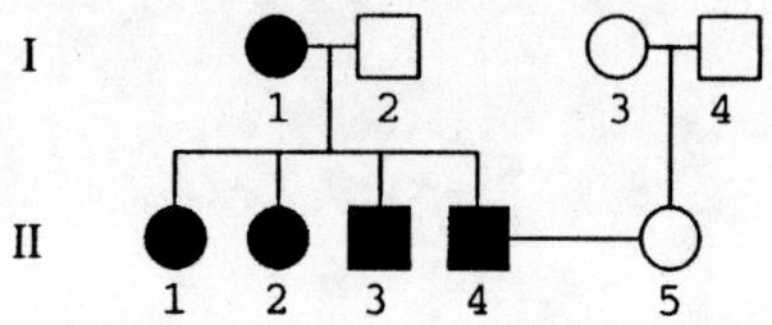

a. Assuming II-4 marries a normal person, what proportion of his offspring should inherit Leber's hereditary optic atrophy?

b. What proportion of the sons of II-2 should be affected?

c. What proportion of the daughters of II-2 should be affected?

Answer:

a. Since II-4 is a male, he will not contribute any of his mutant mitochondria to his offspring. All of his offspring will have normal mitochondria from their mother and be normal.

b. Since II-2 is a female, all of her progeny will obtain her mutant mitochondria, and hence all will be affected. All of her sons will be affected.

c. All of the daughters will be affected, as all will have mutant mitochondria from their mother.

7.14 The inheritance of the shell-coiling direction in the snail *Limnaea peregra* has been studied extensively. A snail produced by a cross between two individuals has a shell with a right-handed twist (dextral coiling). This snail produces only left-handed (sinistral) progeny on selfing. What are the genotypes of the F_1 snail and its parents?

Answer: If the coil direction were controlled by an extranuclear gene, then the progeny would always exhibit the phenotype of the mother, owing to maternal inheritance. If the trait were determined by a nuclear gene that did not show a maternal effect, then one would expect a "true-breeding" individual to give rise to individuals just like itself. Since neither of these are observed, these data are best explained by considering a nuclear gene that shows a maternal effect. In this case, the shell-coiling phenotype is determined by the genotype of the mother. One can infer the maternal genotype from the progeny phenotype. If dextral is *D* and sinistral is *d*, the F_1 genotype must have been *dd*, as it only gives sinistral offspring. Since the F_1 itself has a dextral pattern, its maternal parent must have had a *D* allele. To give the F_1 a *d* allele, the maternal parent must have been *Dd*. The paternal parent must have had a *d* allele and could have been either *dd* or *Dd*.

8

DNA: THE GENETIC MATERIAL

CHAPTER OUTLINE

THE SEARCH FOR GENETIC MATERIAL

Griffith's Transformation Experiment
Avery's Transformation Experiment
The Hershey-Chase Bacteriophage Experiment

THE COMPOSITION AND STRUCTURE OF DNA AND RNA

The Discovery of the DNA Double Helix
Different DNA Structures
DNA in the Cell

THE ORGANIZATION OF DNA IN CHROMOSOMES

Viral Chromosomes
Prokaryotic Chromosomes
Eukaryotic Chromosomes
Chromosomes of Mitochondria and Chloroplasts
Unique-Sequence and Repetitive-Sequence DNA

REVIEW OF KEY TERMS, SYMBOLS, AND CONCEPTS

In your own words, write a brief, precise definition of each term in the groups on the following page. Check your definitions using the text. Then develop a concept map using the terms in each list.

THINKING ANALYTICALLY

This chapter presents a large amount of complex, related information. As such, its information and problems require attention to detail. To comprehend and retain this information, organize it and develop strategies to place it in context. In addition to concept maps, it will help to construct labeled diagrams that display information contextually. Sketch labeled diagrams of nucleic acid structure, chromosome packaging, and sequence organization of chromosomes.

This chapter starts with a presentation of the key experimental evidence for the nature and structure of the genetic material. What we now accept as fact once was hotly debated. These insightful experiments resolved fundamental questions in biology. To organize this information and place it in context, pay close attention to the subtleties of the experiments that are presented, and strive to see how they provided evidence for or against a particular view.

1	2	3
macromolecule	genetic material	Chargaff's rules
chromosome	DNA	% GC
genome	RNA	X-ray diffractions
nucleic acid	transformation	angstrom unit (Å)
DNA, RNA	transforming principle	3′-OH, 5′-P
nucleotide	nuclease	polarity
nucleoside	ribonuclease (RNase)	antiparallel strands
nitrogenous base	deoxyribonuclease (DNase)	complementary base pairs
adenine, guanine	Avery's experiment	double helix
cytosine, thymine	Griffith's experiment	major groove
uracil		minor groove
purine		oligomer
pyrimidine		

4	5	6
viral genome	nuclear matrix	telomere-associated sequences
prokaryotic genome	MARs	transposons
single-, double-stranded	evolutionary conservation	
circular chromosome		
linear chromosome		
T-even phage		
λ, ΦX174 phage		

It is essential to know the chemical structures of nucleic acids to fully understand the material in coming chapters. Start by drawing out the structures, using paper and pencil. Copy the component parts of a nucleotide (the sugar, the bases, the phosphodiester linkages) from the text figures, paying close attention to their polarity. Practice them for a few days in a row so that you can draw them easily from memory.

Having a good sense of the dimensions involved in nucleic acid structure and chromosomal packaging is helpful in grasping a mental image of these physical processes. For this reason, it is important to learn (or review) units of measurement and to be able to make conversions between different dimension units. In solving problems, it is helpful to convert values presented in many different units to a smaller set of units. Particularly important unit conversions are:

Unit/Conversion	Example
1Å = 1 angstrom = 1×10^{-10} meter	distance between 2 protons in H_2 = 0.74 Å
1 nm = 1 nanometer = 1×10^{-9} meter	width of a DNA double helix = 2 nm
1 μm = 1 micrometer = 1×10^{-6} meter	typical red blood cell diameter = 7 μm
1 dalton (unit of mass) = 1.66×10^{-24} grams	1 H atom = 1 dalton

The chromosomal organization and packaging of DNA varies in different organisms. Consequently, while there are general principles that underlie DNA packaging, there are also organism-specific requirements. As you study this chapter, first identify the general principles used to organize and package genetic material. Then relate these principles to specific examples of organization and packaging in viruses, prokaryotes, and eukaryotes.

DNA packaging occurs in three dimensions, and so it helps to visualize DNA packaging by making sketches. You do not have to be a great artist; the point is to convey the different strategies that organisms use to fit a DNA molecule, which if left unpackaged would be larger than most cells, into a cell and have it remain functional.

Not everyone visualizes well in three dimensions, and so it can be especially challenging to visualize supercoiling in a circular DNA molecule. The following exercise might help: Take a spiral telephone cord and hold the ends together to circularize it. Sequentially introduce one, two, three, or four additional twists that tighten the spirals. For each complete twist that is introduced, a supercoil is added. Since the twist tightens the spiral, it introduces a positive supercoil. If you let the cord "relax" and then introduce additional twists that loosen the spirals, you will be introducing negative supercoils. To determine whether your phone cord is a right- or left-handed helix, wrap your right hand around the spiral and trace the direction of the spiral with your right index finger. If, while you hold the spiral with your right hand, your index finger traces up the spiral in the same direction that your right thumb points, you are holding a right-handed helix. If it traces down the spiral in an opposite direction, you are holding a left-handed helix. Notice that the orientation of the helix remains the same regardless of whether your right thumb points up or down.

QUESTIONS FOR PRACTICE

Multiple Choice Questions

1. DNA and RNA are polymers of
   a. nucleosides
   b. nucleotides
   c. pentose sugars connected by phosphodiester bonds
   d. ribonucleotides
2. A molecule consisting of ribose covalently bonded to a purine or pyrimidine base is a
   a. ribonucleoside
   b. ribonucleotide
   c. nuclease
   d. deoxyribonucleotide
3. The transforming principle was found to be
   a. a cellular material that could alter a cell's heritable characteristic
   b. a substance derived from killed viruses
   c. modified RNA that could change a living cell
   d. a transmissible substance that revives dead cells
4. When Griffith injected mice with a mixture of live *R* pneumococcus derived from a *IIS* strain and heat-killed *IIIS* bacteria,
   a. the mice survived and he recovered live type *IIIR* organisms
   b. the mice died, but he recovered live type *IIIR* cells
   c. the mice died, but he recovered live type *IIS* cells
   d. the mice died, but he recovered live type *IIIS* cells

5. In the Hershey-Chase experiment, T2 phage were radioactively labeled with either ^{35}S or ^{32}P and allowed to infect *E. coli*. What results proved that DNA was the genetic material?
   a. The ^{35}S was found in progeny phage, and the ^{32}P was found in phage ghosts.
   b. The ^{35}S was found in phage ghosts, and the ^{32}P was found in progeny phage.
   c. The ^{35}S was found in both progeny phage and in phage ghosts.
   d. The ^{32}P was found in both progeny phage and in phage ghosts.
6. Analysis of the bases of a sample of nucleic acid yielded these percentages: A, 20 percent; G, 30 percent; C, 20 percent; and T, 30 percent. The sample must be
   a. double-stranded RNA
   b. single-stranded RNA
   c. double-stranded DNA
   d. single-stranded DNA
7. Which of the following is not true about a linear molecule of double-stranded DNA?
   a. It is a double helix composed of antiparallel strands.
   b. Bases are paired via hydrogen bonds.
   c. At one end, two 5′ phosphate groups can be found.
   d. Pentose sugars are linked via covalent phosphodiester bonds.
8. Which kind of DNA is not likely to be found in cells?
   a. A-form DNA
   b. B-form DNA
   c. Z-form DNA
   d. none of the above
9. Two double-stranded 25-base-pair DNA fragments are heated in solution. Fragment A has 60 percent GC, and fragment B has 40 percent GC. Which observation(s) might be made as the solution temperature increases?
   a. At a low enough temperature, both fragments will remain double-stranded.
   b. A will separate into single strands at a lower temperature than B.
   c. B will separate into single strands at a lower temperature than A.
   d. At a high enough temperature, both A and B will separate into single strands.
   e. a, c, and d
10. The chromosome of *E. coli* is packaged in the nucleoid region in a
    a. nuclear membrane
    b. semicircular form
    c. relaxed form
    d. supercoiled form
11. Topoisomerases are enzymes that do all of the following except
    a. untwist relaxed DNA
    b. introduce negative supercoils into relaxed DNA
    c. introduce positive supercoils into relaxed DNA
    d. change the topological form of the DNA but not the DNA sequence
12. The total amount of DNA in the haploid genome of any organism is
    a. its karyotype
    b. its C value

c. twice its GC content
d. an indication of its organizational and structural complexity

13. Facultative heterochromatin
    a. is always inactive
    b. is inactive only in certain cells
    c. contains only moderately repetitive DNA
    d. can contain unique-sequence DNA
    e. both b and d
    f. both a and c
14. Which of the following is not true about both centromeres and telomeres?
    a. They are characteristically heterochromatic.
    b. They are associated with *CEN* sequence elements.
    c. They are essential for eukaryotic chromosome function.
    d. They contain a short, species-specific sequence that is repeated hundreds to thousands of times.
15. The fundamental unit of chromatin packaging in eukaryotes is
    a. a histone protein
    b. a nucleosome
    c. a 30-nm chromatin fiber
    d. a looped domain
16. Examples of tandemly repeated DNA sequences include all of the following except
    a. genes for ribosomal RNA
    b. LINEs
    c. genes for histones
    d. simple telomeric sequences
17. Which type of chromatin contains expressed genes?
    a. constitutive heterochromatin
    b. facultative heterochromatin
    c. euchromatin
    d. Barr bodies

Answers: 1b, 2a, 3a, 4d, 5b, 6d, 7c, 8a, 9e, 10d, 11c, 12b, 13e, 14d, 15b, 16b, 17c

Thought Questions

1. The histones, H1, H2A, H2B, H3, and H4, are the most highly conserved of all proteins. It has been proposed that they serve a basic function in all eukaryotes. What are the implications of this assertion for evolutionary theory?
2. The histones are highly basic proteins that interact with DNA (specifically, with the acidic sugar-phosphate backbone). What kinds of proteins are the nonhistones? How might they differ from the histones in their interactions with DNA?
3. How do you account for the presence of both unique and repetitive sequences in the genomes of eukaryotes? Why are the latter mostly lacking in prokaryotes?
4. Explain one of the fundamental puzzles of life: An amphibian that lives in a swamp has 30-fold more DNA than a human that lives on the 48th floor of a luxury apartment complex. In what way is the amphibian richer?

5. Summarize the structural composition and organization of the chromosomes of eukaryotic cells. How do they compare with the chromosomes of viruses and of prokaryotes?
6. Describe the logic behind the series of experiments that conclusively demonstrated that DNA was the genetic material in some cells. Why was this hotly debated for so long? When RNA was found to be the genetic material for some viruses, was the debate rekindled, or was this "the exception that proved the rule"?
7. In a double-stranded DNA molecule, how many strands of DNA exist in a single chromosome during the following stages of the cell cycle (mitosis and meiosis): G_1, G_2, prophase, metaphase, anaphase, pachynema, diplonema, anaphase I, and anaphase II?
8. Describe the various forms of DNA (A, B, Z) that have been identified. What are their most significant differences? Are any of these differences likely to have functional significance? If so, what are they?
9. Watson and Crick reputedly deduced the structure of DNA by applying observations made by others to molecular models, without any direct experimental observations of their own. Is this legitimate scientific procedure? If so, why? If not, why not?
10. Draw out the chemical structure of a double-stranded DNA molecule so that one strand has the sequence 5′-ATG-3′. Indicate the polarity of each strand, the location and the kinds of bonds that exist, and the approximate dimensions of the molecule.
11. Double-stranded DNA molecules have negatively charged sugar-phosphate backbones and a major and minor groove into which the chemical groups of the bases project. Some proteins that interact with DNA do so in a highly sequence-specific manner, while others interact in a largely sequence-nonspecific manner. What different features of the DNA molecule might these two classes of protein be recognizing?

Thought Questions for Media

After reviewing the media at the *Essential iGenetics* website, try to answer these questions.

a

1. In the animation depicting the experiment by Avery and his colleagues that showed that bacteria could be transformed with DNA, how did they show that polysaccharides were not the transforming principle?
2. In the animation depicting the Hershey-Chase experiment, how were the T2 phage initially labeled with either ^{35}S or ^{32}P?
3. The T2 phage labeled with ^{32}P only had their DNA labeled. Why doesn't this information, by itself, provide evidence that DNA is the genetic material of T2?
4. What experiment was done to provide evidence that DNA is the genetic material?
5. In the animation depicting how the *E. coli* chromosome is supercoiled, is the DNA of *E. coli* compacted by adding negative or positive supercoils?
6. Which enzyme is responsible for adding negative supercoils?
7. Which enzyme is responsible for removing negative supercoils?

i

1. We now know that the genetic material of viruses can be either RNA or DNA and that it can be either single- or double-stranded. Why is it still important to determine the type of nucleic acid associated with a new virus?
2. Once you determine the type of nucleic acid of a new virus, how can you determine which one of several known viruses is most closely related to it?

SOLUTIONS TO TEXT PROBLEMS

8.1 Griffith's experiment injecting a mixture of dead and live bacteria into mice demonstrated that (choose the correct answer)

a. DNA is double stranded
b. mRNA of eukaryotes differs from mRNA of prokaryotes
c. a factor was capable of transforming one bacterial cell type to another
d. bacteria can recover from heat treatment if live helper cells are present

Answer: c

8.2 In the 1920s, while working with *Streptococcus pneumoniae,* the agent that causes pneumonia, Griffith injected mice with different types of bacteria. For each of the following bacteria types injected, indicate whether the mice lived or died.

a. type *IIR*
b. type *IIIS*
c. heat-killed *IIIS*
d. type *IIR* + heat-killed *IIIS*

Answer:

a. Lived
b. Died
c. Lived
d. Died (DNA from the *IIIS* bacteria transformed the *IIR* bacteria to a virulent form.)

8.3 Several years after Griffith described the transforming principle, Avery, MacLeod, and McCarty investigated the same phenomenon.

a. List the steps they used to show that DNA from dead *S. pneumoniae* cells was responsible for the change from a nonvirulent to a virulent state.
b. Did their work affirm Griffith's work or disaffirm it, and how?
c. What was the role of the enzymes used in their experiments?

Answer:

a. They made extracts from *S* cells and showed that they could transform *R* cells by exposing them to these extracts. These extracts contained different cellular macromolecules, such as lipids, polysaccharides, proteins, and nucleic acids. They first showed that nucleic acids were the only macromolecular component that could transform the *R* cells. Then, by treating the nucleic acid component with either ribonuclease or deoxyribonuclease, they showed that the transforming principle co-purified with DNA. This strongly suggested that the genetic material was DNA. Although the work suggested that the genetic material was DNA, it could be criticized because the nucleic acids were contaminated by proteins.
b. Their work affirmed and extended Griffith's work in that it identified the probable nature of Griffith's transforming principle as being DNA.
c. Enzymes were used to destroy one of the two components of the nucleic acid mixture and served to identify which of two nucleic acids was the genetic material. Ribonuclease did not destroy the transforming principle, but deoxyribonuclease did

abolish transforming activity. This result supported the contention that DNA was the genetic material.

8.4 Hershey and Chase showed that when phages were labeled with ^{32}P and ^{35}S, the ^{35}S remained outside the cell and could be removed without affecting the course of infection, whereas the ^{32}P entered the cell and could be recovered in progeny phages. What distribution of isotope would you expect to see if parental phages were labeled with isotopes of

a. C?

b. N?

c. H?

Explain your answer.

Answer: a, b, and c. Phage ghosts (supernatant) and progeny would have isotope. Both amino acids and nucleic acids have carbon, nitrogen and hydrogen. If a parental phage was labeled with isotopes of C, N, or H, one would expect the phage to have a labeled protein coat as well as labeled DNA. As a consequence, these isotopes would be recovered in the DNA of the progeny phage, as well as in the phage ghosts left behind in the supernatant after phage infection. This experiment would not be very helpful to determine whether DNA or protein was the genetic material, as neither material would be selectively labeled. The elegance of the Hershey-Chase experiment was that the isotopes used selectively labeled only one component of the phage, so that it could be followed from one generation to the next.

8.5 The X-ray diffraction data obtained by Rosalind Franklin suggested that (choose the correct answer)

a. DNA is a helix with a pattern that repeats every 3.4 nm

b. purines are hydrogen bonded to pyrimidines

c. DNA is a left-handed helix

d. DNA is organized into nucleosomes

Answer: a

8.6 What evidence do we have that in the helical form of the DNA molecule the base pairs are composed of one purine and one pyrimidine?

Answer: Two different lines of evidence support the view that a base pair is composed of one purine and one pyrimidine.

(1) When the chemical components of double-stranded DNA from a wide variety of organisms were analyzed quantitatively by Chargaff, it was found that the amount of purines equaled the amount of pyrimidines. More specifically, it was found that the amount of adenine equaled the amount of thymine and that the amount of cytosine equaled the amount of guanine. The simplest hypothesis to explain these observations was the existence of complementary base pairing, A on one strand paired with T on the other strand, and G paired with C.

(2) More direct physical evidence was provided by X-ray diffraction studies. These established the dimensions of the DNA double helix and allowed for comparison with the known sizes of the bases. The diameter of the double helix is constant throughout its length at 2 nm. This is the right size to accommodate a purine paired with a pyrimidine, but too small for a purine-purine pair and too large for a pyrimidine-pyrimidine pair.

8.7 What exactly is a deoxynucleotide made up of, and how many different deoxynucleotides are there in DNA? Describe the structure of DNA, and describe the bonding mechanism of the molecule (i.e., the kind of bonds are on the sides of the "ladder" and the kind of bonds holding the two complementary strands together). Base pairing in DNA consists of purine-pyrimidine pairs, so why can't A-C and G-T pairs form?

Answer: A deoxynucleotide consists of 2′-deoxyribose plus a phosphate group (PO_4^{-2}) attached to its 5′ carbon, plus a nitrogenous base attached to its 1′ carbon. Since there are four different nitrogenous bases in DNA, adenine (A), thymine (T), guanine (G), and cytosine (C), there are four different nucleotides. In a DNA molecule, one finds monophosphate nucleotides, so that there is deoxyadenosine monophosphate (dAMP), thymidine monophosphate (TMP), deoxyguanosine monophosphate (dGMP) and deoxycytidine monophosphate (dCMP). Along the sides of the "ladder," the 5′ carbon of one deoxyribose is connected by a covalent phosphodiester (O-P-O) bond to the 3′ carbon of another. A phosphate group is found at the 5′ end of a DNA polynucleotide chain, and a hydroxyl (OH) group is found at the 3′ end of a DNA polynucleotide chain. Weaker hydrogen bonds between complementary A-T and G-C bases hold the complementary strands together. A-T pairs have two hydrogen bonds, while G-C pairs have three hydrogen bonds. See text Fig. 8.11.

Although A-C and G-T pairs would be purine-pyrimidine base pairs, they do not form because of their inability to pair using hydrogen bonding. Note that A-T base pairs have two hydrogen bonds, while G-C base pairs have three hydrogen bonds. This is consistent with the evidence provided by Chargaff. Quantitative measurements of the four bases in double-stranded DNA isolated from a wide variety of organisms indicated that in all cases, the amount of A equaled the amount of T, and the amount of G equaled the amount of C. Moreover, different DNAs exhibited different base ratios, so that while (A) = (T) and (G) = (C), in most organisms (A + T) ≠ (G + C). Put another way, the percent GC content of different DNA samples varies. The simplest hypothesis is that there are two base pairs in DNA, A-T and G-C, and the proportion of the two base pairs varies from organism to organism.

8.8 What is the base sequence of the DNA strand that would be complementary to the following single-stranded DNA molecules?

a. `5'-AGTTACCTGATCGTA-3'`

b. `5'-TTCTCAAGAATTCCA-3'`

Answer:

a. `3'-TCAATGGACTAGCAT-5' (or 5'-TACGATCAGGTAACT-3').`

b. `3'-AAGAGTTCTTAAGGT-5' (or 5'-TGGAATTCTTGAGAA-3').`

8.9 Describe the bonding properties of G-C and T-A. Which would be the hardest to break? Why?

Answer: The adenine-thymine base pair has two hydrogen bonds, while the guanine-cytosine base pair has three hydrogen bonds. Thus, the guanine-cytosine base pair requires more energy to break apart and so is harder to break apart.

8.10 The double helix model of DNA, as suggested by Watson and Crick, was based on data gathered on DNA by other researchers. The facts fell into the following two general categories; give two examples of each.

a. chemical composition

b. physical structure

Answer: The double helix model of DNA suggested by Watson and Crick had to incorporate existing information about its chemical composition and physical structure. In terms of its chemical composition, it was known that DNA is composed of polynucleotides, that (A) = (T) and (G) = (C) (Chargaff's rules), and that while the percent GC varies between organisms, the A/T and G/C ratios do not. In terms of its physical structure, the structure and molecular dimensions of the component molecules (the bases, sugars, and phosphates) were known. It was also known from studies of Franklin and Wilkins that the molecule is organized in a highly ordered, helical structure and that there are two distinctive regularities at 0.34 and 3.4 nm along the molecule's axis.

8.11 For double-stranded DNA, which of the following base ratios always equals 1?

a. (A + T)/(G + C)

b. (A + G)/(C + T)

c. C/G

d. (G + T)/(A + C)

e. A/G

Answer: Since (G) = (C) and (A) = (T), it follows that (G + A) = (C + T) and (G + T) = (A + C). Thus, b, c, and d are all equal to 1.

8.12 If the ratio of (A + T) to (G + C) in a particular DNA is 1, does this indicate that the DNA is probably constituted of two complementary strands of DNA or a single strand of DNA, or is more information necessary?

Answer: More information is needed. That (A + T)/(G + C) = 1 indicates only that (A + T) = (G + C). If the DNA were double-stranded, (G) = (C), and (A) = (T). For the observed ratio, there would need to be 25 percent A, 25 percent T, 25 percent C, and 25 percent G. However, if the DNA were single-stranded, one could still observe this ratio. In single-stranded DNA, there are no restrictions on the relative amounts of the different bases. There are many ways in which one could observe (A + T)/(G + C) = 1. For example, suppose there were 35 percent A, 15 percent T, 20 percent G, and 30 percent C.

8.13 The percentage of cytosine in a double-stranded DNA is 17. What is the percentage of adenine in that DNA?

Answer: In double-stranded DNA, if (C) = 17 percent, then (G) = 17 percent. This means that the DNA has 34 percent G-C, and 66 percent A-T base pairs. Hence, the DNA will have $(^{66}/_{2})(100\%) = 33\%$ A.

8.14 A double-stranded DNA polynucleotide contains 80 thymidylic acid and 110 deoxyguanylic acid residues. What is the total nucleotide number in this DNA fragment?

Answer: Since the DNA molecule is double-stranded, (A) = (T) and (G) = (C). If there are 80 T residues, there must be 80 A residues. If there are 110 G residues, there must be 110 C residues. The molecule has (110 + 110 + 80 + 80) = 380 nucleotides, or 190 base pairs.

8.15 Analysis of DNA from a bacterial virus indicates that it contains 33 percent A, 26 percent T, 18 percent G, and 23 percent C. Interpret these data.

Answer: First, notice that (A) ≠ (T) and (G) ≠ (C). Thus, the DNA is not double-stranded. The bacterial virus appears to have a single-stranded DNA genome.

8.16 The genetic material of bacteriophage ΦX174 is single-stranded DNA. What base equalities or inequalities might we expect for single-stranded DNA?

Answer: No predictions can be made regarding the base content of single-stranded DNA. Any base pair equality would depend on the overall sequence of the chromosome: (A) might be equal to (T), but that need not be the case, as that is only one of many possibilities.

8.17 Through X-ray diffraction analysis of crystallized DNA oligomers, different forms of DNA have been identified. These forms include A-DNA, B-DNA, and Z-DNA, and each has unique molecular attributes.

a. Which of these forms is the most common form in living cells?
b. Z-DNA has an unusual conformation resulting in more base pairs per helical turn than B-DNA. What is the conformation? Does this molecule have any function in living cells?
c. Which of these forms is never found in living cells?

Answer:

a. B-DNA is the form most common to living cells.
b. To understand why Z-DNA is considered to have an unusual conformation, it helps to compare it with B-DNA. Z-DNA is a long and thin (about 2 nm wide) double helix, like B-DNA. Unlike the right-handed B-DNA, however, Z-DNA is left-handed. It has an axis that runs through the minor groove and has 12 base pairs per helical turn that are inclined 8.8° from a plane perpendicular to the axis. In contrast, B-DNA has an axis that runs through the base pairs and has 10 base pairs per helical turn that are inclined 2° from a plane perpendicular to the axis. The major groove of Z-DNA is not very distinct, as it is thin and flattened out along the helix surface, while the major groove of B-DNA is wide and intermediate in depth (between that of A-DNA and Z-DNA). The minor groove of Z-DNA is extremely narrow and very deep, while that of B-DNA is narrow and of intermediate depth. It has been proposed that regions with Z-DNA

provide a stretch of left-handed helical turns involved in replication, recombination, and transcription. Stretches of left-handed turns may aid in unwinding right-handed helical turns in B-DNA during these processes. Z-DNA may also be more stable under extreme environmental conditions.

c. A-DNA is found only when the DNA is dehydrated, so it is unlikely that lengthy sections of A-DNA would be found in living cells.

8.18 If a virus particle contains double-stranded DNA with 200,000 bp, how many complete 360° turns occur in this molecule?

Answer: With 10 base pairs per complete 360° turn of a B-form double-stranded DNA molecule, there are 200,000/10 = 20,000 complete turns in the viral DNA.

8.19 A double-stranded DNA molecule is 100,000 bp (100 kb) long.

a. How many nucleotides does it contain?

b. How many complete turns are there in the molecule?

c. How long is the DNA molecule?

Answer:

a. Each base pair has 2 nucleotides, so the molecule has 200,000 nucleotides.

b. There are 10 base pairs per complete 360° turn, so there will be 100,000/10 = 10,000 complete turns in the molecule.

c. There is 0.34 nm between the centers of adjacent base pairs. There will be 100,000 $\times$ 0.34 nm = 3.4×10^4 nm = 34 μm.

8.20 Different organisms have vastly different amounts of genetic material. *E. coli* has about 4.6 million bp of DNA in one circular chromosome; the haploid budding yeast (*S. cerevisiae*) has 12,067,280 bp of DNA in 16 chromosomes; and the gametes of humans have about 2.75 billion bp of DNA in 23 chromosomes.

a. If all of the DNA were B-DNA, what would be the average length of a chromosome in these cells?

b. On average, how many complete turns would be in each chromosome?

c. Would your answers to (a) and (b) be significantly different if the DNA were composed of, say, 20 percent Z-DNA and 80 percent B-DNA?

d. What implications do your answers to these questions have for the packaging of DNA in cells?

Answer:

a. For *E. coli,* the length is (4.6×10^6 bp) $\times$ (0.34 nm/bp) = 1.6×10^6 nm = 1,600 μm (microns). Since *E. coli* has a circular chromosome, this would be a chromosome with a diameter of about 510 μm. For yeast, the average chromosome length is (12,067,280 bp/16 chromosomes) $\times$ (0.34 nm/bp) = 2.56×10^5 nm = 256 μm. For humans, the average chromosome length is (2.75×10^9 bp/23 chromosomes) $\times$ (0.34 nm/bp) = 4.07×10^7 nm = 40,700 μm.

b. For B-DNA, there are 10 bp per helical turn. Therefore, the *E. coli* chromosome would have (4.6×10^6 bp/chromosome) $\times$ (1 turn/10 bp) = 4.6×10^5 turns. The average yeast chromosome would have (12,067,280 bp/16 chromosomes) $\times$ (1 turn/10 bp)

= 7.54×10^4 turns. The average human chromosome would have (2.75×10^9 bp/23 chromosomes) × (1 turn/10 bp) = 1.2×10^7 turns.

c. While Z-DNA and B-DNA differ in the distance between successive base pairs and the number of base pairs per turn, these are not large compared to chromosome length. Thus, the answers to (a) and (b) will not be very different.

Z-DNA has more space between successive base pairs, as there is about 0.57 nm between successive base pairs instead of the 0.34 nm found in B-DNA. Thus, if 20% of each chromosome were Z-DNA, each chromosome would be 20% × [(0.57 – 0.34)/0.34] = 13.5% longer. A full turn of the double helix in Z-DNA utilizes more base pairs, as there are 12 bp/turn in Z-DNA instead of the 10 bp/turn in B-DNA. Thus, if 20% of each chromosome were Z-DNA, each chromosome would have 20% × [(12 – 10)/12] = 3.33% fewer turns.

d. These answers point out the need for DNA to be flexible so that it can be packaged and "fit" into a cell. Other forms of DNA, including Z-DNA, do not necessarily shorten a chromosome but they can affect its flexibility. In the examples considered in this problem, the cells are hundreds to tens of thousands of times smaller than the length of their (average) chromosome. *E. coli* is about 2 μm × 0.5 μm, with a 510-μm diameter chromosome; a yeast cell is about 20 μm wide, with 16 chromosomes that average 256 μm; and a "typical" human cell is about 10 μm wide, with 46 chromosomes that average 40,700 μm. Thus, chromosomes cannot remain as unpackaged molecules inside cells.

8.21 Define topoisomerases, and list the functions of these enzymes.

Answer: Topoisomerases are a class of enzymes that convert one topological form of DNA to another. Topoisomerases untwist relaxed DNA to produce negative supercoils or twist negatively supercoiled DNA to convert it to a more relaxed state. These enzymes allow DNA to be converted between negatively supercoiled, compact DNA and a more relaxed, less compact state.

8.22 In a particular eukaryotic chromosome (choose the best answer),

a. heterochromatin and euchromatin are regions where genes make functional gene products (where they are active)
b. heterochromatin is active, but euchromatin is inactive
c. heterochromatin is inactive, but euchromatin is active
d. both heterochromatin and euchromatin regions are inactive

Answer: c

8.23 Compare and contrast eukaryotic chromosomes and bacterial chromosomes, considering the following features.

a. centromeres
b. hexose sugars
c. amino acids
d. supercoiling
e. telomeres
f. nonhistone proteins
g. DNA
h. nucleosomes
i. circular chromosome
j. looping

Answer:

a. Only eukaryotic chromosomes have centromeres, the sections of the chromosome found near the point of attachment of mitotic or meiotic spindle fibers. In some organisms, such as *S. cerevisiae,* they are associated with specific *CEN* sequences. In other organisms, they have a more complex repetitive structure.
b. Hexose sugars are not found in either eukaryotic or bacterial chromosomes, as *pentose* sugars are found in DNA.
c. Amino acids are found in proteins that are involved in chromosome compaction, such as the proteins that hold the ends of looped domains in prokaryotic chromosomes and the histone and nonhistone proteins in eukaryotic chromatin.
d. Both eukaryotic and bacterial chromosomes share supercoiling.
e. Telomeres are only found at the ends of eukaryotic chromosomes and are required for replication and chromosome stability. They are associated with specific types of sequences: simple telomeric sequences and telomere-associated sequences.
f. Nonhistone proteins are found only in eukaryotic chromosomes and have structural (higher-order packaging) and possibly other functions.
g. DNA is found in both prokaryotic and eukaryotic chromosomes (although some viral chromosomes have RNA as their genetic material).
h. Nucleosomes are the fundamental unit of packaging of DNA in eukaryotic chromosomes and are not found in prokaryotic chromosomes.
i. Circular chromosomes are only found in prokaryotes, not in eukaryotes.
j. Looping is found in both eukaryotic and prokaryotic chromosomes. In eukaryotic chromosomes, the 30-nm nucleofilament is packed into looped domains by nonhistone chromosomal proteins. In bacterial chromosomes, such as that of *E. coli,* there are about 100 looped domains, each containing about 40 kilobases of supercoiled DNA.

8.24 Discuss the structure of nucleosomes, the hierarchical packaging, and the composition of a core particle of a nucleosome. Also discuss the functions of nucleosomes.

Answer: Nucleosomes are the fundamental units of DNA packaging in eukaryotic chromosomes. In a nucleosome core particle, a short segment of DNA is wrapped about 1¾ times around a protein core. This protein core is an octamer consisting of two copies of each of four histone proteins: H2A, H2B, H3, H4. This nucleosome core particle packs the DNA into a flattened disk about 5.7 × 11 nm. This packing effectively condenses the DNA into a smaller size, condensing it by a factor of about seven. The interactions between the histones and the DNA are not sequence-specific, but rather are based on the basic, positively charged histone proteins interacting with the acidic, negatively charged sugar-phosphate backbone of the DNA. Nucleosome core particles serve to package double-stranded DNA at regular intervals, with short stretches of linker DNA between particles. At this level of packaging, DNA is packed into a "beads-on-a-string" form of chromatin known as the 10-nm nucleofilament. This condensation of DNA also forms the basis for subsequent packaging, first through associations between nucleosomes and then by associations with nonhistone chromosomal proteins. Associations between nucleosomes result in further compaction of the 10-nm nucleofilament, producing a 30-nm chromatin fiber that condenses the DNA another sixfold. Histone H1 plays an important role in this coiling. The 30-nm chromatin fiber is further condensed into looped domains, with each loop containing between ten and hundreds of kilobases of DNA. The looped domains are anchored inside the nuclear envelope to a filamentous structural framework of protein, the nuclear matrix.

8.25 Answer the questions below after setting up the following "rope trick." Start with a belt (representing a DNA molecule; imagine the phosphodiester backbones lying along the top and bottom edges of the belt) and a soda can. Holding the belt buckle at the bottom of the can, wrap the belt flat against the side of the can, counterclockwise three times around the can. Now remove the "core" soda can and, holding the ends of the belt, pull the ends of the belt taut. After some reflection, answer the following questions.

a. Did you make a left- or a right-handed helix?
b. How many helical turns were present in the coiled belt before it was pulled taut?
c. How many helical turns were present in the coiled belt after it was pulled taut?
d. Why does the belt appear more twisted when pulled taut?
e. About what percentage of the length of the belt was decreased by this packaging?
f. Is the DNA of a linear chromosome that is coiled around histones supercoiled?
g. Why are topoisomerases necessary to package linear chromosomes?

Answer:

a. The belt forms a right-handed helix. Although you wrapped the belt around the can axis in a counterclockwise direction from your orientation (looking down at the can), the belt was winding up and around the side of can in a clockwise direction from its orientation. With the belt wrapped around the can, curve the fingers of your right hand over it, and use your index finger to trace the direction of the belt's spiral. Your right index finger will trace the spiral upward, the same direction your thumb points when you wrap your hand around the can. Therefore, the belt has formed a right-handed helix. If you do this with your left hand, your thumb will point away from the direction the helix twists, indicating that it has not formed a left-handed helix.
b. Three
c. Three. The number of helical turns is unchanged, although the twist in the belt is changed.
d. The belt appears more twisted because the pitch of the helix was altered and the edges of the belt (positioned much like the complementary base pairs of a double helix) are twisted more tightly.
e. While twisted around the can, the length of the belt decreases by about 70 to 80 percent, depending on the initial length of the belt and the belt diameter.
f. Yes. The DNA of linear chromosomes becomes supercoiled as it is wrapped around histones to form the 10-nm nucleofilament. Just as you must add twists to the belt for it to lie flat on the surface of the can, supercoils must be introduced into the DNA for it to wrap around the histones.
g. Topoisomerases increase or reduce the level of supercoiling in DNA. For linear DNA to be packaged, negative supercoils must be added.

8.26 Arrange the following figures in increasing order of eukaryotic chromosome condensation. Designate the simplest chromosome organization as number 1 and the most complex chromosome organization as number 6.

A. Chromatin fiber of packed nucleosomes

B. Metaphase chromosome

C. DNA double helix

D. Extended section of chromosome

E. "Beads-on-a string" form of chromatin

F. Condensed section of chromosome

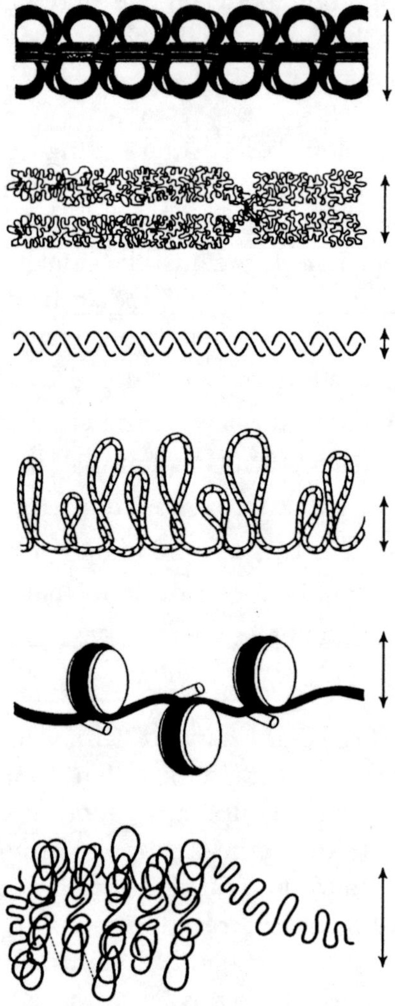

From Alberts et al., *Molecular Biology of the Cell,* 2nd ed., p. 504. New York: Garland Publishing, Inc., 1989.

Answer: 1 = C, DNA double helix

↓

2 = E, "beads-on-a-string" form of chromatin

↓

3 = A, chromatin fiber of packed nucleosomes

↓

4 = D, extended section of chromosome

↓

5 = F, condensed section of chromosome

↓

6 = B, metaphase chromosome

8.27 Telomeres are unique repeated sequences. Where on the DNA strand are they found? Do they serve a function?

Answer: Telomeres are characteristically heterochromatic sequences found at the ends of a linear, eukaryotic chromosome. For most organisms, the telomeric sequences at the extreme end of a chromosome are simple, highly repeated, and species-specific. Nearby, but not at the very end of a chromosome, are telomere-associated sequences. These are repeated, complex sequences and extend for thousands of bases from the simple telomeric sequences. These sequences are replicated, along with other chromosomal DNA sequences, by a set of enzymes that include DNA polymerases. Telomere organization is quite different in some organisms. In organisms such as *Drosophila,* telomeres consist of transposons belonging to the LINE family of repeated sequences. Telomeres function in DNA replication and provide chromosome stability.

8.28 Would you expect to find most protein-coding genes in unique-sequence DNA, in moderately repetitive DNA, or in highly repetitive DNA?

Answer: In unique-sequence DNA

8.29 Both histone and nonhistone proteins are essential for DNA packaging in eukaryotic cells. However, these classes of proteins are fundamentally dissimilar in a number of ways. Describe how these classes of proteins differ in terms of

a. their protein characteristics
b. their presence and abundance in cells
c. their interactions with DNA
d. their role in DNA packaging

Answer:

a. The five histones (H1, H2A, H2B, H3, and H4) are small, basic, positively charged proteins that are rich in arginine and lysine. Histones H2A, H2B, H3, and H4 are among the most highly conserved of all proteins. In contrast to the small number of histones, there are many types of nonhistones. They are variable in size and are usually negatively charged, acidic proteins.

b. The five histones are present in all cells of all eukaryotic organisms. Relative to the total amount of DNA, they are present in fairly constant amounts. In contrast, nonhistones differ in number and type between cell types in an organism, at different times in the same cell type, and from organism to organism. As a class, nonhistones may be between 50 and 100 percent of the mass of DNA.

c. Histones are positively charged and mostly interact with the negatively charged sugar-phosphate backbone of DNA. Nonhistones are typically negatively charged and can interact with the positively charged histones or with DNA. The HMG class of nonhistones binds to the minor groove of the DNA and causes DNA bending.

d. The histones H2A, H2B, H3, and H4 are fundamental to packaging DNA into nucleosome core particles and the formation of the 10-nm nucleofilament. Histone H1 is involved in compaction of this nucleofilament to produce a 30-nm chromatin fiber. Nonhistones are important in higher-order packaging of DNA. They form the protein scaffold that anchors the looped domains of the 30-nm chromatin fiber, and they control the degree of chromosome condensation.

9

DNA REPLICATION

CHAPTER OUTLINE

SEMICONSERVATIVE DNA REPLICATION

- The Meselson-Stahl Experiment
- Semiconservative DNA Replication in Eukaryotes

ENZYMES INVOLVED IN DNA SYNTHESIS

- DNA Polymerase I
- Roles of DNA Polymerases

MOLECULAR MODEL OF DNA REPLICATION

- Initiation of Replication
- Semidiscontinuous DNA Replication

DNA REPLICATION IN EUKARYOTES

- DNA Replication and the Cell Cycle
- Eukaryotic Replication Enzymes
- Replicons
- Origins of Replication
- Replicating the Ends of Chromosomes
- Assembling New DNA into Nucleosomes

DNA RECOMBINATION

REVIEW OF KEY TERMS, SYMBOLS, AND CONCEPTS

In your own words, write a brief, precise definition of each term in the groups on the following page. Check your definitions using the text. Then develop a concept map using the terms in each list.

THINKING ANALYTICALLY

Because of its fundamental importance in genetics and in treating diseases such as cancer, there has been a considerable interest in understanding the mechanism of DNA replication. At this point, we understand the mechanism of DNA replication in considerable detail. For example, we know that DNA replicates semiconservatively, and we know the nature of the complex set of proteins and enzymes that act in a sequential, coordinated fashion to replicate DNA. The best way to

1	2	3
semiconservative DNA replication	semidiscontinuous DNA replication	eukaryote
conservative DNA replication	origin of replication	cell cycle
dispersive DNA replication	replication bubble	checkpoint
Meselson-Stahl experiment	DNA helicase, primase	START
CsCl equilibrium gradient	primosome	cyclins
^{15}N, ^{14}N	replisome	Cdk, MPF
buoyant density	RNA primer	nucleus, mitochondrion
harlequin chromosomes	SSB protein	DNA polymerase α, β, δ, γ, ϵ
BUdR	replication fork	DNA replication
base analogue	Okazaki fragments	DNA repair
DNA polymerase I, II, III	bidirectional replication	proofreading ability
dNTP, dNMP, Mg^{2+}	DNA polymerase I, III	replicon
Kornberg enzyme	$5' \rightarrow 3'$ synthesis	ARS
template strand	proofreading	telomere
nontemplate strand	$3' \rightarrow 5'$, $5' \rightarrow 3'$ exonuclease	telomerase
primer	leading, lagging strand	reverse transcription
	DNA ligase	reverse transcriptase
	polA1, *polA1ex1* mutants	*TLC1*, *EST1*, *TEL1*, *TEL2* genes
	conditional mutant	

understand and retain the details of this complex process is to place them in context. Start by developing a general understanding of the steps of DNA replication. Ask yourself: What must happen first? How is this accomplished? What happens next? How is that accomplished? Then focus on understanding the activities of each of the enzymes and proteins used in each step. To help develop a contextual understanding and establish a mental image of each of the steps, refer to the text figures that summarize the key events of DNA replication.

The following related exercise will help you assess your understanding and retention of the material. After reading each section, close the text and sketch out the aspect of DNA replication that has just been presented. Then relate the figure you have sketched to the previous steps of DNA replication. This will force you to confront unclear concepts and help you identify details you might have missed.

QUESTIONS FOR PRACTICE

Multiple Choice Questions

1. Consider the Meselson-Stahl experiment where *E. coli* were grown in ^{15}N medium. Which model of DNA replication is eliminated by analyzing DNA after exactly one round of replication in ^{14}N medium?
   a. semiconservative
   b. dispersive
   c. conservative

2. The enzymes most directly concerned with catalyzing DNA synthesis are DNA
   a. ligases
   b. exonucleases
   c. polymerases
   d. primases
3. Which of the following is not essential for the in vitro synthesis of DNA?
   a. magnesium ions
   b. DNA polymerase I
   c. DNA primase
   d. a DNA fragment
4. Which enzyme untwists the parent strands of DNA during replication?
   a. DNA primase
   b. DNA helicase
   c. DNA gyrase
   d. DNA ligase
5. Which of the following *do not* provide evidence for semiconservative DNA replication?
   a. The Meselson-Stahl experiment
   b. harlequin chromosomes
   c. the existence of ARS elements
   d. Okazaki fragments
6. Which of the following *E. coli* polymerases have proofreading activity?
   a. DNA polymerase I
   b. DNA polymerase II
   c. DNA primase
   d. DNA polymerase III
   e. a, b, and d
7. During chromosome replication, new histones are synthesized during
   a. G_1 and S, and nucleosomes are re-formed using new and old histones
   b. G_1 and S, and new nucleosomes are formed with only these new histones
   c. G_1, S, and G_2, and nucleosomes are re-formed using new and old histones
   d. G_1, S, and G_2, and new nucleosomes are formed with only these new histones
8. During the eukaryotic cell cycle, checkpoints serve to determine
   a. whether the mitochondria of a cell should replicate their DNA
   b. whether the cell is infected with a virus
   c. whether the cell requires the activity of telomerase
   d. whether the cell is able to or should continue to the next phase of the cell cycle
9. What is the function of the enzyme telomerase in mammals?
   a. to replicate telomeres in all cell types
   b. to replicate and expand simple telomeric repeats in germ cells and tumor cells
   c. to expand the length of simple telomeric repeats in brain cells
   d. to replicate telomere-associated sequences and simple telomeric repeats in all cells

Answers: 1c, 2c, 3c, 4b, 5c, 6e, 7b, 8d, 9b

Thought Questions

1. Describe the mechanism of semidiscontinuous DNA replication.
2. In his investigation of the requirements for in vitro synthesis of DNA, Kornberg found that an absolute minimum of four components was necessary. What were these components, and why in particular were each and all necessary? Also, what limitations did his in vitro method have as compared to in vivo synthesis, which involves other components, such as helicase, primase, and ligase?
3. Describe how eukaryotic DNA is assembled into nucleosomes after DNA replication. Where do the new nucleosomes come from? How might you experimentally determine whether new nucleosomes are composed of all new or some old and some new histone proteins? (Hint: Consider the Meselson-Stahl approach.)
4. What phenotype would you expect to see in a mutant in the *E. coli polA* gene that showed heat-sensitive 3′→5′ exonuclease activity, but normal 5′→3′ exonuclease activity at all temperatures?
5. In humans, how are simple telomeric repeats replicated in a fast-dividing population of cells such as epithelial cells? Are all of them replicated at each cell division?
6. The text describes evidence that telomere length is under genetic control and that in mammals, only immortal cells (e.g., tumor cells, germ cells) have telomerase activity. Why might this be important? In particular, why might it be important for some cells *not* to have telomerase activity?

Thought Questions for Media

After reviewing the media at the *Essential iGenetics* website, try to answer these questions.

a

1. In the animation depicting the experiment by Meselson and Stahl that showed that DNA replication was semiconservative, how did they initially label DNA with ^{15}N?
2. In the animation depicting DNA synthesis using DNA polymerase, when proofreading occurs, what type of molecule is released?
3. In the animation depicting the semidiscontinuous model of DNA replication, in which direction is DNA synthesis on the lagging strand and which direction is this relative to the movement of the replication fork?

i

1. How can you experimentally determine if a replication fork proceeds bidirectionally or unidirectionally?
2. How can you experimentally determine whether DNA primase randomly starts synthesizing an RNA primer or whether it primes DNA synthesis only at specific sites?
3. In the text, the function of *E. coli* DNA polymerase II was not described. Based on the iActivity, what genetic analysis could you perform to demonstrate that it is involved in DNA repair?

SOLUTIONS TO TEXT PROBLEMS

9.1 Describe the Meselson-Stahl experiment, and explain how it showed that DNA replication is semiconservative.

Answer: By growing *E. coli* for many generations in media containing "heavy" nitrogen, or ^{15}N, bacterial DNA was uniformly labeled with heavy nitrogen. Such ^{15}N DNA has a greater buoyant density than ^{14}N DNA and can be differentiated from ^{14}N DNA by its banding position in a CsCl density gradient. Semiconservative DNA replication was demonstrated by placing *E. coli* with ^{15}N DNA into ^{14}N medium and following the density of the DNA that was present in cells after growth proceeded through each of several successive generations. If DNA replication were conservative, after one cell division and one round of DNA replication one would expect to see two distinct, equally dense bands, one corresponding to ^{14}N DNA and one corresponding to ^{15}N DNA. This was not seen. If DNA replication were either semiconservative or dispersive, after one cell division and one round of DNA replication one would expect to see one band consisting of DNA with a density halfway between ^{14}N DNA and ^{15}N DNA. This was seen. To distinguish further between semiconservative and dispersive replication, the consequences of another round of cell division and DNA replication were examined. In semiconservative replication, one would expect two bands, one of density halfway between ^{14}N DNA and ^{15}N DNA and one having ^{14}N DNA density. In dispersive replication, only DNA with both ^{14}N DNA and ^{15}N DNA would be seen. The density of the two bands seen (one at the level of ^{14}N DNA and one at the level of half ^{14}N DNA and half ^{15}N DNA) supported the semiconservative model. See text Figure 9.2.

9.2 In the Meselson-Stahl experiment, ^{15}N-labeled cells were shifted to ^{14}N medium at what we can designate as generation 0.

a. For the semiconservative model of replication, what proportion of ^{15}N-^{15}N, ^{15}N-^{14}N, and ^{14}N-^{14}N would you expect to find after 1, 2, 3, 4, 6, and 8 replication cycles?

b. Answer (a) in terms of the conservative model of DNA replication.

	Semiconservative Model			**Conservative Model**		
Generation	^{15}N-^{15}N	^{15}N-^{14}N	^{14}N-^{14}N	^{15}N-^{15}N	^{15}N-^{14}N	^{14}N-^{14}N
1	0	1	0	1/2	0	1/2
2	0	1/2	1/2	1/4	0	3/4
3	0	1/4	3/4	1/8	0	7/8
4	0	1/8	7/8	1/16	0	15/16
6	0	1/32	31/32	1/64	0	63/64
8	0	1/128	127/128	1/256	0	255/256

9.3 A spaceship lands on Earth and with it is a sample of extraterrestrial bacteria. You are assigned the task of determining the mechanism of DNA replication in this organism.

You grow the bacteria in unlabeled medium for several generations, then grow it in the presence of ^{15}N for exactly one generation. You extract the DNA and subject it to CsCl centrifugation. The banding pattern you find is as follows:

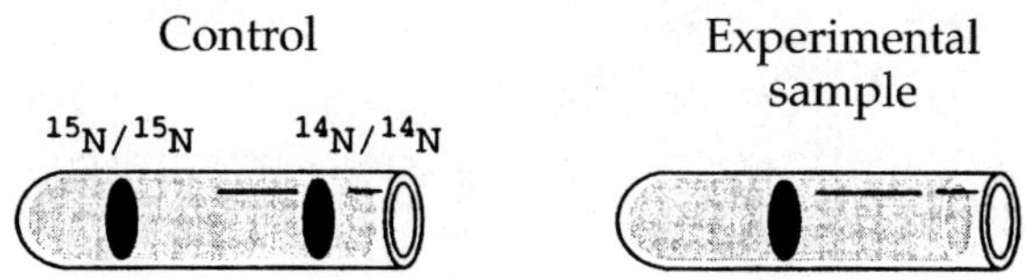

It appears to you that this is evidence that DNA replicates in the semiconservative manner, but you are wrong. Why? What other experiment could you perform (using the same sample and technique of CsCl centrifugation) that would further distinguish between semiconservative and dispersive modes of replication?

Answer: The CsCl centrifugation result eliminates the possibility of the conservative model of replication, but is still consistent with either semiconservative or dispersive models of DNA replication. To distinguish between these two possibilities using the same sample and the technique of CsCl centrifugation, one could denature the DNA and then subject the single-stranded sample to CsCl centrifugation. This could be done in practice by using an alkaline CsCl gradient, as the two DNA strands will denature at high pH. The expected results are shown below.

If semiconservative

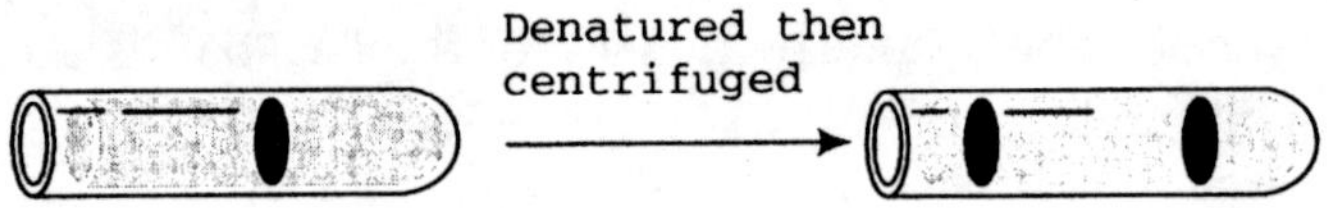

If dispersive

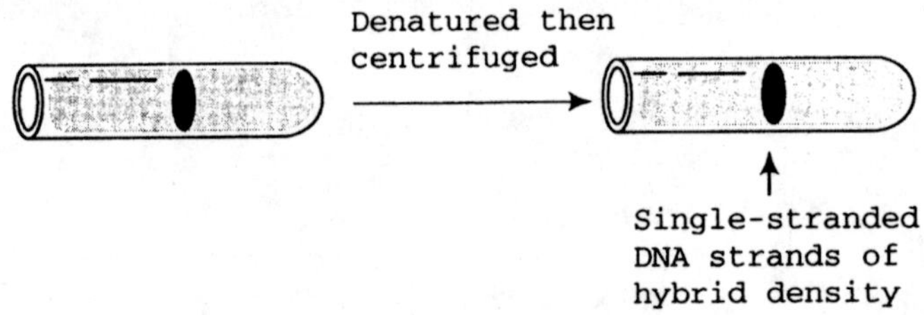

9.4 The elegant Meselson-Stahl experiment was among the first experiments to contribute to what is now a highly detailed understanding of DNA replication. Reconsider this experiment in light of current molecular models by answering the following questions.

a. Does the fact that DNA replication is semiconservative mean that it must be semidiscontinuous?

b. Does the fact that DNA replication is semidiscontinuous ensure that it is also semiconservative?

c. Do any properties of known DNA polymerases ensure that DNA is synthesized semiconservatively?

Answer:

a. That DNA replication is semiconservative does not mean that it also must be semidiscontinuous. For example, if DNA polymerase were able to synthesize DNA in both the 3′→5′ and 5′→3′ directions, DNA replication could proceed continuously on both DNA strands while still being semiconservative. Alternatively, if each of the two old strands were completely (or even substantially) unwound, and replication were initiated from the 3′ end (or 3′ region) of each, it could proceed continuously (or mostly continuously) in a 5′→3′ direction along each strand.

b. That DNA replication is semidiscontinuous does ensure that it is semiconservative. In the semidiscontinuous model, each old, separated strand serves as a template for a new strand. This is the essence of the semiconservative model.

c. That DNA polymerase synthesizes just one new strand from each "old" single-stranded template and can synthesize in only one direction (5′→3′) ensures that replication is semiconservative.

9.5 List the components necessary to make DNA in vitro using the enzyme system isolated by Kornberg.

Answer: DNA can be synthesized in vitro using the Kornberg enzyme (DNA polymerase I), all four dNTPs (dATP, dGTP, dCTP, dTTP), magnesium ions (Mg^{2+}), and a fragment of DNA that will serve as a template.

9.6 Kornberg isolated DNA polymerase I from *E. coli.* DNA polymerase I has an essential function in DNA replication. What are the functions of the enzyme in DNA replication?

Answer: DNA polymerase I functions to remove the RNA primer synthesized during the initiation of DNA replication and replace this RNA with DNA. When DNA polymerase III, the main synthetic enzyme for DNA polymerization, reaches an RNA primer, it dissociates from the DNA. DNA polymerase I functions to continue synthesis of the DNA in a 5′→3′ direction. It simultaneously removes the RNA primer using its 5′→3′ exonuclease activity and replaces the RNA with DNA nucleotides.

9.7 Assume you have a DNA molecule with the base sequence **TATCA** going from the 5′ to the 3′ end of one of the polynucleotide chains. The building blocks of the DNA are drawn as in the following figure. Use this shorthand system to diagram the completed double-stranded DNA molecule, as proposed by Watson and Crick.

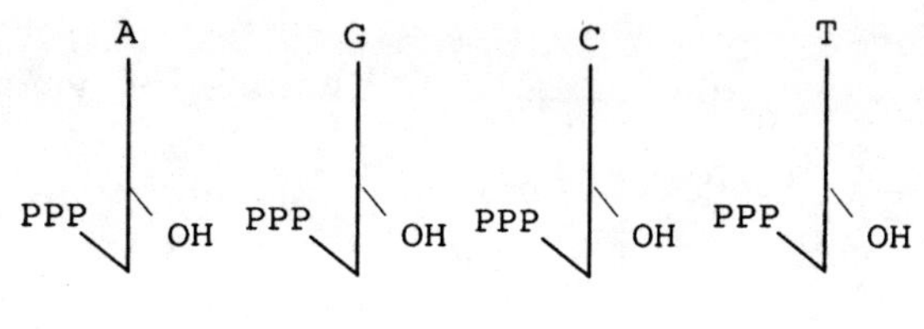

Answer: The deoxyribonucleotides have a triphosphate group at the 5′ end of the molecule and a hydroxyl group at the 3′ end of the molecule. This information establishes the polarity of each of the deoxyribonucleotides (dATP, dGTP, dTTP, and dCTP). Since the phosphodiester bond will form between the 5′ and 3′ ends of adjacent nucleotides, the 5′→3′ bonding of the **TATCA** oligonucleotide can be represented as

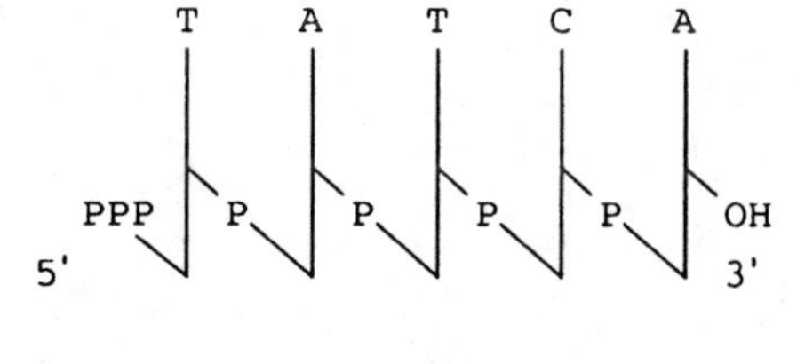

The complementary strand will base-pair with antiparallel orientation:

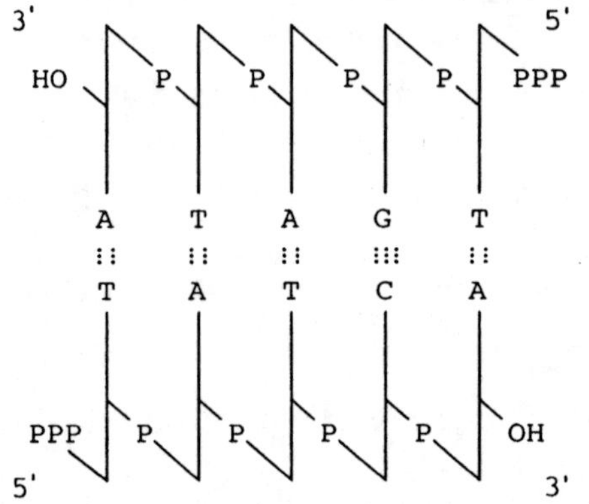

9.8 Base analogues are compounds that resemble the natural bases found in DNA and RNA but are not normally found in those macromolecules. Base analogues can replace their normal counterparts in DNA during in vitro DNA synthesis. Four base analogues were studied for their effects on in vitro DNA synthesis using *E. coli* DNA polymerase. The results were as follows, with the amounts of DNA synthesized expressed as percentages of the DNA synthesized from normal bases only.

	Normal Bases Substituted by the Analogue			
Analogue	**A**	**T**	**C**	**G**
A	0	0	0	25
B	0	54	0	0
C	0	0	100	0
D	0	97	0	0

Which bases are analogues of adenine? of thymine? of cytosine? of guanine?

Answer: None are analogues of adenine, B and D are analogues of thymine, C is an analogue of cytosine, and A is an analogue of guanine.

9.9 a. Describe (draw) models of continuous, semidiscontinuous and discontinuous DNA replication.

b. What was the contribution of Reiji and Tuneko Okazaki and colleagues with regard to those replication models?

Answer:

a. Since DNA replication is known to be semidiscontinuous, it is helpful to first consider the details of this model. As the other models are constructed, reference to the semidiscontinuous model can identify the problems associated with them.

A *semidiscontinuous model* is shown in text Figures 9.5 and 9.6. Here are its main features:

i. DNA replication proceeds only in the 5′→3′ direction.

ii. Because DNA replication proceeds only in one direction, it is continuous on the 3′→5′ template strand (the leading strand), while it is discontinuous on the 5′→3′ template strand (the lagging strand). This is why it is *semi*discontinuous.

iii. Replication initiates at a replication origin, where a complex of proteins and enzymes forms at a replication fork. Initially, helicase untwists the DNA, and SSBs (single-strand DNA-binding proteins) bind the single-stranded DNA to protect it from nuclease attack and keep it unwound.

iv. Since DNA polymerase can only polymerize DNA if a 3′-OH is provided, an RNA primer must be synthesized by a primase-enzyme complex (the primosome). The primosome binds to the single-stranded DNA and synthesizes a short RNA primer in the 5′→3′ direction.

v. Once an RNA primer is synthesized, DNA polymerase III (a complex of proteins) adds nucleotides onto the 3′-OH in the 5′→3′ direction, using the unwound DNA as a template. The newly synthesized DNA is thus synthesized onto an RNA primer. This is referred to as an Okazaki fragment. On the lagging strand, synthesis moves "away" from the fork, and synthesis must be reinitiated frequently. On the leading strand, synthesis proceeds "into" the fork. The proteins involved in replication are associated into a replisome, so that the DNA polymerase III of the lagging strand is complexed with DNA polymerase III on the leading strand. The replisome that forms at a replication fork moves as a unit along the DNA, enabling new DNA to be synthesized efficiently on both the leading and lagging strands.

vi. On the lagging strand, DNA polymerase III will (after about 1,000 to 2,000 bp) encounter the 5′ end of an RNA primer of the previously synthesized Okazaki fragment. At this point, it dissociates and is replaced by DNA polymerase I, which uses its 5′→3′ exonuclease activity to excise the RNA primer and replace it with DNA.

vii. Once the RNA primer is replaced with DNA, the gap between adjacent DNA fragments is sealed by DNA ligase.

A *continuous model* would need to consider what we know about the semiconservative nature of DNA replication, as well as the enzymes known to be involved in DNA replication. In particular, the model would have to

- produce new double helices that each have one old and one new strand,
- use DNA polymerases that synthesize DNA only in the 5′→3′ direction,
- use DNA polymerases requiring a 3′→5′ template,
- use DNA polymerases that need a primer to initiate replication.

One problem to be solved by any model of continuous-only DNA replication is to ensure that the DNA is completely replicated. For example, if DNA replication starts in the middle of a linear chromosome, and proceeds continuously in only a 5′→3′ direction using a 3′→5′ template strand, half of that template strand will remain unreplicated. This problem might be overcome in different ways, depending on whether the chromosome is circular or linear. Each solution, however, presents additional problems.

Suppose DNA replication occurs in a circular chromosome. As in the semidiscontinuous model, a protein or set of proteins would recognize an origin-of-replication site within the DNA molecule. To create a site for initiation of DNA synthesis, an endonuclease would nick one strand. Helicase would then bind the free 5′ end and start unwinding it. DNA primase would not be needed, as the free 3′-OH would serve as the end of a "primer." The complementary (3′→5′) strand would be used as a template for new DNA synthesis. A protein such as SSB would bind to the unwound, nontemplate strand (the strand with an exposed 5′ end) to keep it separate from the template strand. In some unknown manner, this strand would need to be kept separate and perhaps unwound throughout the replication of the template strand. This raises a first problem, as this might be a daunting task given the size of a chromosome compared to the size of a cell. Since DNA replication is semiconservative, this strand (which lacks a priming site) would also need to be replicated. The mechanism by which this would occur is unknown and raises a second problem. In a replication fork proceeding down the length of the template strand, DNA polymerase III would add nucleotides based on the sequence of the 3′→5′ template strand, continuously until the site of the origin was reached. Some mechanism would be needed to mark this site so that DNA polymerase III would stop synthesis. A ligation reaction would be needed to join the two ends of the newly synthesized DNA. There are multiple approaches to solving the two problems posed by this model, but in reality, neither problem occurs due to the semidiscontinuous nature of DNA replication.

For linear eukaryotic chromosomes, it would be difficult to initiate DNA synthesis at the very ends of chromosomes. But this would not always be required, as in germ-line cells of eukaryotes, the simple telomeric repeats at the very ends of chromosomes are replicated by telomerase. However, in somatic cells, a distinct mechanism would be needed to identify origin-of-replication sites very near to, but not at, the ends of the chromosomes. After an endonuclease made a nick near the end of a linear chromosome, a DNA polymerase III–like enzyme could add nucleotides 5′→3′ based on the sequence of the 3′→5′ template strand. A replication fork would proceed down the length of the template strand, replicating just one of the strands. Presumably, the other strand would be bound by an SSB-like protein as it was unwound. (Recall the first problem above.) At the other end of a linear chromosome, a similar process would occur.

In both of these models of continuous DNA replication, no DNA primase was required, and no RNA primer was used. Certainly, other models of continuous DNA replication might be devised that use DNA primase and RNA primers. However, there is then the question of how to remove the primers, and so a DNA polymerase I–like enzyme must be used. One significant problem with these models is that they require one of the two strands of the parental DNA molecule to remain separate from the template strand until it is replicated. Given the sizes of chromosomes in prokaryotes and eukaryotes relative to their cell sizes, this is a considerable problem. It is a problem that is solved by having a replication fork synthesize DNA semidiscontinuously, so that there are only very short stretches of unwound, SSB-bound DNA.

A *discontinuous model* of DNA replication would be nearly identical to the semidiscontinous model [presented in (b)], except that DNA replication would be discontinuous not only on the lagging strand, but also on the leading strand. Presumably, DNA polymerase III would "fall off" the leading-strand template periodically and then come back "on board" to continue synthesizing DNA. No new priming would be required, as there would be a free 3′-OH available for DNA polymerase III to add deoxyribonucleotides onto.

b. Reiji Okazaki provided the primary evidence that DNA synthesis must be discontinuous on at least one strand. He demonstrated the existence of Okazaki fragments, short segments of DNA that have short RNA primers at their 5′ ends. Their existence indicated that DNA synthesis must be periodically reprimed on the lagging strand and therefore that it must be discontinuous on that strand. (See also the answer to problem 9.15.)

9.10 How long would it take *E. coli* to replicate its entire genome (4.2×10^6 bp), assuming a replication rate of 1,000 nucleotides per second at each fork with no pauses?

Answer: In *E. coli,* replication proceeds bidirectionally from the replication origin (*ori*C). Thus, each replication fork proceeds halfway down the length of the circular *E. coli* chromosome, at which point the replication forks meet and replication is complete. Thus, it will take

$$\frac{4.2 \times 10^6 \text{ bp}}{2 \text{ replication forks}} \times \frac{1 \text{ second}}{1{,}000 \text{ bp}} \times \frac{1 \text{ minute}}{60 \text{ seconds}} = 35 \text{ minutes}$$

9.11 A diploid organism has 4.5×10^8 bp in its DNA. This DNA is replicated in 3 minutes. Assuming all replication forks move at a rate of 10^4 bp per minute, how many replicons (replication units) are present in this organism's genome?

Answer: Since a replication fork moves at a rate of 10^4 bp per minute and each replicon has two replication forks moving in opposite directions, in one replicon, replication would be occurring at a rate of 2×10^4 bp/minute. Since all of the organism's DNA is replicated in 3 minutes, the number of replicons in the diploid genome is

$$\frac{4.5 \times 10^8 \text{ bp}}{3 \text{ minutes}} \times \frac{1 \text{ replicon}}{2 \times 10^4 \text{ bp / minute}} = 7{,}500$$

9.12 The following events, steps, or reactions occur during *E. coli* DNA replication. For each entry in column A, select the appropriate entry in column B. Each entry in A may have more than one answer, and each entry in B can be used more than once.

Column A	Column B
____ a. Unwinds the double helix	A. Polymerase I
____ b. Prevents reassociation of complementary bases	B. Polymerase III
____ c. Is an RNA polymerase	C. Helicase
____ d. Is a DNA polymerase	D. Primase
____ e. Is the "repair" enzyme	E. Ligase
____ f. Is the major elongation enzyme	F. SSB protein
____ g. A 5′→3′ polymerase	G. Gyrase
____ h. 3′→5′ polymerase	H. None of these
____ i. Has 5′→3′ exonuclease function	
____ j. Has 3′→5′ exonuclease function	
____ k. Bonds free 3′-OH end of a polynucleotide to a free 5′-monophosphate end of a polynucleotide	
____ l. Bonds 3′-OH end of a polynucleotide to a free 5′ nucleotide triphosphate	
____ m. Separates daughter molecules and causes supercoiling	

Answer: a. C; b. F; c. D; d. A, B; e. A; f. B; g. A, B, D; h. H; i. A; j. A, B, k. E; l. A, B, (D, after the first two bases of the RNA primer are positioned); m. G

9.13 Describe the molecular action of the enzyme DNA ligase. What properties would you expect an *E. coli* cell to have if it had a temperature-sensitive mutation in the gene for DNA ligase?

Answer: DNA ligase catalyzes the formation of a phosphodiester bond between the 3′-OH and the 5′-monophosphate groups on either side of a single-strand DNA gap, sealing the gap. Temperature-sensitive ligase mutants would be unable to seal such gaps at the restrictive (high) temperature, leading to fragmented lagging strands and presumably cell death. If a biochemical analysis were performed on DNA synthesized after *E. coli* were shifted to a restrictive temperature, there would be an accumulation of DNA fragments the size of Okazaki fragments. This would provide additional evidence that DNA replication must be discontinuous on one strand.

9.14 Chromosome replication in *E. coli* commences from a constant point, called the origin of replication. It is known that DNA replication is bidirectional. Devise a biochemical experiment to prove that the *E. coli* chromosome replicates bidirectionally. (Hint: Assume that the amount of gene product is directly proportional to the number of genes.)

Answer: Assume the amount of a gene's product is directly proportional to the number of copies of the gene present in the *E. coli* cell. Assay the enzymatic activity of genes at various positions in the *E. coli* chromosome during the replication period. Then some genes (those immediately adjacent to the origin) will double their activity very shortly after replication begins. Relate the map position of genes having doubled activity to the amount of time that has transpired since replication was initiated. If replication is bidirectional, there should be a doubling of the gene products both clockwise and counterclockwise from the origin.

9.15 Reiji Okazaki concluded that both strands could not replicate continuously. What evidence led him to this conclusion?

Answer: Reiji Okazaki and his colleagues added a pulse of radioactive DNA precursor, ^{3}H-thymidine, to cultures of *E. coli* for a small fraction of its generation time (0.5 percent). To prevent the further incorporation of additional radioactive precursor into the DNA after this period, they chased this pulse with a large amount of nonradioactive thymidine. They then sampled the bacterial DNA at successive intervals as DNA replication continued. When DNA was sampled soon after the labeling period, most of the radioactivity was found in DNA with very low molecular weights, in DNA fragments between 100 and 1,000 nucleotides long. When DNA was sampled at longer intervals, the radioactivity was found in DNA with much higher molecular weight, in much longer DNA fragments. The short DNA fragments seen in the early samplings indicate that DNA replication normally involves the synthesis of short DNA segments—the Okazaki fragments—that are then linked together. This provided evidence that DNA replication occurs in a discontinuous fashion on the lagging strand.

9.16 A space probe returns from Jupiter and brings with it a new microorganism for study. It has double-stranded DNA as its genetic material. However, studies of replication of the alien DNA reveal that although the process is semiconservative, DNA synthesis is continuous on both the leading-strand and the lagging-strand templates. What conclusions can you make from that result?

Answer: Clearly, DNA replication in the Jovian bug does not occur as it does in *E. coli.* Assuming that the double-stranded DNA is antiparallel as it is in *E. coli,* the Jovian DNA polymerases must be able to synthesize DNA in the $5' \rightarrow 3'$ direction (on the leading strand) as well as in the $3' \rightarrow 5'$ direction (on the lagging strand). This is unlike any DNA polymerase on Earth.

9.17 Compare and contrast eukaryotic and prokaryotic DNA polymerases.

Answer: A number of different DNA polymerases have been identified in both prokaryotes and eukaryotes. All DNA polymerases can synthesize DNA using a template and a primed strand. The primer provides a 3′-OH onto which nucleotides are added in the $5' \rightarrow 3'$ direction. There are three DNA polymerases in *E. coli* (I, II, and III) and five in mammalian cells (α, β, γ, δ, and ϵ). All of the enzymes synthesize DNA in a $5' \rightarrow 3'$ direction.

In *E. coli,* at least DNA polymerase I and III are essential for viability. DNA polymerase I and II each consist of one polypeptide coded for by one gene, while DNA polymerase III consists of ten polypeptides coded for by many genes. Both DNA polymerase I and III are involved in DNA replication, with DNA polymerase III being the main synthetic enzyme. Less is known about DNA polymerase II. All three *E. coli* DNA polymerases also have proofreading activity, as they have $3' \rightarrow 5'$ exonuclease activity. However, only DNA polymerase I has $5' \rightarrow 3'$ exonuclease activity, the activity needed to excise the RNA primer at the 5′ end of Okazaki fragments.

In mammalian cells, only two DNA polymerases, α and δ, function in nuclear DNA replication. They function much like *E. coli* DNA polymerase III functions in chromosomal DNA replication. DNA polymerases β and ϵ are nuclear and function in DNA repair. DNA polymerase γ is found in mitochondria and catalyzes mitochondrial DNA replication. In contrast to the *E. coli* polymerases, only a subset of mammalian enzymes (δ, ϵ, and γ), have proofreading $3' \rightarrow 5'$ exonuclease activity. Like the *E. coli* polymerases, the nuclear polymerases α and δ, as well as the mitochondrial polymerase γ, use RNA primers.

9.18 In eukaryotes, cellular proliferation is tightly controlled. Explain why this is important, and outline how key regulatory proteins control the progression of the cell cycle.

Answer: Cellular proliferation is tightly controlled at a series of cell cycle checkpoints. At each checkpoint, a system of checks and balances assesses whether the cell is able to or should be allowed to continue into the next phase of the cell cycle. This is important for many reasons. To proceed through the G_1-S checkpoint, a cell must be large enough, and the environment must be favorable for future growth. This ensures that cells will not expend their energy on DNA replication unless there is a good chance for successful future growth. As will be seen in Chapter 18, G_1-S is also a major checkpoint to ensure that DNA damage is repaired, so that only healthy, undamaged cells progress through the cell cycle. To proceed through the G_2-M checkpoint, a cell must have completed DNA replication, be large enough, and be in a favorable environment. This maximizes the probability that daughter cells will have a euploid genome and that the energy expended on mitosis will result in viable daughter cells. To proceed through the third checkpoint in M, chromosomes must be attached properly to the mitotic spindle, ensuring their correct segregation and the production of euploid daughter cells.

The cyclins and cyclin-dependent kinases (Cdks) are interacting regulatory proteins that control whether cells will proceed past each cell cycle checkpoint. Cyclin-Cdk interactions lead to the activation and inactivation of other cellular proteins, often by altering their phosphorylation state. For example, at the G_1-S (START) checkpoint, one or more G_1 cyclins bind to the Cdk *CDC28/cdc2* kinase and activate it. This Cdk in turn phosphorylates key proteins needed for progression of the cell into S. Following activation of the Cdk, the level of cyclins is decreased through increased proteolysis.

9.19 Autoradiography is a technique that allows radioactive areas of chromosomes to be observed under the microscope. The slide is covered with a photographic emulsion, which is exposed by radioactive decay. In regions of exposure, the emulsion forms silver grains on being developed. The tiny silver grains can be seen on top of the (much larger) chromosomes. Devise a method for finding out which regions in the human karyotype replicate during the last 30 minutes of the S phase. (Assume a cell cycle in which the cell spends 10 hours in G_1, 9 hours in S, 4 hours in G_2, and 1 hour in M.)

Answer: If cells spend 4 hours in G_2, then there will be about 4.5 to 5 hours between the last 30 minutes of S and metaphase in M. To see which chromosomal regions replicate during the last 30 minutes of S, add ^{3}H-thymidine for a 5-minute pulse, chase it with (cold, unlabeled) thymidine, wait 4.5 to 5 hours, and prepare a spread of metaphase chromosomes on a glass slide. (Cells can be blocked at metaphase by adding colchicine, a drug that blocks the separation of the chromatids by interfering with microtubule function.) Immerse the slide in photographic emulsion, and after a suitable exposure period, develop the exposed silver grains and stain the chromosomes. Examine which regions of the chromosomes have silver grains lying over them to infer where ^{3}H-thymidine was incorporated.

9.20 In Figure 9.3, semiconservative DNA replication is visualized in eukaryotic cells using the harlequin chromosome-staining technique.

a. Explain what the harlequin chromosome-staining technique is and how it provides evidence for semiconservative DNA replication in eukaryotes.
b. Propose a hypothesis to explain why, in Figure 9.3, some chromatids appear to contain segments of both DNA containing T and DNA containing BUdR, while others appear to consist entirely of DNA with T or DNA with BUdR.

Answer:

a. Harlequin chromosomes are prepared by allowing tissue culture cells to undergo two rounds of DNA replication in the presence of BUdR. After BUdR (a base analogue that replaces T) becomes incorporated in the DNA, metaphase chromosomes stained with Giemsa stain and a fluorescent dye have one darkly stained and one lightly stained chromatid. Since DNA with BUdR stains less intensely than DNA with T, the presence of two differently stained chromatids indicates that one chromatid has two strands with BUdR while the other has one with BUdR and one with T.

   This supports the semiconservative model of DNA replication. After one round of replication, the DNA molecule contains one new strand with BUdR and one old strand with T. After two rounds of replication, these molecules will each produce a DNA molecule where both strands have BUdR (the light chromatid) and a DNA molecule where one strand has BUdR and one strand has T (the dark chromatid).

b. Chromosomes with chromatids that appear to contain segments with both T and BUdR might have had a sister chromatid exchange (mitotic crossing-over).

9.21 A mutant *Tetrahymena* has an altered repeated sequence in its telomeric DNA. What change in the telomerase enzyme would produce this phenotype?

Answer: Telomerase synthesizes the simple-sequence telomeric repeats at the ends of chromosomes. The enzyme is made up of both protein and RNA, and the RNA component has a base sequence that is complementary to the telomere repeat unit. The RNA component is used as a template for the telomere repeat, so that if the RNA component were to be altered, the telomere repeat would be as well. Thus, the mutant in this question is likely to have an altered RNA component.

10

GENE CONTROL OF PROTEINS

CHAPTER OUTLINE

GENE CONTROL OF ENZYME STRUCTURE
- Garrod's Hypothesis of Inborn Errors of Metabolism
- The One Gene—One Enzyme Hypothesis

GENETICALLY BASED ENZYME DEFICIENCIES IN HUMANS
- Phenylketonuria
- Albinism
- Lesch-Nyhan Syndrome
- Tay-Sachs Disease

GENE CONTROL OF PROTEIN STRUCTURE
- Sickle-Cell Anemia
- Other Hemoglobin Mutants
- Biochemical Genetics of the Human ABO Blood Groups
- Cystic Fibrosis

GENETIC COUNSELING
- Carrier Detection
- Fetal Analysis

REVIEW OF KEY TERMS, SYMBOLS, AND CONCEPTS

In your own words, write a brief, precise definition of each term in the groups on the following page. Check your definitions using the text. Then develop a concept map using the terms in each list.

THINKING ANALYTICALLY

The problems in this chapter require you to make multiple connections. You need to (1) relate the genetic properties of a mutation to its visible phenotype, (2) relate the visible phenotype to a biochemical deficit (a biochemical phenotype), and (3) relate the biochemical deficit of a mutation back to its genetic properties. As you make these connections, layers of complexity will unfold in a problem. Clues at one level (genetic properties of the mutations, mutant phenotypes, biochemical abnormalities found in the mutant) can be used to infer what might be happening at another level. Approach the problems systematically, attempting to connect these levels wherever possible.

1	2
inborn error of metabolism	enzyme
metabolic pathway	nonenzymatic protein
biochemical pathway	ABO blood type
nutritional mutant	antigen, antibody
auxotroph	polysaccharide
prototroph	glycosyltransferase
ascus, asci, ascospore, conidia	cystic fibrosis (CF)
minimal medium	transmembrane conductance regulator
mating type	active transport
one gene–one enzyme hypothesis	sickle-cell anemia, sickle-cell trait
one gene–one polypeptide hypothesis	hemoglobin
amino acid	α, β polypeptide
polypeptide	electrophoresis
protein	amino acid substitution
N, C terminus	hydrophilic
pleiotropic	multiple alleles
alkaptonuria (AKU)	autosomal recessive mutation
phenylketonuria (PKU)	genetic counseling
albinism	pedigree analysis
Lesch-Nyhan syndrome	carrier detection
Tay-Sachs disease	fetal analysis
recessive trait	amniocentesis
OMIM	chorionic villus sampling

To relate biochemical pathways to mutant phenotypes, diagram as much of the biochemical pathway as you can, and then try to understand how specific phenotypic consequences arise when a particular step is blocked. Go slowly and be thorough. It is essential to remember that mutations in different steps of a biochemical pathway do not always cause identical mutant phenotypes. In addition, mutations affecting a particular step can have multiple consequences. They may lead to the absence of the final product and cause one phenotype. They may also lead to the accumulation of intermediate metabolites of the pathway and cause a different phenotype. Furthermore, mutations at different steps of a pathway may be able to be "rescued" if different biochemical intermediates are provided to the mutant.

QUESTIONS FOR PRACTICE

Multiple Choice Questions

1. An inborn error of metabolism is
   a. any biochemical abnormality that results from taking a drug
   b. any heritable biochemical abnormality
   c. any recessive mutation
   d. any dominant mutation

2. Why was the one gene-one enzyme hypothesis recast as the one gene-one polypeptide hypothesis?
   a. Genes can encode proteins that are not enzymes.
   b. Some enzymes are not polypeptides.
   c. Some enzymes have more than one polypeptide subunit.
   d. Some polypeptides are not enzymes.
3. Suppose an individual was diagnosed with PKU as a child and was successfully treated. If that individual and a normal, noncarrier (pku^+/pku^+) partner have children, what is the probability that they will have offspring that are carriers for PKU?
   a. 0.00
   b. 0.25
   c. 0.50
   d. 1.00
4. In individuals affected with AKU,
   a. a block in the pathway leads to the accumulation of homogentisic acid
   b. if homogentisic acid is provided, the pathway can be completed
   c. a block in the pathway leads to the accumulation of phenylalanine
   d. if phenylalanine is provided, the pathway can be completed
5. Two *Neurospora* auxotrophs are unable to grow on minimal medium but are able to grow on minimal medium supplemented with arginine. When each is tested, only one can grow on minimal medium supplemented with ornithine, a biochemical precursor to arginine. Which statement(s) below are supported by these findings?
   a. The auxotrophs are blocked in different biochemical steps.
   b. At least one auxotroph is blocked in a step before ornithine is made.
   c. Both of the auxotrophs are blocked in a step before ornithine is made.
   d. Both of the auxotrophs accumulate ornithine.
   e. a and b
6. In complex metabolic pathways such as the phenylalanine-tyrosine metabolic pathway,
   a. blocks at different steps always result in the same phenotype
   b. a block at one step results solely in the accumulation of the biochemical made just before that step
   c. blocks at different steps can lead to very different phenotypes
   d. a block at one step can lead to the accumulation of potentially toxic derivatives of biochemicals synthesized before that step
   e. a and b
   f. c and d
7. What is the difference between sickle-cell anemia and sickle-cell trait?
   a. None; they refer to individuals equally affected with the same disease.
   b. Individuals with sickle-cell anemia have severe disease and have two abnormal alleles; individuals with the sickle-cell trait have a less severe form of the disease, as they are heterozygotes with one disease allele and one normal allele.
   c. Individuals with sickle-cell anemia have severe disease; the term *sickle-cell trait* refers to the disease allele present in their families, and not to symptomatic status.
   d. Individuals with sickle-cell anemia have severe disease and are homozygous for a severe disease allele; individuals with sickle-cell trait have mild disease and are homozygous for a less severe disease allele.

8. Both Tay-Sachs disease and Lesch-Nyhan syndrome are examples of human diseases that are
   a. pleiotropic
   b. caused by a deficiency in one enzyme activity
   c. recessive
   d. a, b, and c
9. An example of a heritable disease that results from a biochemical defect in a nonenzymatic protein is
   a. PKU
   b. AKU
   c. sickle-cell anemia
   d. Tay-Sachs disease
10. Which of the following is *not* true about both amniocentesis and chorionic villus sampling?
    a. Both can be performed with an acceptably low risk as early as the eighth week of pregnancy.
    b. Each procedure has a risk of fetal loss.
    c. One is more likely than the other to provide an accurate diagnosis.
    d. Both are valid, useful procedures in genetic counseling.

Answers: 1b, 2c, 3d, 4a, 5e, 6f, 7b, 8d, 9c, 10a

Thought Questions

1. If you were a genetic counselor, how would you counsel the following couple who are contemplating having children? The prospective father is apparently normal, except for a nervous twitch of his nose, but his maternal uncle had PKU as a child and his brother, who plays in a heavy metal band, is hard of hearing. The prospective mother also appears normal, but her father has hemophilia and her brother, who plays in the same band as her brother-in-law, is also hard of hearing. You know that hemophilia is often X-linked and that PKU and sometimes deafness are autosomal recessive.
2. Distinguish between the procedures of amniocentesis and chorionic villus sampling. How is each useful in genetic counseling? What reasons are there for or against choosing each procedure?
3. All states currently require testing for PKU in infants. Do you expect the frequency of PKU to decline or increase over time? Justify your answer.
4. In the United States, the frequency of PKU is about 1 in 16,000 newborns. The frequency of cystic fibrosis is about 1 in 4,000 newborns. While all states currently require testing for PKU in infants, not all states require testing for cystic fibrosis. Both are autosomal recessive disorders. What are the risks associated with screening for an inherited disease? What reasons might there be not to routinely test for cystic fibrosis (or PKU) in newborns? What reasons might there be to routinely test for PKU (or cystic fibrosis) in newborns?
5. Over 200 hemoglobin mutants have been detected. Explain why some hemoglobin mutations do not lead to as severe sickle-cell anemia phenotypes as others.
6. Some alleles of sickle-cell anemia, when heterozygous with a normal allele, confer some resistance to malarial infection. What consequences might this have on selection for this allele in different human populations?
7. Keep in mind your answer to questions 5 and 6 as you respond to the following question. Suppose you have the ability to make point mutations in a gene that encodes an enzyme and are able to determine the percentage of normal enzyme activity that remains in the mutant

enzyme. What effects would you expect to see if you systematically mutate the gene for the enzyme, changing only one amino acid at a time?

8. Different diseases are more prevalent in different ethnic groups. For example, within the U.S. population, Tay-Sachs disease is more prevalent in individuals of Ashkenazi Jewish descent, while certain types of sickle-cell anemia are more prevalent in some individuals of Hispanic, Native American, and African American descent. Given this situation and a history of documented discrimination based on ethnicity in the United States, what concerns must be addressed if genetic testing is to be used for carrier testing and prenatal diagnosis with the aim of improving the quality of health care?
9. As you have seen in the study of PKU, Tay-Sachs disease, Lesch-Nyhan syndrome, and sickle-cell anemia, mutations that result in the blockage of a single enzymatic step or the function of a single protein can have pleiotropic phenotypes. Generate a hypothesis about a general principle from these data.
10. What functions do (should) genetic counseling services provide? What do they look for, how do they proceed, what do they recommend, and what, if any, advice do (should) they give?
11. In many states in the United States, legislation is being developed to address the use and privacy of information obtained from genetic testing. What specific protections can be legislated for genetic testing to be performed without fears of stigmatization and discrimination? Under what circumstances, if any, should genetic testing be allowed for diseases for which there is no treatment? No cure? For some diseases with very similar symptoms, genetic tests may be useful to distinguish between diseases and improve therapy. Should laws governing the use of genetic tests as diagnostic tests (in symptomatic individuals) differ from laws governing the use of genetic tests as tests for predicting risk (in asymptomatic individuals)?
12. What is meant by "obtaining informed consent" prior to undergoing a genetic test? What information is needed to give informed consent prior to undergoing a genetic test?

Thought Questions for Media

After reviewing the media at the *Essential iGenetics* website, try to answer these questions.

1. In the animation depicting the experiments by Beadle and Tatum, how can *Neurospora* be propagated asexually?
2. How did Beadle and Tatum obtain nutritional mutations?
3. Why did Beadle and Tatum use two rounds of screening for auxotrophs?

1. Why is the arginine pathway present in *Neurospora* also present in humans?
2. Suppose a block is made in a single enzymatic step of the arginine pathway in either humans or *Neurospora*. What is the consequence to the level of the precursor compound synthesized before the blocked enzymatic step?
3. How might citrullinemia, argininosuccinic aciduria, or OTC deficiency lead to ammonia buildup? How might you implement effective therapeutic treatments?

SOLUTIONS TO TEXT PROBLEMS

10.1 Phenylketonuria (PKU) is an inheritable metabolic disease of humans; its symptoms include mental deficiency. This phenotypic effect results from

a. accumulation of phenylketones in the blood
b. absence of phenylalanine hydroxylase

c. deficiency of phenylketones in the blood
d. deficiency of phenylketones in the diet

Answer: a

10.2 If a person were homozygous for both PKU and alkaptonuria (AKU), would you expect him or her to exhibit the symptoms of PKU, AKU, or both? Refer to the following pathway.

Phenylalanine
↓ (blocked in PKU)
tyrosine → DOPA → melanin
↓
p-Hydroxyphenylpyruvic acid
↓
Homogentisic acid
↓ (blocked in AKU)
Maleylacetoacetic acid

Answer: Given the pathway that is presented, a double homozygote will have PKU, but not AKU. The PKU block will prevent homogentisic acid from being formed, so it could not accumulate to high levels and cause AKU.

10.3 Refer to the pathway shown in problem 10.2. What effect, if any, would you expect PKU and AKU to have on pigment formation? Explain your answer.

Answer: The block in PKU leads to decreased tyrosine levels and so should lead to a decrease in melanin formation and pigmentation. However, some tyrosine will be obtained from food, and so the block in the pathway can be partially circumvented in this way. It has been reported that PKU patients are sometimes lighter in pigmentation than their normal relatives.

Since the block in AKU patients lies after (downstream of) the formation of melanin, pigmentation should not be affected in AKU patients. However, the levels of the products of an enzymatic reaction can affect the efficiency with which that reaction proceeds. If the products of an enzymatic reaction accumulate, the reaction may be inhibited. This in turn can lead to the accumulation of the substrates of the reaction. In this case, the accumulation of homogentisic acid in AKU individuals may ultimately lead to elevation of tyrosine levels. This tyrosine might be available for conversion to melanin. Hence, if AKU has any effect, it may be to increase pigmentation levels.

Using the same logic as in problem 10.2, an individual with both PKU and AKU will show symptoms of PKU, but not AKU. Therefore, it is likely that there will be decreased tyrosine levels and a decrease in melanin formation and pigmentation.

10.4 a^+, b^+, c^+, and d^+ are independently assorting Mendelian genes controlling the production of a black pigment. The alternate alleles that give abnormal functioning of these genes are a, b, c, and d. A black individual of genotype $a^+/a^+\ b^+/b^+\ c^+/c^+\ d^+/d^+$ is crossed with a colorless individual of genotype $a/a\ b/b\ c/c\ d/d$ to produce a black F_1. $F_1 \times F_1$ crosses are then done. Assume that a^+, b^+, c^+, and d^+ act in a pathway as follows:

$$\text{colorless} \xrightarrow{a^+} \text{colorless} \xrightarrow{b^+} \text{colorless} \xrightarrow{c^+} \text{brown} \xrightarrow{d^+} \text{black}$$

a. What proportion of the F_2 are colorless?
b. What proportion of the F_2 are brown?

Answer: The F_1 cross is $a^+/a\ b^+/b\ c^+/c\ d^+/d \times a^+/a\ b^+/b\ c^+/c\ d^+/d$.

a. A colorless F_2 individual would result if an individual has an a/a, b/b, and/or c/c genotype. This would consist of many possible genotypes. Rather than identify all of these combinations, it is possible to use the fact that the proportion of colorless individuals = 1 – the proportion of pigmented individuals. The proportion of pigmented individuals ($a^+/-\ b^+/-\ c^+/-$) is $\frac{3}{4} \times \frac{3}{4} \times \frac{3}{4} = \frac{27}{64}$. The chance of not obtaining this genotype is $1 - \frac{27}{64} = \frac{37}{64}$.
b. A brown individual is $a^+/-\ b^+/-\ c^+/-\ d/d$. The proportion of brown individuals is $\frac{3}{4} \times \frac{3}{4} \times \frac{3}{4} \times \frac{1}{4} = \frac{27}{256}$.

10.5 Using the genetic information given in problem 10.4, now assume that a^+, b^+, and c^+ act in a pathway as follows:

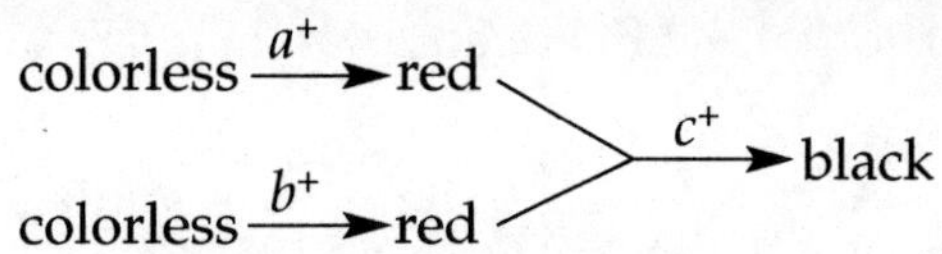

Black can be produced only if both red pigments are present; that is, c^+ converts the two red pigments together into a black pigment.

a. What proportion of the F_2 are colorless?
b. What proportion of the F_2 are red?
c. What proportion of the F_2 are black?

Answer:

a. Colorless individuals are $a/a\ b/b\ -/-$. The chance of obtaining such an individual from a cross of $a^+/a\ b^+/b\ c^+/c \times a^+/a\ b^+/b\ c^+/c$ is $\frac{1}{4} \times \frac{1}{4} = \frac{1}{16}$.
b. There are two ways to determine the proportion of red individuals. First, notice that red individuals are obtained either when only one of the a^+ or b^+ functions is present (no matter what genotype is at c) or when both a^+ and b^+ functions are present but c^+ is not. Thus, these phenotypes are obtained from the following genotypes: $a^+/-b/b\ -/-$, $a/a\ b^+/-\ -/-$, or $a^+/-\ b^+/-\ c/c$. The probability of obtaining one of these genotypes is

$$(\tfrac{3}{4} \times \tfrac{1}{4} \times 1) + (\tfrac{1}{4} \times \tfrac{3}{4} \times 1) + (\tfrac{3}{4} \times \tfrac{3}{4} \times \tfrac{1}{4}) = \tfrac{33}{64}$$

A second way is to use the information from parts (a) and (c). From part (c), one has that $\frac{27}{64}$ individuals are black, so the remaining $\frac{37}{64}$ are either red or colorless. From part (a), $\frac{1}{16} = \frac{4}{64}$ are colorless. Thus, there are $(37 - 4)/64 = \frac{33}{64}$ that are red.

c. Black individuals are formed only if both red pigments are available and are then converted to black pigment. Thus, an individual must be $a^+/-\ b^+/-\ c^+/-$. The chance of obtaining such an individual is $\frac{3}{4} \times \frac{3}{4} \times \frac{3}{4} = \frac{27}{64}$.

10.6 a. Three genes on different chromosomes are responsible for three enzymes that catalyze the same reaction in corn:

$$\text{colorless compound} \xrightarrow{a^+,\ b^+,\ c^+} \text{red compound}$$

The normal functioning of any one of these genes is sufficient to convert the colorless compound to the red compound. The abnormal functioning of these genes is designated by a, b, and c. A red $a^+/a^+\ b^+/b^+\ c^+/c^+$ is crossed with a colorless $a/a\ b/b\ c/c$ to give a red F_1, $a^+/a\ b^+/b\ c^+/c$. The F_1 is selfed. What proportion of the F_2 are colorless? b. It turns out that another step is involved in the pathway. It is controlled by gene d^+, which assorts independently of a^+, b^+, and c^+:

$$\text{colorless compound 1} \xrightarrow{d^+} \text{colorless compound 2} \xrightarrow{a^+,\ b^+,\ c^+} \text{red compound}$$

The inability to convert colorless 1 to colorless 2 is designated d. A red $a^+/a^+\ b^+/b^+\ c^+/c^+\ d^+/d^+$ is crossed with a colorless $a/a\ b/b\ c/c\ d/d$. The F_1 are all red. The red F_1s are now selfed. What proportion of the F_2 are colorless?

Answer:

a. Since any of the normal alleles a^+, b^+, or c^+ is sufficient to catalyze the reaction leading to color, in order for color to fail to develop, all three normal alleles must be missing. That is, the colorless F_2 must be $a/a\ b/b\ c/c$. The chance of obtaining such an individual is $\frac{1}{4} \times \frac{1}{4} \times \frac{1}{4} = \frac{1}{64}$.

b. Now, colorless F_2 are obtained if *either* one or both steps of the pathway are blocked. That is, colorless F_2 are obtained in either of the following genotypes: $d/d\ -/-\ -/-\ -/-$ (the first or both steps blocked) or $d^+/-\ a/a\ b/b\ c/c$ (second step blocked). The chance of obtaining such individuals is

$$(\tfrac{1}{4} \times 1 \times 1 \times 1) + (\tfrac{3}{4} \times \tfrac{1}{4} \times \tfrac{1}{4} \times \tfrac{1}{4}) = (64 + 3)/256 = \tfrac{67}{256}$$

10.7 In *Drosophila*, the recessive allele *bw* causes a brown eye, and the (unlinked) recessive allele *st* causes a scarlet eye. Flies homozygous for both recessives have white eyes. The genotypes and corresponding phenotypes, then, are as follows:

$bw^+/-$	$st^+/-$	red eye
bw/bw	$st^+/-$	brown
$bw^+/-$	st/st	scarlet
bw/bw	st/st	white

Outline a hypothetical biochemical pathway that would produce this type of gene interaction. Demonstrate why each genotype shows its specific phenotype.

Answer: Since the *bw/bw st*$^+$/–individuals have brown pigmentation but only have *st*$^+$ function, the *st*$^+$ gene must make brown pigment. Similarly, the scarlet pigmentation of *bw*$^+$/– *st/st* individuals indicates that the *bw*$^+$ gene must make scarlet pigment. The combination of the brown and scarlet pigments in *bw*$^+$/– *st*$^+$/– individuals leads to red eyes. The absence of either pigment in *bw/bw st/st* individuals leads to white (colorless) eyes.

The production of two different colored pigments, with each affected by a separate gene, suggests that there are two distinct biochemical pathways. Each of these pathways might have a number of steps. One pathway produces a brown pigment and the other produces a scarlet pigment. This can be illustrated in the following diagram:

Two pathways, each with an unknown number of steps:

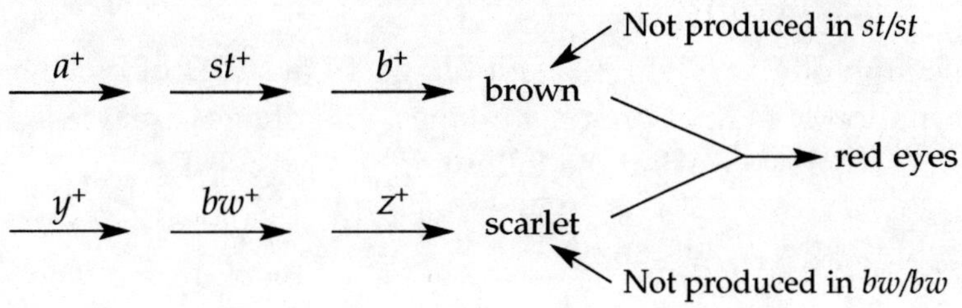

10.8 In J. R. R. Tolkien's *The Lord of the Rings,* the Black Riders of Mordor ride steeds with eyes of fire. As a geneticist, you are very interested in the inheritance of the fire-red eye color. You discover that the eyes contain two types of pigments, brown and red, that are usually bound to core granules in the eye. In wild-type steeds, precursors are converted by these granules to the above pigments, but in steeds homozygous for the recessive X-linked gene *w* (white eye), the granules remain unconverted and a white eye results. The metabolic pathways for the synthesis of the two pigments are shown in Figure 10.A. Each step of the pathway is controlled by a gene: A mutation *v* results in vermilion eyes; *cn* results in cinnabar eyes; *st* results in scarlet eyes; *bw* results in brown eyes; and *se* results in black eyes. All these mutations are recessive to their wild-type alleles and all are unlinked. For the following genotypes, show the proportions of steed phenotypes that would be obtained in the F_1 of the given matings.

a. *w/w bw*$^+$*/bw*$^+$ *st/st* × *w*$^+$*/Y bw/bw st*$^+$*/st*$^+$
b. *w*$^+$*/w*$^+$ *se/se bw/bw* × *w/Y se*$^+$*/se*$^+$ *bw*$^+$*/bw*$^+$
c. *w*$^+$*/w*$^+$ *v*$^+$*/v*$^+$ *bw/bw* × *w/Y v/v bw/bw*
d. *w*$^+$*/w*$^+$ *bw*$^+$*/bw st*$^+$*/st* × *w/Y bw/bw st/st*

Figure 10.A

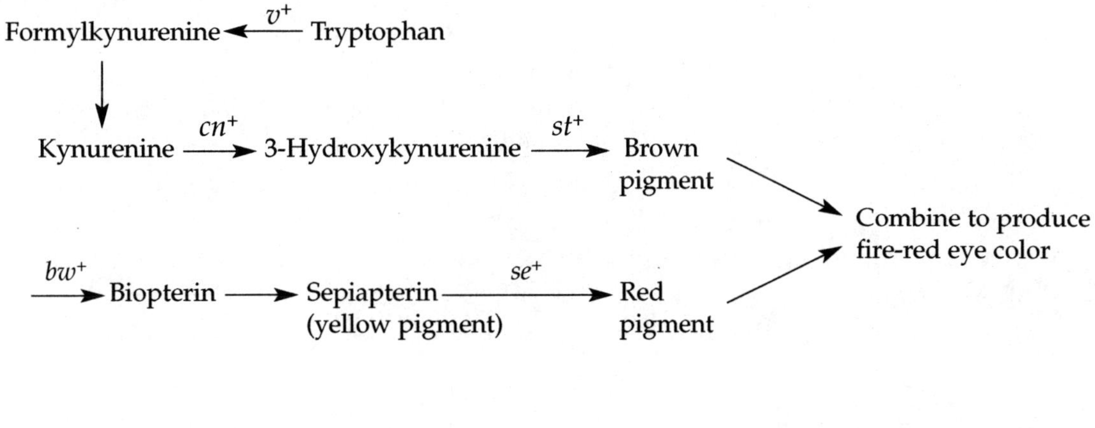

Answer:

a. ½ w/w^+ bw/bw st/st^+, fire-red-eyed daughters;
   ½ w/Y bw/bw^+ st/st^+, white-eyed sons

b. w/w^+ se/se^+ bw/bw^+ and w^+/Y se/se^+ bw/bw^+, all fire-red eyes

c. w/w^+ v/v^+ bw/bw and w^+/Y v/v^+ bw/bw, all brown eyes

d. ¼ w^+/w or w^+/Y, bw/bw^+ st/st^+, fire-red eyes;
   ¼ w^+/w or w^+/Y, bw/bw st/st^+, brown eyes;
   ¼ w^+/w or w^+/Y, bw/bw^+ st/st, scarlet eyes;
   ¼ w^+/w or w^+/Y, bw/bw st/st (the color of 3-hydroxykynurenine plus the color of the precursor to biopterin, or colorless = white)

10.9 Upon infection of *E. coli* with bacteriophage T4, a series of biochemical pathways results in the formation of mature progeny phages. The phages are released after lysis of the bacterial host cells. Suppose that the following pathway exists:

$$\text{A} \xrightarrow{\text{enzyme}} \text{B} \xrightarrow{\text{enzyme}} \text{mature phage}$$

Also suppose that we have two temperature-sensitive mutants that involve the two enzymes catalyzing these sequential steps. One of the mutations is cold-sensitive (*cs*) in that no mature phages are produced at 17°C. The other is heat-sensitive (*hs*) in that no mature phages are produced at 42°C. Normal progeny phages are produced when phages are produced when phages carrying either of the mutations infect bacteria at 30°C. However, let us assume that we do not know the sequence of the two mutations. Two models are therefore possible:

$$(1)\quad \text{A} \xrightarrow{hs} \text{B} \xrightarrow{cs} \text{phage}$$

$$(2)\quad \text{A} \xrightarrow{cs} \text{B} \xrightarrow{hs} \text{phage}$$

Outline how you would experimentally determine which model is the correct model without artificially lysing phage-infected bacteria.

Answer: Consider what will happen under each model if *E. coli* is infected with a doubly mutant phage (one step is cold-sensitive, one step is heat-sensitive), and the growth temperature is shifted between 17° and 42° during phage growth.

Suppose model (1) is correct and cells infected with the double mutant are first incubated at 17° and then shifted to 42°. Progeny phages will be produced and the cells will lyse, as each step of the pathway can be completed in the correct order. In model (1), the first step, *A* to *B*, is controlled by a gene whose product is heat-sensitive, but not cold-sensitive. At 17°, the enzyme works, and *A* will be converted to *B*. While at 17°, the second, cold-sensitive step of the pathway prevents the production of mature phage. However, when the temperature is shifted to 42°, the accumulated *B* product can be used to make mature phage, so that lysis will occur.

Under model (1), a temperature shift performed in the reverse direction does not allow for growth. When *E. coli* cells are infected with a doubly mutant phage and placed at 42°, the heat-sensitive first step precludes the accumulation of *B*. When the culture is shifted to 17°, B can accumulate, but now, the second step cannot occur, so that no progeny phage can be produced. Therefore, if model (1) is correct, lysis will be seen only in a temperature shift from 17° to 42°.

If model (2) is correct, growth will be seen only in a temperature shift from 42° to 17°, and not in a temperature shift from 17° to 42°. Hence, the correct model can be deduced by performing the temperature-shift experiment in each direction and observing which direction allows progeny phage to be produced.

10.10 Four mutant strains of *E. coli* (*a, b, c,* and *d*) all require substance X to grow. Four plates were prepared, as shown in the following figure. In each case, the medium was minimal, with just a trace amount of substance X to allow a small amount of growth of the mutant cells. On plate (a), cells of mutant strain *a* were spread over the entire surface of the agar and grew to form a thin lawn (continuous bacterial growth over the plate). On plate (b), the lawn is composed of mutant *b* cells. And so on. On each plate, cells of the four mutant types were inoculated over the lawn, as indicated by the circles. Dark circles indicate luxuriant growth. This experiment tests whether the bacterial strain spread on the plate can feed the four strains inoculated on the plate, allowing them to grow. What do these results show about the relationship of the four mutants to the metabolic pathway leading to substance X?

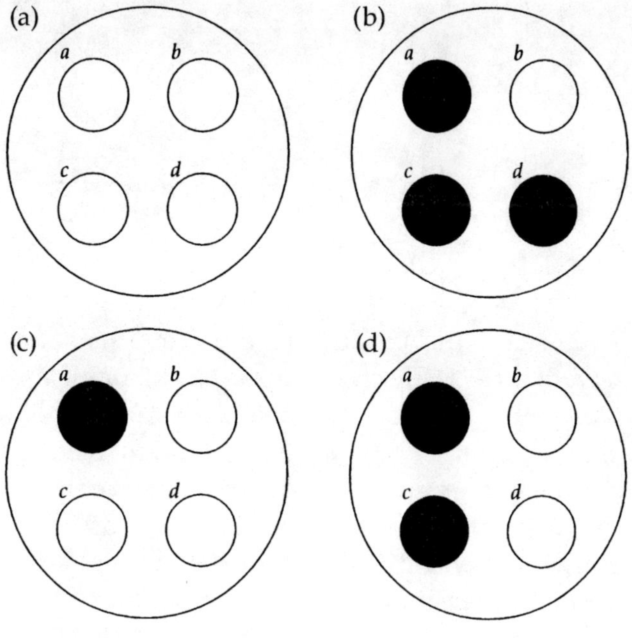

Answer: A strain blocked at a later step in the pathway accumulates a metabolic intermediate that can "feed" a strain blocked at an earlier step. It will secrete the metabolic intermediate into the medium, thereby providing a nutrient to bypass the earlier block of another strain. Consequently, a strain that feeds all others (but itself) is blocked in the last step of the pathway, while a strain that feeds no others is blocked in the first step of the pathway. Mutant *a* is blocked in the earliest step in the pathway, as it cannot feed any of the others. Mutant *c* is next, as it can supply the substance *a* needs, but cannot feed *b* or *d*. Mutant *d* is next, and mutant *b* is last in the pathway, as it can feed all the others. The pathway is $a \rightarrow c \rightarrow d \rightarrow b$.

10.11 The following growth responses (where + = growth and 0 = no growth) of mutants 1–4 were seen on the related biosynthetic intermediates A, B, C, D, and E. Assume that all intermediates are able to enter the cell, that each mutant carries only one mutation, and that all mutants affect steps after B in the pathway.

	Growth On				
Mutant	**A**	**B**	**C**	**D**	**E**
1	+	0	0	0	0
2	0	0	0	+	0
3	0	0	+	0	0
4	0	0	0	+	+

Which of the schemes in the following figure best fits the data with regard to the biosynthetic pathway?

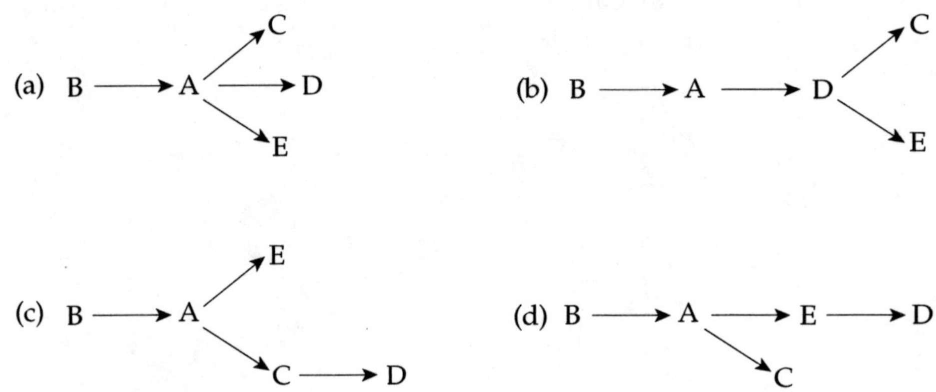

Answer: One approach to this problem is to try to sequentially fit the data to each pathway, as if each were correct. Check where each mutant *could* be blocked (remember, each mutant carries only one mutation), whether the mutant would be able to grow if supplemented with the *single* nutrient that is listed, and whether the mutant would *not* be able to grow if supplemented with the "no growth" intermediate. It will not be possible to fit the data for mutant 4 to pathway a, the data for mutants 1 and 4 to pathway b, or data for mutants 3 and 4 to pathway c. The data for all mutants can be fit only to pathway d. Thus, d must be the correct pathway.

A second approach to this problem is to realize that in any linear segment of a biochemical pathway (a segment without a branch), a block early in the segment can be circumvented by *any* metabolites that normally appear later in the same segment. Consequently, if two (or more) intermediates can support growth of a mutant, they normally are made after the blocked step in the same linear segment of a pathway. From the data given, compounds D and E both circumvent the single block in mutant 4. This means that compounds D and E lie after the block in mutant 4 on a linear segment of the metabolic pathway. The only pathway where D and E lie in an unbranched linear segment is pathway d. Mutant 4 could be blocked between A and E in this pathway. Mutant 4 cannot be fit to a *single* block in any of the other pathways that are shown, so the correct pathway is d.

10.12 Four strains of the haploid fungus *Neurospora,* all of which require arginine but have unknown genetic constitution, have the following nutrition and accumulation characteristics:

	Growth On				
Strain	**Minimal Medium**	**Ornithine**	**Citrulline**	**Arginine**	**Accumulates**
1	–	–	+	+	Ornithine
2	–	–	–	+	Citrulline
3	–	–	–	+	Citrulline
4	–	–	–	+	Ornithine

The pairwise complementation tests of the four strains gave the following results (+ = growth on minimal medium and 0 = no growth on minimal medium):

	4	3	2	1
1	0	+	+	0
2	0	0	0	
3	0	0		
4	0			

Crosses among mutants yielded prototrophs in the following percentages:

1×2: 25 percent
1×3: 25 percent
1×4: none detected among 1 million progeny (ascospores)
2×3: 0.002 percent
2×4: 0.001 percent
3×4: none detected among 1 million progeny (ascospores)

Analyze the data and answer the following questions.

a. How many distinct mutational sites are represented among these four strains?
b. In this collection of strains, how many types of polypeptide chains (normally found in the wild type) are affected by mutations?
c. Write the genotypes of the four strains, using a consistent and informative set of symbols.
d. Determine the map distances between all pairs of linked mutations.
e. Determine the percentage of prototrophs that would be expected among ascospores of the following types: (1) strain $1 \times$ wild type; (2) strain $2 \times$ wild type; (3) strain $3 \times$ wild type; (4) strain $4 \times$ wild type.

Answer: Use the nutrition and accumulation data to infer where each mutant is blocked. 1 accumulates ornithine, and can grow only if citrulline or arginine is added. Hence, 1 is blocked in the conversion of ornithine to citrulline. 2 and 3 are blocked in the conversion of citrulline to arginine. Mutant 4 is more complex. It can grow only if supplemented with arginine, but accumulates ornithine. 4 must be a double mutant and is blocked in the conversion of ornithine to citrulline as well as from citrulline to arginine. This gives the following pathway:

$$\text{precursor} \longrightarrow \text{ornithine} \xrightarrow{1,\,4} \text{citrulline} \xrightarrow{2,\,3,\,4} \text{arginine}$$

Enzyme: A B C

Use the complementation test data to determine whether two mutants affect the same function. Two mutants able to grow on minimal media complement each other: One provides a function missing in the other. Failure to grow indicates no complementation: Both are missing the same function. 2 and 3 complement 1, so 2 and 3 affect a different step than 1. 2 and 3 fail to complement each other, so they affect the same step. 4 fails to complement 1, 2, or 3 and affects both steps. This is consistent with the above diagram: 1 and 4 affect enzyme B, and 2, 3, and 4 affect enzyme C.

Start to analyze the recombination data by assigning symbols to the genes and writing out the genotypes and crosses. Let b^+ encode enzyme B and c^+ encode enzyme C. Then $1 = b^1\ c^+$, $2 = b^+\ c^2$, $3 = b^+\ c^3$, and $4 = b^4\ c^4$. The prototrophic ascospores produced in the crosses 1×2, 1×3, and 2×3 are $b^+\ c^+$ and are half of the recombinant progeny (e.g., in 1×2, parental types are $b^1\ c^+$ and $b^+\ c^2$, recombinant types are $b^+\ c^+$ and $b^1\ c^2$). The results can be tabulated as:

Cross	Genotypes	$b^+\ c^+$ Spores
1×2	$b^1\ c^+ \times b^+\ c^2$	25.0%
1×3	$b^1\ c^+ \times b^+\ c^3$	25.0%
1×4	$b^1\ c^+ \times b^4\ c^4$	$< 1 \times 10^{-6}$
2×3	$b^+\ c^2 \times b^+\ c^3$	0.002%
2×4	$b^+\ c^2 \times b^4\ c^4$	0.001%
3×4	$b^+\ c^3 \times b^4\ c^4$	$< 1 \times 10^{-6}$

Use the recombination frequencies to infer whether the b and c genes are linked and whether two alleles at one gene are identical or can be separated by recombination. 1×2 and 1×3 give 25 percent prototrophs (show 50 percent recombinants), so b and c are unlinked. 2×3 produces very few prototrophs, as they can only arise from rare, intragenic recombination. The 0.002% prototrophs (= half the recombinants) indicate that c^2 and c^3 are 0.004 mu apart. 4 fails to recombine with 1 or 3, indicating that $b^4 = b^1$ and $c^4 = c^3$ (they occupy the same site). The 0.001% prototrophs from 2×4 arise when a c^+ allele is obtained after intragenic recombination between c^2 and c^4 *and* the b^+ allele (in mutant 2) assorts with it into the ascospore. When c^+ assorts with b^4, (half of the time), a prototrophic ascospore is not obtained. This is why only 0.001% prototrophs are recovered from 2×4, instead of the 0.002% prototrophs recovered from 2×3. This results in the following map:

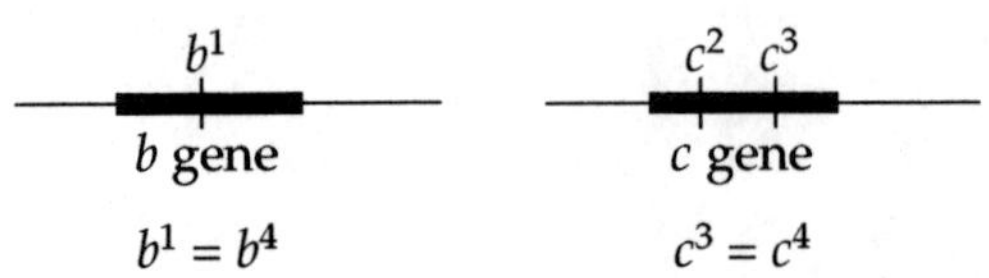

a. Three distinct mutational sites exist based on the recombination analysis: one in the *b* gene and two in the *c* gene.

b. Two polypeptide chains are affected based on the complementation analysis: one encoded by the *b* gene and one encoded by the *c* gene.

c. The genotypes are as given in the table above.

d. There are 0.004 mu between the two mutant sites in the *c* gene identified by c^2 and c^3. The *b* and *c* genes are unlinked.

e. The *b* and *c* genes assort independently of one another. Thus, strain 1, 2, or 3, when mated with the wild type, will give 50 percent prototrophs and 50 percent auxotrophs. As strain 4 is a double mutant, it will only give 25 percent prototrophs. Four equally frequent genotypes would be expected from the cross $b^4\ c^4 \times + +$. Only one of the four will be + +.

10.13 A breeder of Irish setters has a particularly valuable show dog that he knows is descended from the famous bitch Rheona Didona, who carried a recessive gene for atrophy of the retina. Before he puts the dog to stud, he must ensure that it is not a carrier for this allele. How should he proceed?

Answer: One method is to testcross the male to a retinal atrophic female. Retinal atrophic pups (expected half of the time) would indicate that the male is heterozygous. If the retinal atrophy is associated with a known biochemical phenotype (e.g., altered enzyme activity), a preferred method is to take a tissue biopsy or blood sample from the male and assess whether normal levels of enzyme activity are present in the male. Recessive mutations in genes for many enzymes, when heterozygous with a normal allele, lead to reduced levels of enzyme activity even though no visible phenotype is observed.

10.14 Suppose you were on a jury to decide the following case:

The Jones family claims that Baby Jane, given to them at the hospital, belongs not to them but to the Smith family and that the Smiths' baby Joan really belongs to the Jones family. It is alleged that the two babies were accidentally exchanged soon after birth. The Smiths deny that such an exchange has been made. Blood group determinations show the following results:

Mrs. Jones, AB

Mr. Jones, O

Mrs. Smith, A

Mr. Smith, O

Baby Jane, A

Baby Joan, O

Which baby belongs to which family?

Answer: Since Mrs. Jones is type AB and Mr. Jones is type O, the cross can be diagrammed as $I^AI^B \times ii$, which would give I^Ai (type A) or I^Bi (type B) children. Since Mrs. Smith is type A and Mr. Smith is type O, the cross can be diagrammed as $I^A\text{–} \times ii$, which could give either I^Ai (type A) or potentially type O children. In any case, Mr. and Mrs. Jones could not have a type O child, so Baby Jane must be theirs.

10.15 Glutathione (GSH) is important for a number of biological functions, including prevention of oxidative damage in red blood cells, synthesis of deoxyribonucleotides from ribonucleotides, transport of some amino acids into cells, and maintenance of protein conformation. Mutations that have lowered levels of glutathione synthetase (GSS), a key enzyme in the synthesis of glutathione, result in one of two clinically distinguishable disorders. The severe form is characterized by massive urinary excretion of 5-oxoproline (a chemical derived from a synthetic precursor to glutathione), metabolic acidosis (an inability to regulate physiological pH appropriately), anemia, and central nervous system damage. The mild form is characterized solely by anemia. The characterization of GSS activity and the GSS protein in two affected patients, each with normal parents, is shown here.

Patient	Disease Form	GSS Activity in Fibroblasts (percent of Normal)	Effect of Mutation on GSS Protein
1	severe	9%	Arginine at position 267 replaced by tryptophan
2	mild	50%	Aspartate at position 219 replaced by glycine

a. What pattern of inheritance do you expect these disorders to exhibit?

b. Explain the relationship of the form of the disease to the level of GSS activity.

c. How can two different amino acid substitutions lead to dramatically different phenotypes?

d. Why is 5-oxoproline produced in significant amounts only in the severe form of the disorder?

e. Is there evidence that the mutations causing the severe and mild forms of the disease are allelic (in the same gene)?

f. How might you design a test to aid in prenatal diagnosis of this disease?

Answer:

a. Since normal parents have affected offspring, the disease appears to be recessive. However, since patients with 50 percent of GSS activity have a mild form of the disease, individuals may show mild symptoms if they are heterozygous (mutant/+) for a mutation that eliminates GSS activity. In a population, individuals having the disease may not all show identical symptoms, and some may have more severe disease than others. The severity of the disease in an individual will depend on the nature of the person's GSS mutation and, possibly, whether the person is heterozygous or homozygous for the disease allele. The alleles discussed here appear to be recessive.

b. Patient 1, with 9 percent of normal GSS activity, has a more severe form of the disease, while patient 2, with 50 percent of normal GSS activity, has a less severe form of the disease. Thus, increased disease severity is associated with less GSS enzyme activity.

c. The two different amino acid substitutions may disrupt different regions of the enzyme's structure (consider the effect of different amino acid substitutions on hemoglobin function, discussed in the text). As amino acids vary in their polarity and charge, different amino acid substitutions within the same structural region could have different chemical effects on protein structure. This, too, could lead to different levels of enzymatic function. (For a discussion of the chemical differences between amino acids, see Chapter 12).

d. By analogy with the disease PKU discussed in the text, 5-oxoproline is produced only when a precursor to glutathione accumulates in large amounts due to a block in a biosynthetic pathway. When GSS levels are 9 percent of normal, this occurs. When GSS levels are 50 percent of normal, sufficient GSS enzyme is present to partially complete the pathway and prevent high levels of 5-oxoproline.

e. The mutations are allelic (in the same gene), since both the severe and the mild forms of the disease are associated with alterations in the same polypeptide that is a component of the GSS enzyme. (Note that while the data in this problem suggest that the GSS enzyme is composed of a single polypeptide, they do not exclude the possibility that GSS has multiple polypeptide subunits encoded by different genes.)

f. If GSS is normally found in fetal fibroblasts, one could, in principle, measure GSS activity in fibroblasts obtained via amniocentesis. The GSS enzyme level in cells from at-risk fetuses could be compared with that in normal control samples to predict disease due to inadequate GSS levels. Some variation in GSS level might be seen, depending on the allele(s) present. Since more than one mutation is present in the population, it is important to devise a functional test that assesses GSS activity, rather than a test that identifies a single mutant allele.

10.16 Some methods used to gather fetal material for prenatal diagnosis are invasive and therefore pose a small but very real risk to the fetus.

a. What specific risks and problems are associated with chorionic villus sampling and amniocentesis?

b. How are these risks balanced with the benefits of each procedure?

c. Fetal cells are reportedly present in maternal circulation after about 8 weeks of pregnancy. However, the number of cells is very low, perhaps no more than one in several million maternal cells. To date, it has not been possible to isolate fetal cells from maternal blood in sufficient quantities for routine genetic analysis. If the problems associated with isolating fetal cells from maternal blood were overcome, and sufficiently sensitive methods were developed to perform genetic tests on a small number of cells, what would be the benefits of performing genetic tests on these fetal cells?

Answer:

a. Amniocentesis is typically done after 12 weeks of pregnancy (after the first trimester), while chorionic villus sampling can be done between 8 and 12 weeks of pregnancy. In either sampling method, there is a risk of fetal loss. While chorionic villus sampling can allow parents to learn if the fetus has a genetic defect earlier than amniocentesis, there is more risk of fetal loss and more chance of an inaccurate diagnosis due to the contaminating presence of maternal cells.

b. These sampling methods offer parents an opportunity to learn about the genetic constitution and health of the fetus early in a pregnancy. If the fetus is diseased, early prenatal diagnosis allows parents to make an informed decision about the pregnancy. They can seek genetic counseling and consider what options might be available to them: to treat the diseased fetus in utero, to plan for the care of a diseased child, or to terminate the pregnancy.

c. Such a method would allow for early, *noninvasive* prenatal diagnosis. This would substantially reduce the risk to the fetus and the discomfort to the mother. Since a blood sample could be obtained by many health professionals and shipped off-site for processing, it would also eliminate the need for a highly skilled physician and sophisticated on-site support services to perform the sampling procedure. Consequently, it would allow for more accessible prenatal testing at lower cost.

10.17 In evaluating my teacher, my sincere opinion is that

a. he/she is a swell person whom I would be glad to have as a brother-in-law/sister-in-law
b. he/she is an excellent example of how tough it is when you do not have either genetics or environment going for you
c. he/she may have okay DNA to start with, but somehow all the important genes got turned off
d. he/she ought to be preserved in tissue culture for the benefit of other generations

Answer: Your choice!

10.18 A *Neurospora* mutant has been isolated in the laboratory in which you are working. This mutant cannot make an amino acid we will call Y. Wild-type *Neurospora* cells make Y from a cellular product X through a biochemical pathway involving three intermediates called c, d, and e. How would you establish that your mutant contains a defective gene for the enzyme that catalyzes the d-e reaction?

Answer: If the enzyme that catalyzes the d $\rightarrow$ e reaction is missing, the mutant strain should accumulate d and be able to grow on minimal medium to which e is added. In addition, it should not be able to grow on minimal medium or on minimal medium to which X, c, or d is added, but should grow if Y is added. Therefore, plate the strain on these media, and test which allow for growth of the mutant strain and which intermediate is accumulated if the strain is plated on minimal medium.

10.19 You have been introduced to the functions and levels of proteins and their organization. List as many protein functions as you can, and give an example of each.

Answer: Proteins serve a multitude of different functions. Here are some possibilities and examples:

Protein Function	Examples and Their Roles
Enzyme	DNA polymerase I—synthesizes DNA, proofreads, excises an RNA primer
	DNA ligase—seals a nick in a DNA sugar-phosphate backbone
	Phenylalanine hydroxylase—converts phenylalanine to tyrosine
	Topoisomerase—introduces/removes negative supercoils in DNA

(continued)

Protein Function	Examples and Their Roles
Structural protein	Tubulin—polymerizes to form microtubules, used in chromosome segregation at anaphase Myosin—a muscle protein used in muscle contraction Actin—polymerizes to form actin filaments, which provide structural integrity to the cell
Viral protein	Proteins making up the coat of T2, TMV
Receptor	Acetylcholine receptor—binds the neurotransmitter acetylcholine, found on the surface of muscle cells
Transporter	Hemoglobin—transports oxygen
Ion channel	CFTR—transports chloride ions
Regulatory molecule	Transcription factors—determine where mRNA transcripts are produced (see Chapter 11)
Bind DNA	Single-stranded DNA-binding protein—keeps single-stranded DNA strands intact and from renaturing within a replication bubble during DNA synthesis
Antibody	Immunoglobulins—immune function
Blood type	A, B, O glycoproteins—specify blood type

10.20 We have discovered that most enzymes are proteins, but not all proteins are enzymes. What are the functions of enzymes, and why are they essential for living organisms to carry out their biological functions?

Answer: Enzymes are macromolecules that catalyze chemical reactions. These reactions can be quite simple, such as the addition of a hydroxyl (–OH) group onto a simple compound, or they can be exceedingly complex, such as the replication of an entire chromosome. Enzymes are essential to biological systems because they provide catalytic power, usually increasing reaction rates (the rate of production of the product of the chemical reaction) by at least a millionfold. Nearly all of the chemical transformations made in the body proceed at very low rates in the absence of enzymes. Therefore, enzymes are necessary to catalyze reactions that produce needed compounds and perform essential processes.

10.21 We know that the function of any protein is tied to its structure. Give an example of how a disruption of structure by mutation can lead to a distinctive phenotypic effect.

Answer: Many examples are given in the text. Here are two.

Hb-S mutations in the β-globin gene change the sixth amino acid from the N terminus of the β-globin polypeptide. This amino acid is changed from a negatively charged, acidic amino acid (glutamic acid) to an uncharged, neutral amino acid (valine). This single amino acid change causes the β-globin polypeptide to fold in an abnormal way, which in turn causes sickling of red blood cells and sickle-cell anemia.

The most common mutation causing cystic fibrosis (CF) is a mutation in the CFTR (cystic fibrosis transmembrane regulator) gene called ΔF508. This mutation deletes three consecutive base pairs of DNA and results in the deletion of a single amino acid in the CFTR protein. This amino acid is deleted in a region of the CFTR protein important for binding ATP. Since this region contributes to the function of the protein—to transport chloride ions across cell membranes—this single amino acid deletion affects the function of the protein and causes CF.

10.22 Define *autosomal recessive mutation*, and give some examples of diseases that are caused by autosomal recessive mutations. Explain how two parents who do not display any symptoms of a given disease (any of the diseases you have named) can have two or even three children who have the disease. How can these same parents have no children with the disease?

Answer: Autosomes are chromosomes that are found in two copies in both males and females. That is, an autosome is any chromosome except the X and Y chromosome. Since individuals have two of each type of autosome, they have two copies of each gene on an autosome. The forms of the gene, or the alleles at the gene, can be the same or different. They can have either two normal alleles (homozygous for the normal allele), one normal allele and one mutant allele (heterozygous for the normal and mutant alleles), or two mutant alleles (homozygous for the mutant allele). A recessive mutation is one that exhibits a phenotype only when it is homozygous. Therefore, an autosomal recessive mutation is a mutation that occurs on any chromosome except the X or Y and that causes a phenotype only when homozygous. Heterozygotes exhibit a normal phenotype.

Of the diseases discussed in this chapter, many are autosomal recessive diseases. For example, phenylketonuria, albinism, Tay-Sachs disease, and cystic fibrosis are all autosomal recessive diseases. Heterozygotes for the disease allele are normal, but homozygotes with the disease allele are affected. For phenylketonuria and albinism, homozygotes are affected because they lack a required enzymatic function. In these cases, heterozygotes have a normal phenotype because their single normal allele provides sufficient enzyme function.

Parents contribute one of their two autosomes to their gametes, so that each offspring of a couple receives an autosome from each parent. If, in a particular conception, each of two heterozygous parents contributes a chromosome with the normal allele, the offspring will be homozygous for the normal allele and be normal. If, in a particular conception, one of the two heterozygous parents contributes a chromosome with the normal allele and the other parent contributes a chromosome with the mutant allele, the offspring will be heterozygous but be normal. If, in a particular conception, each parent contributes the chromosome with the mutant allele, the offspring will be homozygous for the mutant allele and develop the disease. Therefore, heterozygous parents can have both normal and affected children. Since each conception is independent, two heterozygous parents can have all normal, all affected, or any mix of normal and affected children.

10.23 What can prospective parents do to reduce the risk of having offspring who have genetically based enzyme deficiencies?

Answer: At one level, the answer to this question is straightforward. Couples who are prospective parents can undergo genetic testing for mutations that result in enzyme deficiencies. If both members of a couple are heterozygous for mutations in the same gene, then they may elect to undergo genetic counseling to assess the risk of having an affected child and to discuss the options available to them concerning prenatal screening.

In practice, however, this question raises very complex issues. First, a decision to undergo genetic testing is associated with certain risks. While the risks associated with drawing blood for the testing procedure itself can be minimal, there are additional risks with very real, significant consequences. These include a risk of emotional discomfort upon learning that one is (or is not) a carrier, risks to personal and familial relationships, and potential risks to privacy, insurability, employment, discrimination, and stigmatization. Second, it is difficult to know which mutations to test for. Sometimes this issue can be rationally addressed if there is a clear family history of a particular inherited disease. However, this may not be the case. For example, for autosomal recessive diseases, family members may show no signs of a disease because they are heterozygous, and some individuals will not know their familial history. In addition, because the genes we have are shared within our ethnic group, the risks we have for different inherited diseases are related to our ethnicity. Given this, genetic screening and testing can raise issues related to cultural and societal values, historical (and perhaps current) levels of discrimination, and the availability and quality of health-care services. Third, this question raises personal, philosophical, and religious issues. Some individuals may not be willing to undergo genetic testing out of deeply held personal beliefs. While some couples may be willing to undergo genetic testing and prenatal testing so that they can use this information in planning, they may not be willing to electively terminate a fetus who has a genetically based enzyme deficiency. With current and anticipated advances in genomics (see Chapter 15), these and other policy, societal, and ethical issues related to genetic screening and genetic testing are becoming more significant.

11

GENE EXPRESSION: TRANSCRIPTION

CHAPTER OUTLINE

GENE EXPRESSION: AN OVERVIEW

THE TRANSCRIPTION PROCESS

RNA Synthesis
Initiation of Transcription at Promoters
Elongation and Termination of an RNA Chain

TRANSCRIPTION IN EUKARYOTES

Eukaryotic RNA Polymerases
Transcription of Protein-Coding Genes by RNA Polymerase II
Eukaryotic mRNAs
Transcription of Other Genes

REVIEW OF KEY TERMS, SYMBOLS, AND CONCEPTS

In your own words, write a brief, precise definition of each term in the groups on the following two pages. Check your definitions using the text. Then develop a concept map using the terms in each list.

THINKING ANALYTICALLY

As transcription and RNA processing are central to gene expression, they have been investigated intensely. Our substantial understanding of these processes is reflected in the wealth of detailed information presented in this chapter. As was the case for understanding DNA replication, the best way to understand and retain this information is to organize the details and place them in context. As you read the chapter, continually refer to the text figures that summarize the key events of transcription and RNA processing. Try explaining the figures to yourself out loud to enhance your understanding. To clarify and solidify your understanding of the material after you have read the chapter once, go through it again and categorize its detailed information.

To start categorizing the information, develop a list of terms and identify their chemical nature. Is the term a region of DNA, of RNA, or part of a protein? After this, relate the term to its function in the overall process of gene expression. If it pertains to DNA, how does it relate to the structure of a gene or a transcription unit? If it pertains to RNA, how does it relate to the RNA being synthesized or being processed? If it pertains to a part of a protein, how is it involved in transcription or RNA processing?

1

central dogma
replication
transcription
translation
primary transcript
precursor RNA
mRNA
tRNA
rRNA
snRNA
RNA polymerase
RNA polymerases I, II, III
prokaryotic genes
eukaryotic genes
structural gene
protein-coding gene
α-amanitin
nucleolus
nucleus
gene regulatory elements
5′→3′, 3′→5′
coding strand
template strand
nontemplate strand
NTPs

2

prokaryotic RNA polymerase
transcription initiation
promoter
closed, open promoter complex
holoenzyme
core enzyme
sigma factor
σ^{70}, σ^{32}, σ^{54}, σ^{23}
heat shock, stress
consensus sequence
–10, –35 regions
Pribnow box
elongation
transcription bubble
rho-dependent termination
type I, II terminators
hairpin loop
twofold symmetry
coupled transcription and translation
transcription unit
leader, coding, trailer sequences
5′, 3′ UTR
polycistronic

3

RNA polymerase II
protein-coding genes
mRNA, snRNA
regulatory elements
basal transcription factors
regulatory factors
basal promoter elements
promoter proximal elements
Goldberg-Hogness, TATA, CAAT, GC boxes
enhancer element
silencer elements
activator, repressor
adapter
upstream activator sequence (UAS)
TBP, TFIIB, D, E, F, H
initial committed complex
minimal transcription initiation complex
complete transcription initiation complex
preinitiation complex

4

RNA polymerase II
transcription unit
leader, trailer, coding sequences
5′, 3′ UTR
precursor mRNA
pre-, mature mRNA
posttranscriptional modification
monocistronic
5′ cap
5′ 7-methyl guanosine
5′-5′, 2′-5′ bonds
capping enzyme
methylation
3′ polyadenylation
poly(A) site, tail
poly(A) polymerase
RNA endonuclease
CPSF, CstF, CFI, CFII
introns, exons
hnRNA, gRNA
mRNA splicing
branch-point sequence
3′, 5′ splice sites
snRNA, snRNP
spliceosome
U1, U2, U4, U5, U6
RNA lariat structure
RNA editing

To help you relate a term to its function, construct sketches of the processes of transcription and transcript processing. Diagram how the structure of a gene at the DNA level relates to the structure of its primary transcript at the RNA level. Then diagram how the structure of the primary transcript relates to the structure of the processed, mature transcript.

Next, distinguish between processes and characteristics in prokaryotes and eukaryotes. While there are some similarities between prokaryotes and eukaryotes, there are many, many differences. The fact that three RNA polymerases are active in eukaryotes results in three sets of data. For each RNA polymerase, there are distinct requirements for initiation and termination. Different DNA sequences are used, and different protein factors are involved in each case. It may help to prepare a list of analogous sites and factors, pairing prokaryotic elements with those doing similar jobs in eukaryotes.

5

E. coli RNA polymerase
RNA polymerase I
sucrose density gradient
Svedberg unit
S value
30S, 50S subunits
70S ribosome
23S, 16S, 5S rRNA
40S, 60S subunits
80S ribosome
28S, 18S, 5.8S, 5S rRNA
ribosomal protein
rDNA
rRNA transcription unit (*rrn*)
precursor rRNA
RNase III
rDNA repeat unit
spacer sequence
nontranscribed spacer
external transcribed spacer (ETS)

6

RNA polymerase III
5S rRNA genes
tRNA genes
tDNA
internal control region (ICR)
pre-tRNA molecule
codon
anticodon
posttranscriptional modification
TFIIIA, B, C,
transcription initiation factor
cloverleaf stem-loop structure
RNA ligase

7

spliceosome
snRNPs ("snurps")
self-splicing
group I introns
protein-independent reaction
secondary structure
lariat molecule
Tetrahymena pre-mRNA
28S spacer sequence removal
28S intron removal
RNA enzyme
ribozyme
RNA ligase

QUESTIONS FOR PRACTICE

Multiple Choice Questions

1. Each row of the table below lists an item composed of DNA, RNA, and/or protein. Indicate an item's chemical composition by placing an X in the appropriate column(s).

Item	DNA	RNA	Protein
sigma factor			
promoter			
terminator			
transcription factor			
regulatory factor			
Pribnow box			
transcription initiation complex			
spacer sequences			
NTS			
snRNP			
spliceosome			

(continued)

Item	DNA	RNA	Protein
rho			
promoter element			
RNA polymerase core enzyme			
ribozyme			
consensus sequence			
U2			
enhancer			

2. Now indicate whether each item is found only in prokaryotes, only in eukaryotes, or in both prokaryotes and eukaryotes by placing an X in the appropriate column.

Item	Only in Prokaryotes	Only in Eukaryotes	Found in Both
sigma factor			
promoter			
terminator			
transcription factor			
regulatory factor			
Pribnow box			
transcription initiation complex			
spacer sequences			
NTS			
snRNP			
spliceosome			
rho			
promoter element			
RNA polymerase core enzyme			
ribozyme			
consensus sequence			
U2			
enhancer			

3. Which one statement is correct?
   a. During transcription, the template strand is read in a $3'\rightarrow5'$ direction.
   b. During transcription, the template strand is read in a $5'\rightarrow3'$ direction.
   c. During transcription, the nontemplate strand is read in a $3'\rightarrow5'$ direction.
   d. During transcription, the nontemplate strand is read in a $5'\rightarrow3'$ direction.

4. Which one statement is correct?
   a. During transcription, an RNA is synthesized in the 5′→3′ direction.
   b. During transcription, an RNA is synthesized in the 3′→5′ direction.
   c. During transcription, DNA is synthesized in the 5′→3′ direction.
   d. During transcription, DNA is synthesized in the 3′→5′ direction.
5. Which one of the following is incorrect?
   a. DNA polymerases can proofread but RNA polymerases cannot.
   b. Some RNA polymerases, but no DNA polymerases, are sensitive to α-amanitin.
   c. Only RNA polymerases can initiate the formation of a polynucleotide.
   d. RNA polymerases synthesize in a 3′→5′ direction while DNA polymerases synthesize in a 5′→3′ direction.
6. Which one of the following is essential for RNA polymerase to bind an *E. coli* promoter?
   a. sigma factor
   b. snRNP
   c. TFIID
   d. U1
7. Which one of the following is essential for RNA polymerase II to bind a eukaryotic promoter?
   a. sigma factor
   b. *rho*
   c. TFIID
   d. U1
8. Which one of the following is incorrect?
   a. Prokaryotic and eukaryotic mRNAs have 5′ untranslated leader sequences.
   b. Only prokaryotic mRNAs are polyadenylated at their 3′ ends.
   c. In prokaryotes, transcription and translation are coupled.
   d. Consensus sequences can be found in both DNA and RNA.
   e. In eukaryotes, RNA splicing occurs in the nucleus.

Answers:

1. A promoter, terminator, Pribnow box, NTS, promoter element, and enhancer are composed of DNA. U2 and a ribozyme are composed of RNA. A sigma factor, transcription factor, regulatory factor, *rho,* and the RNA polymerase core enzyme are composed of protein. Spacer and consensus sequences can be found in both RNA and DNA. A transcription initiation complex contains proteins that interact with DNA. Spliceosomes and snRNPs contain both RNA and proteins.
2. Sigma factor, the Pribnow box, *rho,* and the RNA polymerase core enzyme are only found in prokaryotes (*E. coli*). An NTS, snRNP, spliceosome, ribozyme, enhancer, and U2 are only found in eukaryotes. A promoter, terminator, transcription factor, regulatory factors, transcription initiation complex, spacer sequence, promoter element, and consensus sequence can be found in both eukaryotes and prokaryotes.

3.a, 4.a, 5.d, 6.a, 7.c, 8.b

Thought Questions

1. Compare and contrast each of the following groups of terms with respect to location, function, and host organism (prokaryote or eukaryote).
   a. Pribnow box and TATA element
   b. promoter elements, enhancer elements, silencer elements
   c. the promoters bound by RNA polymerase I, II, III
   d. rRNA genes in prokaryotes and eukaryotes
2. Do prokaryotes or eukaryotes have more RNA polymerase (per unit total protein)? Why should the two groups of organisms differ?
3. How does *rho*-dependent termination differ from *rho*-independent termination? What is the role of the hairpin loop in each of these events? Is the hairpin loop in DNA or RNA?
4. What takes the place of sigma factor in eukaryotes?
5. Recall that there was no simple relationship between the C value and the structural and organizational complexity of an organism. Do you expect a relationship to exist between the number of rDNA repeat units and the structural or organizational complexity of the organism? The amount of hnRNA? The amount of cytoplasmic mRNA? Why or why not?
6. The promoters for RNA polymerase II have been highly conserved and thus differ little among eukaryotes. However, the promoters for RNA polymerase I show significant differences among eukaryotes. Why might this be the case? (Hint: Consider the number of genes read by each polymerase.)
7. Consider what will happen to an mRNA once it is processed, and then address the following questions: How precise does RNA splicing need to be? How precise does transcription initiation need to be? How precise does polyadenylation need to be?
8. What posttranscriptional modifications are routinely made to a eukaryotic mRNA? Where do they occur in the cell?

Thought Questions for Media

After reviewing the media at the *Essential iGenetics* website, try to answer these questions.

1. At which end of the template strand does transcription begin?
2. When trascription is complete, which strand of DNA is the final RNA molecule identical to if the thymine in DNA is replaced with uracil in RNA?
3. What purpose does the sequence AAUAAA serve in a pre-mRNA's 3′ UTR?

1. Suppose a mutation in the β-globin gene introduced a new 3′ splice site within its first intron.
   a. In what different ways could the mutation affect the reading frame of the mature mRNA?
   b. If the mature mutant mRNA were translated in the same reading frame after the mutant site, what would its effect be on the β-globin protein?
   c. If the mutation caused the mRNA to be translated in a different reading frame, what would its effect be on the β-globin protein? (Remember that when mRNAs are "read" out of their correct reading frames, stop codons are often introduced.)
2. Suppose a point mutation occurred in the first intron of the β-globin gene, but did not change either the branch point, the 5′ splice site or the 3′ splice site sequence. Could such a mutation ever affect the level of mRNA transcripts available for translation?

How might your answer change if the mutation were a deletion or insertion of 100 base pairs of DNA? (Recall that the first intron is 130 base pairs long.)

3. In the iActivity, the first multiple-choice question describes two forms of beta-thalassemia: beta-plus and beta-zero. Beta-plus thalassemia occurs when a reduced amount of β-globin protein is produced, while beta-zero thalassemia occurs when no β-globin protein is produced.
   a. Do you expect the phenotypes of these forms of thalassemia to be identical in heterozygotes having a disease allele and a normal allele? Why or why not?
   b. Will they be identical in homozygotes for a disease allele? Why or why not?

SOLUTIONS TO TEXT PROBLEMS

11.1 Compare DNA and RNA with regard to structure, function, location, and activity. Also, how do these molecules differ with regard to the polymerases used to synthesize them?

Answer: While both DNA and RNA are composed of linear polymers of nucleotides, their bases and sugars differ. DNA contains deoxyribose and thymine, while RNA contains ribose and uracil. Their structures also differ. DNA is frequently double-stranded, while RNA is usually single-stranded. Single-stranded RNAs are capable of forming stable, functional, and complex stem-loop structures, such as those seen in tRNAs. Double-stranded DNA is wound in a double helix and packaged by proteins into chromosomes, either as a nucleoid body in prokaryotes or within the eukaryotic nucleus. RNA, after being transcribed from DNA, can be exported into the cytoplasm. If it is mRNA, it can be bound by ribosomes and translated. Eukaryotic RNAs are highly processed before being transported out of the nucleus. DNA functions as a storage molecule, while RNA functions either as a messenger (mRNA carries information to the ribosome) or in the processes of translation (rRNA functions as part of the ribosome, tRNA brings amino acids to the ribosome) or in eukaryotic RNA processing (snRNA functions within the spliceosome).

Both DNA polymerases and RNA polymerases catalyze the synthesis of nucleic acids in the 5′→3′ direction. Both use a DNA template and synthesize a nucleic acid polynucleotide that is complementary to the template. However, DNA polymerases require a 3′-OH to add on to, while RNA polymerases do not. That is, RNA polymerases can initiate chains without primers, while DNA polymerases cannot. Furthermore, RNA polymerases usually require specific base pair sequences as signals to initiate transcription.

11.2 All base pairs in the genome are replicated during the DNA synthesis phase of the cell cycle, but only some of the base pairs are transcribed into RNA. How is it determined which base pairs of the genome are transcribed into RNA?

Answer: The DNA sequences that are transcribed into RNA are determined using two general principles. First, signals in the DNA base sequences identify the specific region to be transcribed. Only regions bounded by transcription initiation and transcription termination signals are transcribed. In regions bounded by these signals, only one strand is ordinarily transcribed, so that transcripts are formed in a 5′→3′ direction using a single DNA template strand. Second, transcription within a defined region occurs only if additional transcription-inducing molecules are present. Some of these molecules recognize the signals within the DNA that define the transcription unit and are used in the transcription process.

11.3 Discuss the similarities and differences between the *E. coli* RNA polymerase and eukaryotic RNA polymerases.

Answer: Both eukaryotic and *E. coli* RNA polymerases transcribe RNA in a 5′→3′ direction using a 3′→5′ DNA template strand. There are many differences between the enzymes, however. In *E. coli,* a single RNA polymerase core enzyme is used to transcribe genes. In eukaryotes, there are three types of RNA polymerase molecules, RNA polymerase I, II, and III. RNA polymerase I synthesizes 28S, 18S, and 5.8S rRNA and is found in the nucleolus. RNA polymerase II synthesizes hnRNA, mRNA, and some snRNAs and is nuclear. RNA polymerase III synthesizes tRNA, 5S rRNA, and some snRNAs and is nuclear.

Each RNA polymerase uses a unique mechanism to identify those promoters at which it initiates transcription. In prokaryotes such as *E. coli,* a sigma factor provides specificity to the sites bound by the four-polypeptide core enzyme, so that it binds to promoter sequences. The holoenzyme loosely binds a sequence of about 35 bp before transcription initiation (the –35 region), changes configuration, and then tightly binds a region of about 10 bp before transcription initiation (the –10 region) and melts about 17 bp of DNA around this region. The two-step binding to the promoter orients the polymerase on the DNA and facilitates transcription initiation in the 5′→3′ direction. After about 8 or 9 bases are formed in a new transcript, sigma factor dissociates from the holoenzyme, and the core enzyme completes the transcription process. Although the principles by which eukaryotic RNA polymerases bind their promoters are similar in that they use a set of ancillary protein factors—transcription factors—the details are quite different. In eukaryotes, each of the three types of RNA polymerases recognizes a different set of promoters by using a polymerase-specific set of transcription factors, and the mechanisms of interaction are different.

11.4 What are the most significant differences between the organization and expression of prokaryotic genes and eukaryotic genes?

Answer: Prokaryotic and eukaryotic genes differ in their gene structure, in their RNA processing, and in how transcription is coupled to translation.

Structure: Prokaryotic genes are defined by an upstream promoter, an RNA-coding sequence, and a downstream terminator. While these three features also exist in eukaryotic genes, eukaryotic promoters are more complex, and nearby or distant enhancer and silencer elements can strongly affect the level of transcription of eukaryotic genes. The RNA-coding sequences of eukaryotes can be interrupted with introns. Finally, prokaryotic mRNAs are often polycistronic, containing the amino acid–coding information for more than one gene. In contrast, eukaryotic mRNAs are monocistronic, containing the amino acid–coding information from just one gene.

Processing: Prokaryotic genes lack introns, while eukaryotic genes typically have one or more introns, and therefore a transcribed region can be larger than the size of a mature mRNA. The excision of introns from primary mRNAs is only one aspect of the processing of eukaryotic RNAs. In addition, eukaryotic mRNAs are modified at both their 5′ and 3′ ends: they are capped and polyadenylated.

Coupling of transcription and translation: Since prokaryotes lack a nucleus, transcription is directly coupled to translation. In eukaryotes, mRNAs must be processed and then transported out of a nucleus before they are translated by ribosomes in the cytoplasm.

These three differences provide cell types in a multicellular eukaryote three means to regulate gene expression. Gene expression can be regulated at each of these three levels: transcription, mRNA processing, and translation.

11.5 Discuss the molecular events involved in the termination of RNA transcription in prokaryotes. In what ways is this process fundamentally different in eukaryotes?

Answer: Termination of transcription in *E. coli* is signaled by controlling elements (sequences within the DNA) called terminators. Two classes of terminators exist, *rho*-independent and *rho*-dependent. Both types of termination events lead to the cessation of RNA synthesis and the release of both the RNA chain and the RNA polymerase from the DNA.

In eukaryotes, transcription termination differs, depending on the RNA polymerase under consideration. RNA polymerase I transcribes repeated 18S, 5.8S, and 28S rDNA clusters as single transcription units. Transcription of each unit is terminated at a specific site. The termination site lies in the nontranscribed spacer (NTS) between adjacent rDNA transcription units.

RNA polymerase II transcribes genes for mRNAs and some snRNAs. These genes lack specific transcription termination sequences at their 3′ ends, so that mRNA transcription can continue for hundreds or thousands of nucleotides downstream of the protein-coding sequence until it is past a poly(A) site in the RNA. The poly(A) site is recognized and cleaved by a complex set of proteins. It is positioned 10 to 30 nucleotides after an AAUAAA sequence and is followed by a GU-rich or U-rich sequence. Once the RNA is cleaved, poly(A) polymerase adds A nucleotides onto the 3′-OH of the RNA to produce a poly(A) tail.

RNA polymerase III transcribes genes for 5S rRNA, tRNA, and some snRNAs. Termination events for RNA polymerase III were not discussed in this chapter.

11.6 Three different RNA polymerases are found in all eukaryotic cells, and each is responsible for synthesizing a different class of RNA molecules. How do characteristics of the eukaryotic RNA polymerases differ in terms of their cellular location and products?

RNA Polymerase	Cellular Location	Product and Other Characteristics
I	nucleolus	28S, 18S, 5.8S rRNA: These are structural and functional components of the ribosome, which functions during translation.
II	nucleoplasm	(1) mRNAs: These provide protein-coding information to ribosome. (2) Some snRNAs (small nuclear RNAs): These function in a variety of nuclear processes, including RNA splicing and processing.
III	nucleoplasm	(1) tRNAs (transfer RNAs): These carry amino acids to the ribosome during translation. (2) Some small nuclear RNAs: These function in a variety of nuclear processes. (3) 5S rRNA: This is a functional and structural component of the ribosomes's 60S (large) subunit.

11.7 *E. coli* RNA polymerase can transcribe all the genes of *E. coli*, but the three eukaryotic RNA polymerases transcribe only specific, nonoverlapping subsets of eukaryotic genes. What mechanisms are used to restrict transcribing each of the three eukaryotic polymerases to a particular subset of eukaryotic genes?

Answer: The initiation of transcription at different genes by RNA polymerases I, II, and III is restricted through the use of three distinct sets of transcription factors that recognize three distinct classes of promoters. Since the genes transcribed by each polymerase have uniquely organized controlling regions and the transcription factors that interact with each polymerase recognize only the regulatory elements in one type of controlling region, transcription by each polymerase is specific. See Figure 11.5 for initiation by RNA polymerase II.

11.8 More than 100 promoters in prokaryotes have been sequenced. One element of these promoters is sometimes called the Pribnow box, named after the investigator who compared several *E. coli* and phage promoters and discovered a region they held in common. Discuss the nature of this sequence (where is it located and why is it important?). Another consensus sequence appears a short distance from the Pribnow box. Diagram the positions of the two prokaryotic promoter elements relative to the start of transcription for a typical *E. coli* promoter.

Answer: The Pribnow box is the –10 element of the prokaryotic promoter. It is located at –10 bp relative to the starting point of transcription and has the consensus sequence 5′-TATAAT-3′ (in the coding strand). A second consensus sequence is located about 35 bp before transcription initiation (the –35 region). Each of these sequences is important for the binding of RNA polymerase and initiation of RNA transcription. At the start of transcription initiation, the holoenzyme loosely binds the –35 region, then changes configuration and tightly binds the –10 region. It melts about 17 bp of DNA around this region. The two-step binding to the promoter orients the polymerase on the DNA and facilitates transcription initiation in the 5′→3′ direction, as shown in Figure 11.4 and the cartoon below:

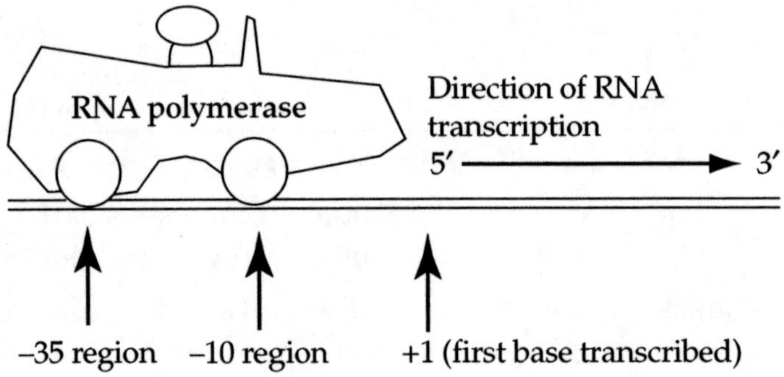

11.9 Eukaryotic promoters have two major types of regulatory regions, one of which consists of nucleotide sequences that serve as the recognition point for RNA polymerase binding. They function somewhat similarly in both eukaryotes and prokaryotes. What do you know about these sequences with regard to location, base makeup, and organization? A diagram with labels and a short explanation would suffice.

Answer: Eukaryotic promoters consist of two general types of elements: basal promoter elements and promoter proximal elements. The basal promoter elements are the TATA box (TATA element), and a pyrimidine-rich sequence near the transcription start site called the initiator element (Inr). Like the Pribnow box in prokaryotes, the AT-rich TATA box, being easier to denature than GC-rich sequences, facilitates strand separation for the initiation of transcription. The TATA box determines the exact start point for eukaryotic transcription. After it is recognized and bound by the TATA-binding protein (TBP), a subunit of TFIID, TFIIB is recruited, which in turn recruits RNA polymerase II and TFIIF to produce a minimal transcription initiation complex. Additional TFIIs (TFIIE, TFIIH) bind to produce the complete transcription initiation complex or preinitiation complex (PIC).

Promoter proximal elements are upstream of the TATA box, between 50 and 200 nucleotides prior to the +1 transcription start site. These elements, which include CAAT and GC boxes, can work in either orientation and are present in one or multiple copies.

Both promoter proximal and basal promoter elements have only general activities in transcription initiation and allow for low levels of transcription. For high levels of transcription to occur, other transcription factors called activators must bind to distinct DNA elements called enhancers. The activator proteins interact through adapter proteins with the preinitiation complex and stimulate transcription. See text Figure 11.5.

11.10 In addition to promoter regions, transcription of most eukaryotic genes is regulated by other DNA sequences called enhancers. How do these sequences differ from promoters?

Answer: An enhancer element is a DNA sequence that increases the amount of transcription of a gene lying in its vicinity. Enhancers are distinct from promoters, both in terms of location and in terms of the proteins that bind them. Promoters, if bound by a preinitiation complex, can provide only a low level of transcription. To obtain maximal transcription, enhancers are bound by activator proteins, (which are not basal transcription factors). The activator proteins interact with adapters to form bridges to the preinitiation complex, thereby stimulating transcription. Specific enhancer elements have consensus sequences and are bound by particular activators. Since the activators are cell- and tissue-specific, enhancers contribute to tissue-specific gene expression. Enhancer elements are found singly or in multiple copies and do not have a fixed position relative to the start of transcription. They can lie over a thousand base pairs in either direction from the start of transcription. In addition, they do not have to be in a particular orientation relative to the direction of transcription, although most lie upstream of the gene.

11.11 How do the structures of mRNA, rRNA and tRNA differ? Hypothesize a reason for the difference.

Answer: Although mRNA, rRNA, and tRNA are single-stranded, they can differ substantially in the secondary and tertiary structures achieved by intramolecular base-pairing interactions. These interactions, in the case of rRNA and tRNA, are essential for them to fulfill their different functions.

mRNA typically has a 5′ untranslated leader sequence, a protein-coding sequence, and a 3′ untranslated trailer sequence (see Figure 11.6). In eukaryotes, mRNAs are processed to have 5′ caps and 3′ poly(A) tails and to remove intronic sequences (see Figures 11.8, 11.9, 11.10, and 11.11).

rRNAs are of three sizes in prokaryotes (5S, 16S, 23S) and four sizes in eukaryotes (5S, 5.8S, 18S, and 28S). They are single-stranded, and three rRNAs are produced by processing a single rRNA precursor transcript (exception is eukaryotic 5S rRNA) (see Figures 11.15 and 11.16). They fold in a complex manner and have both structural and functional roles in the ribosome.

tRNA molecules are small, about 4S, and consist of a single chain of 75 to 90 nucleotides. There is extensive base pairing between different parts of the tRNA molecule, allowing it to form a cloverleaf structure having three or four loops. The tRNA molecule functions to carry amino acids to the ribosome. An amino acid is attached to its 3′ end in the cytoplasm for delivery to the ribosome. After an anticodon in loop II binds to the mRNA, the ribosome transfers the amino acid to a growing polypeptide chain. There is extensive posttranscriptional modification of tRNA bases, so that tRNAs contain modified bases such as inosine, ribothymidine, pseudouridine, dihydrouridine, methylguanosine, dimethylguanosine, and methylinosine. See Figure 11.18.

11.12 Many eukaryotic mRNAs, but not prokaryotic mRNAs, contain introns. Describe how these sequences are removed during the production of mature mRNA.

Answer: Introns are removed from a pre-mRNA molecule by the action of a spliceosome. Consensus sequences at the intron-exon boundaries and within the intron itself are recognized by spliceosome components as the spliceosome assembles around the intronic sequence. First, U1 snRNP binds to the 5′ splice site by base-pairing interactions between the U1 snRNA and the 5′ splice site sequence. Then U2 snRNP binds to a branch-point region inside of the intron. Following the association of a preassembled U4/U6/U5 particle with the bound U1 and U2 snRNPs, U4 snRNP dissociates from the complex, allowing the formation of an active spliceosome. Splicing proceeds via cleavage of the 5′ exon-intron junction, attaching the 5′ end of the intron sequence to a branch-point sequence via an unusual 2′-5′ bond, and then cleavage at the 3′ exon-intron junction. When the two exons are ligated together, a lariat structure composed of intron sequences is released. See Figures 11.10 and 11.11.

11.13 Distinguish between leader sequence, trailer sequence, coding sequence, intron, spacer sequence, nontranscribed spacer sequence, external transcribed spacer sequence, and internal transcribed sequence. Give examples of actual molecules in your answer.

Answer: Leader and *trailer sequences* are terms used to denote untranslated sequences in mRNA not containing protein-coding information that lie at the 5′ end (for leader) or 3′ end (for trailer). See text Figures 11.6 and 11.10.

Coding sequences are sequences in mRNA that are translated by the ribosome. These sequences encode information (in the form of a triplet code—see Chapters 1 and 12) that can be read by the ribosome to direct the synthesis of a polypeptide. *Introns* are sequences present in eukaryotic precursor RNA molecules that are spliced out in the mature, functional molecule. Introns are present in pre-mRNAs as well as in pre-rRNAs and pre-tRNAs.

Nontranscribed spacer sequences are sequences, such as those in the repeated rDNA genes, that lie between adjacent, repeating transcription units. *External transcribed spacer sequences* are found in pre-rRNA transcripts in eukaryotes (from transcription of 18S, 28S, and 5.8S pre-rRNAs). They are located immediately upstream of the 5′ end of the 18S sequence and downstream of the 3′ end of the 28S sequence. The *internal transcribed spacer sequences* are located on either side of the 5.8S sequence. See text Figure 11.16.

11.14 Discuss the posttranscriptional modifications and processing events that take place on the primary transcripts of eukaryotic rRNA and protein-coding genes.

Answer: In eukaryotes, 18S, 28S, and 5.8S rRNA genes are transcribed from the rDNA into a single pre-rRNA molecule. The pre-rRNA is processed by removing the internal and external spacer sequences, leading to the production of the mature rRNAs. See text Figure 11.16. The eukaryotic 5S rRNA is transcribed separately to produce mature rRNA molecules that need no further processing. Eukaryotic 5S rRNAs are imported into the nucleolus, where they are assembled with the other mature rRNAs and the ribosomal protein subunits to produce functional ribosomal subunits. Some *Tetrahymena* pre-rRNAs have self-splicing introns (group I) in the 28S rRNA. It is important to remember that the removal of the intron in the 28S rRNA via self-splicing is a separate process from the cleavage of spacer sequences from the mature rRNAs. See text Figure 11.17.

Protein-coding eukaryotic transcripts synthesized by RNA polymerase II are extensively processed. A 5′-5′ bonded m^7Gppp cap is added to the 5′ end of the nascent transcript when the chain is 20 to 30 nucleotides long. A poly(A) tail of variable length can be added to the 3′ end of the transcript, and spliceosomes remove intronic sequences. The site of poly(A) addition, as well as splice site selection, can be regulated, so that more than one alternatively processed mature mRNA is sometimes produced from a single precursor mRNA transcript.

11.15 Describe the organization of the ribosomal DNA repeating unit of a higher eukaryotic cell.

Answer: rDNA in a eukaryotic cell is organized into tandem arrays of repeating units, each of which have an 18S, 5.8S, and 28S rRNA gene. In between each repeating unit is a nontranscribed spacer. The three genes within each unit are transcribed by RNA polymerase I as a 45S precursor rRNA molecule in the nucleolus. 5′ and 3′ flanking spacer sequences (external transcribed spacer sequences) as well as internal spacer sequences are removed, leading to the production of 18S, 5.8S, and 28S rRNAs. See text Figure 5.18.

11.16 Which of the following kinds of mutations would likely be recessive lethals in humans? Explain your reasoning.

a. deletion of the U1 genes
b. deletion within intron 2 of β-globin
c. deletion of four bases at the end of intron 2 and three bases at the beginning of exon 3 in β-globin

Answer: A recessive lethal is a mutation that results in lethality (death!) when the mutation is homozygous, that is, when only mutant alleles are present. Heterozygotes for such mutations can be viable. One way to think about recessive lethals is that they result in death because some essential function is missing. Neither copy of the gene functions, and so the organism dies.

a. Deletion of the U1 genes will be recessive lethal, as U1 snRNA is essential for the identification of the 5′ splice site in RNA splicing. Incorrect splicing would lead to nonfunctional gene products for many genes, a nonviable situation.
b. If a deletion within intron 2 did not affect a region important for its splicing (e.g., the branch point, the regions near the 5′ or 3′ splice sites), it would have no effect on the

mature mRNA produced. Consequently, such a mutation would lack a phenotype if it were homozygous. However, if splicing of intron 2 were affected and the mRNA altered, such a mutation, if homozygous, could result in the production of only nonfunctional hemoglobin, leading to severe anemia and death.

c. The deletion described would affect the 3′ splice site of intron 2, leading to, at best, aberrant splicing of intron 2. If it were homozygous, only a nonfunctional protein would be produced, resulting in severe anemia and death.

11.17 The figure below shows the transcribed region of a typical eukaryotic protein-coding gene. What is the size (in bases) of the fully processed, mature mRNA? Assume in your calculations a poly(A) tail of 200 As.

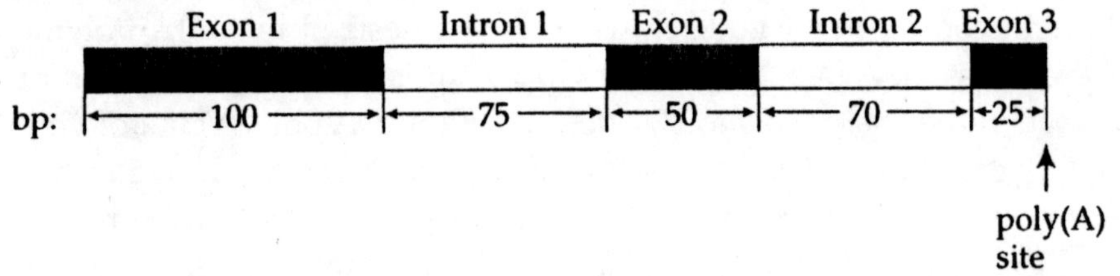

Answer: 1 (5′ m^{7}G cap) + 100 (exon 1) + 50 (exon 2) + 25 (exon 3) + 200 (poly(A) tail) = 376 bases

11.18 Most human obesity does not follow Mendelian inheritance patterns because body fat content is determined by a number of interacting genes and environmental variables. Insights into how specific genes function to regulate body fat content have come from studies of mutant, obese mice. In one mutant strain, *tubby* (*tub*), obesity is inherited as a recessive trait. Comparison of the DNA sequence of the *tub*$^+$ and *tub* alleles has revealed a single base pair change: Within the transcribed region, a 5′ GC base pair has been mutated to a TA base pair. The mutation causes an alteration of the initial 5′ base of the first intron. Therefore, in the homozygous *tub/tub* mutant, a longer transcript is found. Propose a molecularly based explanation for how a single base change causes a nonfunctional gene product to be produced, why a longer transcript is found in *tub/tub* mutants, and why the *tub* mutant is recessive.

Answer: The first two bases of an intron are typically 5′-GU-3′, which are essential for base pairing with the U1 snRNA during spliceosome assembly. A GC-to-TA mutation at the initial base pair of the first intron impairs base pairing with the U1 snRNA, so that the 5′ splice site of the first intron is not identified. This causes the retention of the first intron in the *tub* mRNA and a longer mRNA transcript in *tub/tub* mutants. When the mutant *tub* mRNA is translated, the retention of the first intron could result in either the introduction of amino acids not present in the *tub*$^+$ protein or, if the intron contained a chain termination (stop) codon, premature translation termination and the production of a truncated protein. In either case, a nonfunctional gene product is produced.

The *tub* mutation is recessive because the single *tub*$^+$ allele in a *tub/tub*$^+$ heterozygote produces mRNAs that are processed normally, and when these are translated, enough normal (*tub*$^+$) product is produced to obtain a wild-type phenotype. Only the *tub* allele produces abnormal transcripts. When both copies of the gene are mutated in *tub/tub* homozygotes, no functional product is made and a mutant, obese phenotype results.

11.19 Which of the following could occur in a single mutational event in a human? Explain.

a. deletion of 10 copies of the 5S ribosomal RNA genes only
b. deletion of 10 copies of the 18S rRNA genes only
c. simultaneous deletion of 10 copies of the 18S, 5.8S, and 28S rRNA genes only
d. simultaneous deletion of 10 copies each of the 18S, 5.8S, 28S, and 5S rRNA genes

Answer:

a. This could occur. Since the 5S genes are clustered, 10 copies could be removed in one deletion.
b. This could not occur. The 18S genes are part of a cluster of 18S, 5.8S, and 28S genes. The 18S genes cannot be deleted without also deleting the others.
c. This could occur. The genes are clustered.
d. This could not occur. The 18S, 5.8S, and 28S genes are clustered at a locus separate from the 5S genes.

11.20 During DNA replication in a mammalian cell a mistake occurs: 10 wrong nucleotides are inserted into a 28S rRNA gene. This mistake is not corrected. What will probably be the effect on the cell?

Answer: The cell will most likely be just fine. The remaining 1,000 copies of the 28S rRNA genes are normal, so this error will have little phenotypic consequence.

11.21 Choose the correct answers, noting that each blank may have more than one correct answer and that each answer (1–4) may be used more than once. Answers:

1. eukaryotic mRNAs
2. prokaryotic mRNAs
3. transfer RNAs
4. ribosomal RNAs

a. ________ have a cloverleaf structure
b. ________ are synthesized by RNA polymerases
c. ________ display an anticodon each
d. ________ are the template of genetic information during protein synthesis
e. ________ contain exons and introns
f. ________ are of four types in eukaryotes and only three types in *E. coli*
g. ________ are capped on their 5′ end and polyadenylated on their 3′ end

Answer:

a. 3
b. 1, 2, 3, 4
c. 3
d. 1 and 2
e. 1 (note that some tRNAs and rRNAs have introns)
f. 4
g. 1

12

GENE EXPRESSION: TRANSLATION

CHAPTER OUTLINE

PROTEINS

- Chemical Structure of Proteins
- Molecular Structure of Proteins

THE NATURE OF THE GENETIC CODE

- The Genetic Code Is a Triplet Code
- Deciphering the Genetic Code
- Characteristics of the Genetic Code

TRANSLATION: THE PROCESS OF PROTEIN SYNTHESIS

- Charging tRNA
- Initiation of Translation
- Elongation of the Polypeptide Chain
- Termination of Translation

PROTEIN SORTING IN THE CELL

REVIEW OF KEY TERMS, SYMBOLS, AND CONCEPTS

In your own words, write a brief, precise definition of each term in the groups on the following page. Check your definitions using the text. Then develop a concept map using the terms in each list.

THINKING ANALYTICALLY

Much of the material presented in this chapter is detailed and descriptive. The initiation, elongation, and termination of translation are all complex events. Each involves ribosomal subunits, RNAs, and stage-specific factors that differ in prokaryotes and eukaryotes. As in the last two chapters, placing the information in context will help you learn and retain it.

There are several levels in which to organize this chapter's information. First, consider the level of spatial organization. Associate the different types of RNA (rRNA, mRNA, tRNA) and factors involved in translation with the different components of the translation machinery in the cell. Then construct a temporal framework to follow these components through the stages of translation. For prokaryotes, and then for eukaryotes, what RNAs and factors are components of

1	2	3	4
protein	genetic code	translation	endoplasmic reticulum
polypeptide	triplet code	aminoacyl-tRNA synthetase	membrane-bound ribosome
amino acid	codon	codon recognition	signal hypothesis
α-carbon	frameshift mutation	initiation	signal sequence
peptide bond	revertant	elongation	signal recognition particle (SRP)
acidic, basic, neutral, polar, nonpolar amino acid	reversion	termination	docking protein
C, N terminus	random copolymers	charged tRNA	cotranslational transport
primary stucture	ribosome-binding assay	transformylase	signal peptidase
secondary structure	cell-free, protein-synthesizing system	fMet, tRNA.fMet	Golgi apparatus
tertiary structure	reading frame	tRNA.Met	cisternal space
quarternary structure	comma-free code	30S initiation complex	glycoprotein
conformation	nonoverlapping code	ribosome-binding site	
α-helix	universal, degenerate code	Shine-Dalgarno sequence	
β-pleated sheet	start, stop codon	IF1, IF2, IF3; eIFs	
	sense, nonsense codon	cap-binding protein	
	wobble hypothesis	peptidyl transferase	
	context effect	P, A, E sites	
	site-specific variation	EF-Tu, EF-Ts, EF-G	
		translocation	
		polyribosome	
		polysome	
		termination factor	
		RF1, RF2, RF3	
		UGA, UAG, UAA codon	

the first initiation complex? What RNAs and factors are present during elongation? What new factors are required for termination? Then come back to the level of spatial organization. For eukaryotes, once proteins are synthesized, what factors are used to shuttle them to the appropriate location?

Keep in mind the polarity of DNA and RNA when analyzing DNA sequences, inferring transcripts that are produced, and identifying translation reading frames. Transcripts are always synthesized in the 5′ → 3′ direction using the 3′ → 5′ DNA template strand, and polypeptides are always synthesized starting at their N terminus using codons that start relatively nearer to the 5′ end of the mRNA. The rhyme "5 to 3, N to C" may help you remember the orientations of mRNA and polypeptides. Pay close attention to maintain the correct mRNA, tRNA anticodon, and DNA polarity.

One of the highlights of this chapter is the presentation of a set of elegant experiments used to decipher the genetic code. These technically demanding experiments addressed one of the fundamental questions in genetics: How is information in DNA used to code for proteins? Consider them carefully, and consider how interpreting them often uses probabilistic thinking. The product rule, which states that the frequency of two independent events occurring together is the product of their separate probabilities, allows one to deduce the appropriate codons that specify particular amino acids. Consider the following related problem:

(a) A synthetic mRNA is generated by randomly polymerizing adenine and guanine ribonucleotides. What codons would be formed, and what would be their relative frequency, if the synthesis began with a limitless mixture of 3 adenine to 1 guanine?

(b) When the mRNAs generated in part (a) are added to a cell-free translation system, polypeptides are formed that, on average, have about 55 percent lysine, about 19 percent arginine, about 19 percent glutamic acid, and about 7 percent glycine. What can you infer about codon usage from these data (without consulting a table of the genetic code)?

Answer: First, determine what possible codons (three-nucleotide combinations) could be generated. Then determine the frequency of that codon by multiplying together the probability of obtaining each base in one codon. Since there is a 3 A : 1 G ratio, the chance of finding an A is ¾, and the chance of finding a G is ¼.

$P(AAA) = (\frac{3}{4})^3 = 0.421875 \sim 42\%$
$P(AAG) = (\frac{3}{4})^2(\frac{1}{4}) = 0.140625 \sim 14\%$
$P(AGA) = (\frac{3}{4})^2(\frac{1}{4}) = 0.140625 \sim 14\%$
$P(AGG) = (\frac{3}{4})(\frac{1}{4})^2 = 0.046875 \sim 5\%$
$P(GAA) = (\frac{1}{4})(\frac{3}{4})^2 = 0.140625 \sim 14\%$
$P(GAG) = (\frac{1}{4})^2(\frac{3}{4}) = 0.046875 \sim 5\%$
$P(GGA) = (\frac{1}{4})^2(\frac{3}{4}) = 0.046875 \sim 5\%$
$P(GGG) = (\frac{1}{4})^3 = 0.015625 \sim 2\%$

Since AAA accounts for about 42 percent of the total codons that are translated, and only lysine constitutes at least 42 percent of the amino acid content of the polypeptides, AAA must encode lysine. By similar reasoning, GGG must encode glycine. Since there is only 7 percent glycine, and no codon alone accounts for 7 percent of the total codon usage, glycine must be encoded by at least two codons. GGG must be one of these, since if any other two codons encoded glycine, the amount of glycine would exceed 7 percent. The remaining glycine must be encoded by one of the GAG, GGA, or AGG codons, since they are represented 5 percent of the time. The remaining lysine must be encoded by either AAG, AGA, or GAA, as no combination of other codons would give 55 percent lysine. This leaves arginine to be encoded by one of the AAG, AGA, GAA codons and one of the AGG, GAG, GGA codons. Similarly, glutamic acid must be encoded by one of the AAG, AGA, GAA codons and one of the AGG, GAG, GGA codons. To make additional inferences, more experimentation (e.g., with different copolymers) must be done.

QUESTIONS FOR PRACTICE

Multiple Choice Questions

1. The term *secondary structure* refers to a polypeptide's
   a. sequence of amino acids
   b. structure that results from local interactions between the residues of different amino acids
   c. folding that results from local and distant interactions between the residues of different amino acids
   d. interactions with a second polypeptide chain

2. The eukaryotic equivalent of the Shine-Dalgarno sequence is the
   a. 60S large ribosomal subunit
   b. 40S small ribosomal subunit
   c. 5′ cap of the mRNA
   d. poly(A) tail
3. In which stage of translation is GTP not used?
   a. initiation
   b. elongation
   c. translocation
   d. termination
4. Which of the following are the same in both prokaryotes and eukaryotes?
   a. elongation factors
   b. stop codons
   c. the use of fMet.tRNA
   d. AUG initiation codons
   e. both b and d
5. Specificity in translation is obtained by all of the following except
   a. using specific tRNA synthetases for each amino acid
   b. using an anticodon
   c. using a degenerate triplet code having codons for each amino acid
   d. using RNA polymerase
6. Where can coding sequences not be found?
   a. introns
   b. exons
   c. mRNA
   d. hnRNA
7. In an unprocessed, newly synthesized eukaryotic protein,
   a. the N terminus is encoded by sequences in the 5′ region of the mRNA
   b. the C terminus is encoded by sequences in the 5′ region of the mRNA
   c. the first amino acid is always f-methionine
   d. the last amino acid is always encoded by the codon AAA
8. During translation in *E. coli*,
   a. an intact 70S ribosome binds the mRNA during initiation
   b. Shine-Dalgarno sequences present in the mRNA align it with the 23S rRNA
   c. both initiation and elongation require energy in the form of GTP
   d. termination occurs when the ribosome recognizes a chain termination codon with the aid of EF-Tu
9. The RNAs that are used in eukaryotic translation include all of the following except
   a. 5S rRNA
   b. 28S rRNA
   c. mRNA
   d. tRNA
   e. snRNA

10. In eukaryotes, proteins destined to remain in the cell are distributed by all of the following except
    a. packaging into secretory vesicles
    b. after complexing with an SRP, using an N-terminal signal sequence
    c. using posttranslational transport

Answers: 1b, 2c, 3d, 4e, 5d, 6a, 7a, 8c, 9e, 10a

Thought Questions

1. The following sequence of bases in *E. coli* DNA is part of a gene. Mark the region that, if transcribed, will form the first codon to be translated. Mark the region that, if transcribed, will form a Shine-Dalgarno sequence. Is the strand shown the template or nontemplate strand? (Hint: Find the initiation codon. Be careful about polarity!)

   5'-CCGCATTCCTCCGGCGGGACCTACT-3'

2. Why is the code for methionine not degenerate? (Hint: What special role does methionine play in translation?)
3. What are the components of a cell-free, protein-synthesizing system? What is the function of each?
4. Describe three complementary methods by which the code was deciphered. Is any single method sufficient to completely decipher the code? What advantages and disadvantages does each method have?
5. The genetic code is nearly universal. To what extent does this indicate that the code was "settled on" very early in the evolution of cells?
6. Consider which bases can wobble to base-pair with each other. What structural similarities exist between bases at the 3′ end of a codon that can wobble to base-pair with the base at the 5′ end of the anticodon?
7. What percentage of 70S ribosomes do you expect to be associated with mRNA in prokaryotes? What percentage of 80S ribosomes do you expect to be associated with mRNA in eukaryotes?
8. Discuss the various ways in which proteins within a cell are targeted for secretion. How can a genetic approach be used to dissect the secretory pathway?
9. How do prokaryotes and eukaryotes differ in the mechanisms they use to identify the first codon in a reading frame? In what ways are the mechanisms similar?

Thought Questions for Media

After reviewing the media at the *Essential iGenetics* website, try to answer these questions.

1. How does the ribosome IF-GTP complex identify where it should bind to the mRNA?
2. What are the functions of EF-Tu and EF-G in translation elongation?
3. At what points in translation is GTP hydrolysis required?
4. Review the A, E, and P sites on the ribosome. Which site never has a charged tRNA in it? Which site is only bound by a charged tRNA or an RF? Which site is almost always occupied by a tRNA with an attached polypeptide?
5. Peptidyl transferase provides two functions, one during translation elongation and one during translation termination. Are they the same enzymatic function?

1. Based on viewing the iActivity, what different features of a coding region mutation would cause it to be associated with a mutant phenotype?
2. What type of coding region mutations have no effect whatsoever on protein function?
3. What type of coding region mutations are likely to have little, if any, effect on protein function?

SOLUTIONS TO TEXT PROBLEMS

12.1 The form of genetic information used directly in protein synthesis is (choose the correct answer)

a. DNA
b. mRNA
c. rRNA
d. ribosomes

Answer: b

12.2 Most genes encode proteins. What exactly is a protein, structurally speaking? List some of the functions of proteins.

Answer: A protein is composed of one or more polypeptides, each of which is composed of a folded, linear chain of amino acids. Proteins function in many different ways, including as enzymes to catalyze biochemical reactions, as structural molecules (e.g., actin, myosin, tubulin), as receptors (e.g., on the cell surface, for small signaling molecules such as neurotransmitters), as transporters (e.g., a protein that transports sugars across a membrane), as ion channels (e.g., the protein defective in cystic fibrosis, CFTR, transports Cl^- ions), and as regulatory molecules (e.g., transcription factors).

12.3 Several researchers have studied the structure and makeup of ribosomes because the best biochemical tradition of separating and reconstituting allows us to identify the components necessary for ribosomes to be active. Contrast prokaryotic ribosomes with eukaryotic ribosomes. What exactly is the function of ribosomes, and how do they carry out this function?

Answer: Prokaryotic and eukaryotic ribosomes each have two subunits with parallel functions and components. However, as presented in Chapter 11, the detailed constituents differ. The 70S prokaryotic ribosome has a 30S small subunit, containing a 16S rRNA and 20 different proteins, and a 50S large subunit, containing a 23S rRNA and 5S rRNA and 34 different proteins. The larger 80S eukaryotic ribosome has a 40S small subunit, containing an 18S rRNA and about 35 proteins, and a 60S large subunit, containing 28S, 5.8S, and 5S rRNAs and about 50 ribosomal proteins. In both eukaryotic and prokaryotic ribosomes, the proteins serve as structural units that help to organize the rRNA into key ribozyme elements.

Ribosomes function in translation, which proceeds through three phases: initiation, elongation, and termination. During the initiation phase of translation, the small subunit, with the aid of initiation factors, is bound to the mRNA and the initiating aminoacyl-tRNA (tRNA.fMet in prokaryotes, an initiator Met-tRNA in eukaryotes). The large subunit binds, forming an initiation complex with the initiator tRNA in the P site of the ribosome. During elongation, a second aminoacyl-tRNA is bound to the A site in the ribosome, the peptidyl transferase activity of the ribosome catalyzes the formation of the peptide bond, a now-uncharged tRNA is released (from the E site of the large subunit), and the ribosome translocates one codon down the mRNA. Translation elongation continues, so that codons are matched to amino acids in a growing polypeptide chain, until a stop (nonsense) codon is reached. Stop codons are recognized by release factors, which initiate three processes: the release of the polypeptide from the tRNA in the P site, the release of the tRNA from the ribosome, and the dissociation of the two ribosomal subunits and the release factor from the mRNA.

12.4 The term *genetic code* refers to the set of three-base code words (codons) in mRNA that stand for the 20 amino acids in proteins. What are some of the characteristics of the code?

Answer:

1. Triplet: Three nucleotides specify either the insertion of an amino acid into a polypeptide chain or a chain termination event.
2. Continuous: An mRNA-coding region is read in a continuous fashion without skipping nucleotides.
3. Nonoverlapping: A coding region is read in a nonoverlapping manner. Triplets are read sequentially.
4. Nearly universal: Nearly all organisms use the same code, with exceptions found in mammalian mitochondria and the nuclear genomes of some protozoa.
5. Degenerate: More than one codon typically codes for a single amino acid.
6. Signals start and stop: The code has translation stop (chain termination) and translation start signals. AUG, which encodes methionine within a coding region, is usually used as a start codon. One of three nonsense (*not sensing* an amino acid) codons are used as chain termination codons.
7. Wobble: The 5′ base of some anticodons wobbles, so that some tRNAs can recognize multiple codons (coding for the same amino acid).

12.5 If the codon were four bases long, how many codons would exist in a genetic code?

Answer: At each position of a codon four bases long, there will be one of four possible bases. So, there are $4 \times 4 \times 4 \times 4 = 4^4 = 256$ possible codons. This is far more than is needed to code for 20 amino acids. On the other hand, if there were only two bases in a codon, there would only be $4 \times 4 = 4^2 = 16$ possible codons, too few codons to uniquely specify 20 amino acids. Codons three bases long provide for $4 \times 4 \times 4 = 4^3 = 64$ possible codons.

12.6 Base-pairing wobble occurs in the interaction between the anticodon of the tRNAs and the codons. On a theoretical level, determine the minimum number of tRNAs needed to read the 61 sense codons.

Answer: Consider Figure 12.7 in answering this question.

Amino Acid	tRNAs Needed	Rationale
Ile	1	3 codons can use 1 tRNA (wobble)
Phe	1	2 codons can use 1 tRNA (wobble)
Tyr	1	2 codons can use 1 tRNA (wobble)
His	1	2 codons can use 1 tRNA (wobble)
Gln	1	2 codons can use 1 tRNA (wobble)
Asn	1	2 codons can use 1 tRNA (wobble)
Lys	1	2 codons can use 1 tRNA (wobble)
Asp	1	2 codons can use 1 tRNA (wobble)
Glu	1	2 codons can use 1 tRNA (wobble)
Cys	1	2 codons can use 1 tRNA (wobble)
Trp	1	1 codon
Met	2	Single codon, but need one tRNA for initiation and one tRNA for elongation
Val	2	4 codons: 2 can use 1 tRNA (wobble)
Pro	2	4 codons: 2 can use 1 tRNA (wobble)
Thr	2	4 codons: 2 can use 1 tRNA (wobble)
Ala	2	4 codons: 2 can use 1 tRNA (wobble)
Gly	2	4 codons: 2 can use 1 tRNA (wobble)
Leu	3	6 codons: 2 can use 1 tRNA (wobble)
Arg	3	6 codons: 2 can use 1 tRNA (wobble)
Ser	3	6 codons: 2 can use 1 tRNA (wobble)
Total	**32**	**61 codons**

12.7 Antibiotics have been very useful in elucidating the steps of protein synthesis. If you have an artificial messenger RNA with the sequence AUGUUUUUUUUUUUUU . . . , it will produce the following polypeptide in a cell-free, protein-synthesizing system: fMet-Phe-Phe-Phe . . . In your search for new antibiotics, you find one called putyermycin, which blocks protein synthesis. When you try it with your artificial mRNA in a cell-free system, the product is fMet-Phe. What step in protein synthesis does putyermycin affect? Why?

Answer: Since a dipeptide is formed, translation initiation is not affected, nor is the first step of elongation—the binding of a charged tRNA in the A site and the formation of a peptide bond. However, since *only* a dipeptide is formed, it appears that translocation is inhibited.

12.8 Describe the reactions involved in the aminoacylation (charging) of a tRNA molecule.

Answer: Aminoacylation of tRNAs occurs enzymatically using aminoacyl-tRNA synthetases. To provide for specificity in charging a tRNA, a separate aminoacyl-tRNA synthetase exists for each amino acid. This prevents an inappropriate amino acid from becoming attached to a tRNA. If an inappropriate amino acid were attached to a tRNA, the wrong amino acid would be inserted into a growing polypeptide chain, even with correct codon-anticodon base pairing between the mRNA and tRNA.

Aminoacyl-tRNA synthetases catalyze two sequential reactions to attach an amino acid to its appropriate tRNA. The first reaction results in the formation of an aminoacyl-AMP complex from the amino acid and ATP. The second reaction results in a covalent bond between the carboxyl group of the amino acid and the last ribose at the 3′ end of the tRNA. See text Figure 12.9.

12.9 If the initiating codon of an mRNA became mutated, what might be the effect on the transcript?

Answer: In both prokaryotes and eukaryotes, the initiating codon of a transcript is AUG, recognized by a special charged tRNA (tRNA.fMet in prokaryotes, initiator Met-tRNA in eukaryotes). This first codon is recognized in part because it is in a particular context: In prokaryotes, it is near the Shine-Dalgarno sequence, while in eukaryotes, it is embedded in a Kozak sequence. It is possible that if it were mutated, the 30S (in prokaryotes) or 40S (in eukaryotes) initiation complex would not form correctly, since the initiator tRNA would not find the correct initiator codon. The consequence of this would be that the mRNA would not be translated. On the other hand, if another AUG codon was very close by (and in eukaryotic mRNAs, embedded in a Kozak-type sequence), it might be used as an initiation codon, but the mRNA might be translated at a lower efficiency.

12.10 What differences would we find in protein synthesis initiation between prokaryotes and eukaryotes? What differences would we find in protein synthesis termination between prokaryotes and eukaryotes?

Answer: In both prokaryotes and eukaryotes, protein synthesis is initiated using special initiator tRNAs. In prokaryotes, methionine is attached to tRNA.fMet by methionine-tRNA synthetase, and transformylase catalyzes the addition of a formyl group to the methionine. The resulting fMet-tRNA.fMet is used at the AUG initiation codon. In eukaryotes, the N-terminal methionine is not formylated, but a special initiator methionine-tRNA is still used.

In both prokaryotes and eukaryotes, an initiation complex containing the small ribosomal subunit, initiator factors, GTP, and the mRNA forms sequentially. In prokaryotes, after IF1, IF2, IF3, and GTP bind to the 30S ribosomal subunit, fMet-tRNA.fMet and the mRNA attach to form the 30S initiation complex. The mRNA is aligned using an internal ribosome-binding site (the Shine-Dalgarno sequence) lying near the AUG initiation codon that base-pairs with sequences at the 3′ end of the 16S rRNA. After the 50S ribosomal subunit binds, GTP is hydrolyzed and IF1 and IF2 are released, leaving a 70S initiation complex. In eukaryotes, after eIF4A binds to the 5′ cap of the mRNA, a complex of the 40S ribosomal subunit, the initiator Met-tRNA.Met, GTP, and additional eIFs bind. Only after the 40S ribosomal subunit scans and finds the initiator codon does the 60S ribosomal subunit bind. This displaces the eIFs and produces the 80S initiation complex.

In both prokaryotes and eukaryotes, chain termination is signaled by one of three stop codons (UAG, UAA, UGA). These stop codons do not code for an amino acid, but are recognized with the help of termination factors or release factors. Of the three RFs in *E. coli*, RF1 recognizes UAA and UAG, RF2 recognizes UAA and UGA, and RF3 stimulates chain termination. Eukaryotes have a single release factor, eRF, that recognizes all three stop codons. The release factors trigger release of the polypeptide from the tRNA in the P site of the ribosome, release of the tRNA from the ribosome, and dissociation of the two ribosomal subunits from the mRNA. The initiating amino acid is usually cleaved from the completed peptide.

12.11 Random copolymers were used in some of the experiments that revealed the characteristics of the genetic code. For each of the following ribonucleotide mixtures, give the expected codons and their frequencies, and give the expected proportions of the amino acids that would be found in a polypeptide directed by the copolymer in a cell-free, protein-synthesizing system.

a. 4 A : 6 C

b. 1 A : 3 U : 1 C

Answer: Determine the expected amino acids in each case by calculating the expected frequency of each kind of triplet codon that might be formed, and inferring from these what types and frequencies of amino acids would be used during translation.

a. 4 A : 6 C gives $2^3 = 8$ codons, specifically AAA, AAC, ACC, ACA, CCC, ACA, CAC, CAA. Since there is 40% A and 60% C,

   $P(\text{AAA}) = 0.4 \times 0.4 \times 0.4 = 0.064$, or 6.4% Lys

   $P(\text{AAC}) = 0.4 \times 0.4 \times 0.6 = 0.096$, or 9.6% Asn

   $P(\text{ACC}) = 0.4 \times 0.6 \times 0.6 = 0.144$, or 14.4% Thr

   $P(\text{ACA}) = 0.4 \times 0.6 \times 0.4 = 0.096$, or 9.6% Thr (24% Thr total)

   $P(\text{CCC}) = 0.6 \times 0.6 \times 0.6 = 0.216$, or 21.6% Pro

   $P(\text{CCA}) = 0.6 \times 0.6 \times 0.4 = 0.144$, or 14.4% Pro (36% Pro total)

   $P(\text{CAC}) = 0.6 \times 0.4 \times 0.6 = 0.144$, or 14.4% His

   $P(\text{CAA}) = 0.6 \times 0.4 \times 0.4 = 0.096$, or 9.6% Gln

b. 1 A : 3 U : 1 C gives $3^3 = 27$ different possible codons. Of these, one will be UAA, a chain-terminating codon. Since there is 20% A, 60% U, and 20% C, the probability of finding this codon is $0.6 \times 0.2 \times 0.2 = 0.024$, or 2.4%. All of the remaining 26 (97.6%) codons will be sense codons. Proceed in the same manner as in (a) to determine their frequency, and determine the kinds of amino acids expected. To take the frequency of nonsense codons into account, divide the frequency of obtaining a particular amino acid considering all 27 possible codons by the frequency of obtaining a sense codon. This gives

   (0.8/0.976)% = 0.82% Lys

   (3.2/0.976)% = 3.28% As n

   (12.0/0.976)% = 12.3% Ile

   (9.6/0.976)% = 9.84% Tyr

   (19.2/0.976)% = 19.67% Leu

   (28.8/0.976)% = 29.5% Phe

   (4.0/0.976)% = 4.1% Thr

   (0.8/0.976)% = 0.82% Gln

   (3.2/0.976)% = 3.28% His

   (4.0/0.976)% = 4.1% Pro

   (12.0/0.976)% = 12.3% Ser

It is likely that the chains produced would be relatively short due to the chain-terminating codon.

c. 1 A : 1 U : 1 G : 1 C will produce $4^3 = 64$ different codons, all of those possible in the genetic code. The probability of each codon is $\frac{1}{64}$, and so there will be a $\frac{3}{64}$ chance of a codon being chain terminating. With those exceptions, the relative proportion of amino acid incorporation is directly dependent on the codon degeneracy for each amino acid. Inspecting the table of the genetic code in Figure 12.7, and taking the frequency of nonsense codons into account, there will be:

Amino Acid	# Codons	Frequency
Trp	1	1/61 = 1.64%
Met	1	1.64%
Phe	2	2/61 = 3.28%
Try	2	3.28%
His	2	3.28%
Gln	2	3.28%
Asn	2	3.28%
Lys	2	3.28%
Asp	2	3.28%
Glu	2	3.28%
Cys	2	3.28%
Ile	3	3/61 = 4.92%
Val	4	4/61 = 6.56%
Pro	4	6.56%
Thr	4	6.56%
Ala	4	6.56%
Gly	4	6.56%
Leu	6	6/61 = 9.84%
Arg	6	9.84%
Ser	6	9.84%

12.12 What would the minimum word (codon) size need to be if the number of different bases in the mRNA were, instead of four,

a. two

b. three

c. five

Answer:

a. If only two bases existed in mRNA and codons were four bases long, there would be $2^4 = 16$ possible codons, not enough to code for 20 amino acids. To encode 20 different amino acids, codons would need to be at least five bases long. For five-base codons, there would be $2^5 = 32$ possible codons.

b. If only three bases existed in mRNA, codons would need to be at least three bases long ($3^3 = 27$ possible combinations).

c. If five bases existed in mRNA, codons would need to be at least two bases long ($5^2 = 25$ possible combinations).

12.13 Suppose that at stage A in the evolution of the genetic code, only the first two nucleotides in the coding triplets led to unique differences and that any nucleotide could occupy the third position. Then suppose there was a stage B in which differences in meaning arose, depending on whether a purine (A or G) or pyrimidine (C or U) was present at the third position. Without reference to the number of amino acids or multiplicity of tRNA molecules, how many triplets of different meaning could be constructed out of the code at stage A? At stage B?

Answer: Stage A: $4^2 = 16$ different meaningful triplets
Stage B: $4^2 \times 2 = 32$ different meaningful triplets

12.14 A gene encodes a polypeptide 30 amino acids long containing an alternating sequence of phenylalanine and tyrosine. What are the sequences of nucleotides corresponding to this sequence in

a. the DNA strand that is read to produce the mRNA, assuming Phe = UUU and Tyr = UAU in mRNA

b. the DNA strand that is not read

c. tRNAs

Answer:

a. `3′-TAC AAA ATA AAA ATA AAA ATA AAA ATA . . . -5′` (The first fMet or Met is removed following translation of the mRNA.)

b. `5′-ATG TTT TAT TTT TAT TTT TAT TTT TAT . . . -3′`

c. 3′-AAA-5′ is the anticodon for Phe, and 3′-AUA-5′ is the anticodon for Tyr

12.15 A segment of a polypeptide chain is Arg-Gly-Ser-Phe-Val-Asp-Arg. It is encoded by the following segment of DNA:

```
------GGCTAGCTGCTTCCTTGGGGA------
      |||||||||||||||||||||
------GCGATCGACGAAGGAACCCCT------
```

Which strand is the template strand? Label each strand with its correct polarity (5′ and 3′).

Answer: The template strand is the one read to produce the mRNA. It is complementary to the mRNA, and has an opposite polarity. The nontemplate strand has the same 5′-to-3′ polarity as the mRNA, and if U is replaced by T, the same sequence. Use the polypeptide segment and a table of the genetic code to determine what possibilities exist for the first three codons of the mRNA (read 5′-to-3′) and the sequence of the nontemplate DNA strand. Let N = any nucleotide, Y = a pyrimidine (C or U), and R = a purine (G or A).

```
                  amino acids:      Arg  -  Gly  -  Ser

             potential codons:  5'  AGR     GGN     AGY  3'
                                    or              or
                                    CGN             UCN

  nontemplate strand sequence:  5'  AGR     GGN     AGY  3'
                                    or              or
                                    CGN             TCN
```

By comparing the nontemplate strand sequence to the one given, the top strand is the nontemplate strand, with the 5′ end on the right side. The template strand is the bottom strand, and transcription occurs from right to left. The C-terminus and the N-terminus of the polypeptide segment can also be determined from this information.

```
----3'-GGCTAGCTGCTTCCTTGGGGA-5'----
       |||||||||||||||||||||
----5'-CCGATCGACGAAGGAACCCCT-3'----

Transcribed into:  3' GGC UAG CUG CUU CCU UGG GGA 5'
  Translated into: C-Arg-Asp-Val-Phe-Ser-Gly-Arg-N
```

12.16 Two populations of RNAs are made by the random combination of nucleotides. In population A the RNAs contain only A and G nucleotides (3A : 1G), while in population B the RNAs contain only A and U nucleotides (3A : 1U). In what ways *other than amino acid content* will the proteins produced by translating the population A RNAs differ from those produced by translating the population B RNAs?

Answer: In population A, the codons that can be produced encode Lys (AAA, AAG), Arg (AGG, AGA), Glu (GAG, GAA) and Gly (GGA, GGG). All of these are sense codons, and so long polypeptide chains will be synthesized containing these amino acids. In population B, the codons that can be produced encode Lys (AAA), Asn (AAU), Ile (AUA, AUU), Tyr (UAU), Leu (UUA), Phe (UUU) and stop (UAA). The frequency of the stop codon will be $(\frac{1}{4} \times \frac{3}{4} \times \frac{3}{4}) = \frac{9}{64} = 0.14$ or 14%. Thus, the polypeptides formed in population B will, on average, be shorter than those formed in population A. If a stop codon appears about 14 percent of the time, there will be, on average, $1/0.14 = 7.14$ codons from one stop codon to the next. On average, six sense codons will lie in between the stop codons, so polypeptides will be synthesized that are about six amino acids long.

12.17 In *E. coli* a particular tRNA normally has the anticodon 5′-GGG-3′, but because of a mutation in the tRNA gene, the mutant tRNA has the anticodon 5′-GGA-3′.

a. What amino acid would this tRNA carry?

b. What codon would the normal tRNA recognize?

c. What codon would the mutant tRNA recognize?

d. What would be the effect of the mutation on the proteins in the cell?

Answer:

a. The normal anticodon 5′-GGG-3′ binds to the codons 5′-CCC-3′ and (because of wobble, see Table 12.1) 5′-CCU-3′. Both of these codons encode Pro (proline). Since the amino acid that is attached to the tRNA is unaffected by a mutant anticodon, the mutant tRNA will continue to carry proline to the ribosome.
b. The normal tRNA would recognize 5′-CCC-3′ and 5′-CCU-3′.
c. The mutant tRNA has the anticodon 5′-GGA-3′, which recognizes the codons 5′-UCC-3′ and (because of wobble) 5′-UCU-3′.
d. There are two consequences of the anticodon mutation. First, since 5′-UCC-3′ and 5′-UCU-3′ encode Ser (serine), the mutant tRNA will compete with the normal tRNA.Ser in the cell for binding to UCC and UCU codons during the translation of an mRNA. In a percentage of UCC codons, this leads to the insertion of proline instead of serine into a polypeptide chain. Second, unless the cell normally has an additional tRNA.Pro able to bind to the codons 5′-CCC-3′ and 5′-CCU-3′ (e.g., a tRNA with the anticodon 5′-IGG-3′), these codons would no longer be able to be read as sense codons. Consequently, when the ribosome encounters these codons in an mRNA, it will stall and chain termination will occur. This would lead to truncated proteins.

12.18 A particular protein found in *E. coli* normally has the N-terminal sequence Met-Val-Ser-Ser-Pro-Met-Gly-Ala-Ala-Met-Ser. . . . In a particular cell a mutation alters the anticodon of a particular tRNA from 5′-GAU-3′ to 5′-CAU-3′. What would be the N-terminal amino acid sequence of this protein in the mutant cell? Explain your reasoning.

Answer: The anticodon 5′-GAU-3′ recognizes the codon 5′-AUC-3′, which encodes Ile. The mutant tRNA anticodon 5′-CAU-3′ would recognize the codon 5′-AUG-3′, which normally encodes Met. The mutant tRNA would therefore compete with tRNA.Met for the recognition of the 5′-AUG-3′ codon, and if successful, insert Ile into a protein where Met should be. Since a special tRNA.Met is used for initiation, only AUG codons other than the initiation AUG will be affected. Thus, this protein will have four different N-terminal sequences, depending on which tRNA occupies the A site in the ribosome when the codon AUG is present there:

Met-Val-Ser-Ser-Pro-**Ile**-Gly-Ala-Ala-**Ile**-Ser
Met-Val-Ser-Ser-Pro-**Met**-Gly-Ala-Ala-**Ile**-Ser
Met-Val-Ser-Ser-Pro-**Ile**-Gly-Ala-Ala-**Met**-Ser
Met-Val-Ser-Ser-Pro-**Met**-Gly-Ala-Ala-**Met**-Ser

12.19 The gene encoding an *E. coli* tRNA containing the anticodon 5′-GUA-3′ mutates so that the anticodon now is 5′-UUA-3′. What will be the effect of this mutation? Explain your reasoning.

Answer: The normal tRNA recognizes the codon 5′-UAC-3′ and so must have carried the amino acid tyrosine. The altered anticodon will recognize the codon 5′-UAA-3′, a chain termination codon. Consequently, a tyrosine will be inserted when the nonsense codon UAA in an mRNA is positioned in the A site of the ribosome. This will result in read-through of the mRNA some of the time (when the termination factor does not compete for binding to the chain termination codon) and the addition of amino acids onto the C terminus of the protein. mRNAs having UAG and UGA chain termination codons will not be affected.

12.20 The normal sequence of the coding region of an mRNA is shown here, along with six mutant versions of the same mRNA. Indicate what protein would be formed in each case. [(. . .) = a multiple of three unspecified bases.]

Normal: AUGUUCUCUAAUUAC(...)AUGGGGUGGGUGUAG
Mutant *a*: AUGUUCUCUAAUUAG(...)AUGGGGUGGGUGUAG
Mutant *b*: AGGUUCUCUAAUUAC(...)AUGGGGUGGGUGUAG
Mutant *c*: AUGUUCUCGAAUUAC(...)AUGGGGUGGGUGUAG
Mutant *d*: AUGUUCUCUAAAUAC(...)AUGGGGUGGGUGUAG
Mutant *e*: AUGUUCUCUAAUUC(...)AUGGGGUGGGUGUAG
Mutant *f*: AUGUUCUCUAAUUAC(...)AUGGGGUGGGUGUGG

Answer: First rewrite the sequences so that the codons can be readily seen, noting the mutations (underlined):

normal: AUG UUC UCU AAU UAC (...) AUG GGG UGG GUG UAG
a: AUG UUC UCU AAU UAG (...) AUG GGG UGG GUG UAG
b: AGG UUC UCU AAU UAC (...) AUG GGG UGG GUG UAG
c: AUG UUC UCG AAU UAC (...) AUG GGG UGG GUG UAG
d: AUG UUC UCU AAA UAC (...) AUG GGG UGG GUG UAG
e: AUG UUC UCU AAU UC. ..)A UGG GGU GGG UGU AG.
f: AUG UUC UCU AAU UAC (...) AUG GGG UGG GUG UGG

Mutants *a, b, c, d,* and *f* are point mutations, in which one base has been substituted for another. Mutant *e* is a deletion of a single base that results in a shift in the reading frame of the mRNA (a frameshift mutation). Translating each sequence using the genetic code, determine the proteins that would be formed if these sequences were translated:

normal: AUG UUC UCU AAU UAC ... AUG GGG UGG GUG UAG
Met Phe Ser Asn Tyr ... Met Gly Trp Val Stop

a: AUG UUC UCU AAU UAG ... AUG GGG UGG GUG UAG
Met Phe Ser Asn Stop

Mutant *a* is a nonsense mutation and results in premature chain termination.

b: AGG UUC UCU AAU UAC ... AUG GGG UGG GUG UAG
Met Gly Trp Val Stop

Mutant *b* mutates the initiation codon, so that a polypeptide would be formed (if formed at all) using a downstream initiation codon. It results in a polypeptide missing amino acids at its N terminus.

```
c:  AUG UUC UCG AAU UAC ... AUG GGG UGG GUG UAG
    Met Phe Ser Asn Tyr ... Met Gly Trp Val Stop
```

Mutant *c* changes a base in the 3′ end of the codon. This does not alter the amino acid that is inserted. It will be "silent" and have no phenotypic effect.

```
d:  AUG UUC UCU AAA UAC ... AUG GGG UGG GUG UAG
    Met Phe Ser Lys Tyr ... Met Gly Trp Val Stop
```

Mutant *d* changes a base in the 3′ end of the codon. This does alter the amino acid that is inserted. It is a missense mutation, resulting in the insertion of a Lys instead of an Asn.

```
e:  AUG UUC UCU AAU UC. ..A UGG GGU GGG UGU AG.
    Met Phe Ser Asn Ser ... Trp Gly Gly Cys ?
```

Mutant *e* is a single base pair deletion and results in a frameshift mutation that alters the reading frame of the protein. All amino acids inserted following Asn are likely to be incorrect. It is conceivable that a stop codon could be read in the region that is indicated by the . . . , leading to premature chain termination.

```
f:  AUG UUC UCU AAU UAC ... AUG GGG UGG GUG UGG
    Met Phe Ser Asn Tyr ... Met Gly Trp Val Trp..
```

Mutant *f* changes a base in the chain-terminating UAG codon so that the amino acid Trp will now be inserted. It will result in the placement of additional amino acids onto the C terminus of the protein.

12.21 The normal sequence of a particular protein is given here, along with several mutant versions of it. For each mutant, explain what mutation occurred in the coding sequence of the gene. [(. . .) = a multiple of three unspecified bases.]

Normal: Met-Gly-Glu-Thr-Lys-Val-Val-...-Pro
Mutant 1: Met-Gly
Mutant 2: Met-Gly-Glu-Asp
Mutant 3: Met-Gly-Arg-Leu-Lys
Mutant 4: Met-Arg-Glu-Thr-Lys-Val-Val-...-Pro

Answer: One approach to this problem is to infer the possible coding sequence(s) that could be used for the normal protein (using N = any nucleotide, R = purine, Y = pyrimidine) and then examine this sequence to deduce what possible mutations could have resulted in the mutant proteins.

Based on the normal amino acid sequence, the mRNA sequence is:

amino acid sequence: Met-Gly-Glu-Thr-Lys-Val-Val-...-Pro
potential mRNA: 5′-AUG GGN GAR ACN AAR GUN GUN ... CCN-3′

In mutant 1, premature chain termination has occurred. This could have occurred if, in the DNA transcribed into the third (GAR (Glu)) codon, a GC base pair was changed to a TA base pair. This would lead to a UAR (stop) codon.

normal sequence: Met-Gly-Glu-Thr-Lys-Val-Val-...-Pro
normal mRNA: 5′-AUG GGN GAR ACN AAR GUN GUN ... CCN-3′
mutant mRNA: 5′-AUG GGN UAR ACN AAR GUN GUN ... CCN-3′
mutant sequence: Met-Gly-Stop

It could also have occurred if a TA base pair insertion mutation occurred in the DNA so that a U was transcribed in between the normal second and third codons, resulting in a UGA (stop) third codon.

amino acid sequence: Met-Gly-Glu-Thr-Lys-Val-Val-...-Pro
potential mRNA: 5′-AUG GGN GAR ACN AAR GUN GUN ... CCN-3′
mutant mRNA: 5′-AUG GGN UGA RAC NAA RGU NGU N.. .CC-3′
mutant sequence: Met-Gly-Stop

In mutant 2, premature chain termination has occurred after a wrong amino acid has been inserted. To explain both of these results as a consequence of a single mutational event, try either insertion or deletion mutations that would alter the reading frame. One possible explanation is that a GC base pair insertional mutation in the DNA resulted in a G being inserted after the third codon. If the N of the fourth codon were a U, then such a frameshifting insertion would change the Thr to Asp and also introduce a chain termination codon into the fifth codon position. If the R in the GAR coding for Glu (the third amino acid) is a G, this mutation could also be caused by a GC or AT base pair insertional mutation after the first two nucleotides coding for Glu. As before, assume the N of the fourth codon is a U.

amino acid sequence: Met-Gly-Glu-Thr-Lys-Val-Val-...-Pro
potential mRNA: 5′-AUG GGN GAR ACN AAR GUN GUN ... CCN-3′
mutant mRNA: 5′-AUG GGN GAR GAC UAA RGU NGU N.. .CC-3′
mutant sequence: Met-Gly-Glu-Asp-Stop

In mutant 3, the situation is similar to that with mutant 2. This time, however, several wrong amino acids are inserted before chain termination. To explain all of these consequences as the result of a single mutational event, check for the consequences of deletions or insertions in the region of the second and third codons. One possible explanation is that a deletion mutation in the DNA resulted in the N of the second codon being deleted. If, as in mutant 2, the N of the fourth codon is a U, the R of the third codon is a G, the R of the fifth codon is an A, and the N of the sixth codon is an A, the mutant sequence would be obtained.

```
normal sequence:     Met-Gly-Glu-Thr-Lys-Val-Val ... Pro
  normal mRNA: 5'-AUG GGN GAR ACN AAR GUN GUN ... CCN-3'
  normal mRNA: 5'-AUG GGN GAG ACU AAA GUA GUN ... CCN-3'
  mutant mRNA: 5'-AUG GGG AGA CUA AAG UAG UN. ..C CN-3'
mutant sequence:     Met-Gly-Arg-Leu-Lys-Stop
```

In mutant 4, the normal second amino acid (Gly) has been replaced with Arg. Arg is encoded by AGR or CGN, while Gly is encoded by GGN. If a GC base pair were substituted for a CG base pair in the DNA so that the first G of the second codon were replaced by a C, Arg would be inserted as the second amino acid.

```
normal sequence:     Met-Gly-Glu-Thr-Lys-Val-Val ... Pro
  normal mRNA: 5'-AUG GGN GAR ACN AAR GUN GUN ... CCN-3'
  mutant mRNA: 5'-AUG CGN GAR ACN AAR GUN GUN ... CCN-3'
mutant sequence:     Met-Arg-Glu-Thr-Lys-Val-Val ... Pro
```

12.22 The N terminus of a protein has the sequence Met-His-Arg-Arg-Lys-Val-His-Gly-Gly. A molecular biologist wants to synthesize a DNA chain that could encode this portion of the protein. How many possible DNA sequences could encode this polypeptide?

Answer: Because of the redundancy in the genetic code, a number of different DNA strands could serve as coding (nontemplate) strands for this sequence. If N = any of the four nucleotides, Y = either pyrimidine nucleotide (C, T), and R = either purine nucleotide (A, G), then the amino acid sequence could be encoded by any of the following strands:

```
Met-His-    Arg     -    Arg     -Lys-Val-His-Gly-Gly
ATG CAY (AGR or CGN) (AGR or CGN) AAR GTN CAY GGN GGN
```

Calculate the number of different sequences by multiplying the possibilities (using the product and sum rules):

```
ATG CAY (AGR or CGN) (AGR or CGN) AAR GTN CAY GGN GGN
 1 × 2 ×  (2 + 4)   ×  (2 + 4)    × 2 × 4 × 2 × 4 × 4 = 18,432
```

12.23 In the recessive condition in humans known as sickle-cell anemia, the β-globin polypeptide of hemoglobin is found to be abnormal. The only difference between it and the normal β-globin is that the sixth amino acid from the N-terminal end is valine, whereas the normal β-globin has glutamic acid at this position. Explain how this occurred in terms of differences in the DNA and the mRNA.

Answer: Both GAA and GAG code for glutamic acid, while GUU, GUC, GUA, and GUG code for valine. The simplest explanation is that there was an AT-to-TA change in the DNA, at the 17th base pair in the coding region of the gene. In this event, the 6th codon, instead of being GAA or GAG, would be GUA or GUG and encode valine.

12.24 Cystic fibrosis is an autosomal recessive disease in which the cystic fibrosis transmembrane conductance regulator (CFTR) protein is abnormal. The transcribed portion of the cystic fibrosis gene spans about 250,000 base pairs of DNA. The CFTR protein, with 1,480 amino acids, is translated from an mRNA of about 6,500 bases. The most common mutation in this gene results in a protein that is missing a phenylalanine at position 508 (ΔF508).

a. Why is the RNA-coding sequence of this gene so much larger than the mRNA from which the CFTR protein is translated?
b. About what percentage of the mRNA together makes up 5′ untranslated leader and 3′ untranslated trailer sequences?
c. At the DNA level, what alteration would you expect to find in the ΔF508 mutation?
d. What consequences might you expect if the DNA alteration you described in (c) occurred at random in the protein-coding region of the cystic fibrosis gene?

Answer:

a. If the primary mRNA for this gene is 250 kb, it must be substantially processed by RNA splicing (removal of introns) and polyadenylation to a smaller mature mRNA.
b. A 1,480-amino-acid protein requires $1{,}480 \times 3 = 4{,}440$ bases of protein-coding sequence. This leaves $6{,}500 - 4{,}440 = 2{,}060$ bases of 5′ untranslated leader and 3′ untranslated trailer sequence in the mature mRNA, about 32 percent.
c. The ΔF508 mutation could be caused by a DNA deletion for the three base pairs encoding the mRNA codon for phenylalanine. This codon is UUY (Y = U or C), and the DNA sequence of the nontemplate strand is 5′-TTY-3′. The segment of DNA containing these bases would be deleted in the appropriate region of the gene.
d. If positioned at random and solely within a gene's coding region (i.e., not in 3′ or 5′ untranslated sequences or in intronic sequences), a deletion of three base pairs results in either an mRNA missing a single codon or an mRNA missing bases from two adjacent codons. If three of the six bases from two adjacent codons were deleted, the remaining three bases would form a single codon. In this case, an incorrect amino acid might be inserted into the polypeptide at the site of the left codon, and the amino acid encoded by the right codon would be deleted. If the 3′ base of the left codon were deleted, it would be replaced by the 3′ base of the right codon. Since the code is degenerate, and wobble occurs in the 3′ base, this type of deletion might not alter the amino acid specified by the left codon. The adjacent amino acid would still be deleted, however.

13

RECOMBINANT DNA CLONING TECHNOLOGY

CHAPTER OUTLINE

REVIEW OF KEY TERMS, SYMBOLS, AND CONCEPTS

In your own words, write a brief, precise definition of each term in the groups on the following page. Check your definitions using the text. Then develop a concept map using the terms in each list.

THINKING ANALYTICALLY

The material in this chapter introduces core concepts in molecular genetics. It presents many techniques and ideas that are widely used throughout genetics. Many of these have been developed by coupling fundamental genetic principles with technological advances. For example, the development of the ability to clone DNA required the synthesis of information from bacterial genetics (e.g., how to transform cells, how to design and use selectable markers, how to perform bacterial crosses, how to design plasmids, an understanding of the λ phage life cycle and genetics); DNA replication (the discovery, characterization and purification of DNA polymerases and ligases); and the discovery, characterization and purification of restriction enzymes. As you learn

1	2	3	4
recombinant DNA genetic engineering cloning vector molecular cloning restriction enzyme, site endonuclease twofold rotational symmetry palindrome sticky, staggered, blunt ends 5′, 3′ overhang plasmid; shuttle vectors polylinker multiple cloning site selectable marker unique restriction site blue/white selection	genomic library complementary DNA cDNA cDNA library partial digestion insert size plasmid, shuttle vectors oligo(dT) chains reverse transcriptase RNase H restriction site linker adapter subcloning *TRP1*, *URA3* markers	library screening expression vector autoradiogram autoradiography replica plating nonradioactive labeling complementation GenBank	restriction map preparative agarose gel electrophoresis Southern blot Technique northern blot analysis Sanger (dideoxy) sequencing dideoxynucleotide computer database ORF motif polymerase chain reaction (PCR) thermal cycler amplification primer ancient DNA thermostable DNA polymerase *Taq* DNA polymerase DNA typing

about molecular genetic methods, observe how they rely on a synthesis of information from different areas of genetics and biochemistry.

Learning this material requires you to understand the logic and fundamental genetic principles behind many different techniques. While it would be helpful to perform all of the techniques, this may not be an option. Indeed, the current repertoire of techniques is so great and so rapidly advancing that no one scientist has firsthand experience in all of them. Still, scientists learn to think effectively about them and apply them to their research questions. As you approach a new method, think through the steps of the method and consider the rationale for each step. Then consider how different methods can be used together or interchangeably.

This chapter discusses manipulations of DNA that cannot be directly sensed (with the eyes, ears, nose, or touch). Like material from the physical sciences—chemistry and physics—that considers molecular interactions, it requires visualization and abstract conceptualization. While the results of techniques presented in this chapter cannot be directly sensed, they are based on understanding and developing a mental image of the molecular structure, function, and activity of nucleic acids and proteins in test tubes, cells, and viruses under defined environmental conditions. When the techniques produce reasonable, expected results, they validate and sometimes expand our understanding of the structure and function of these molecules. For example, we have never *seen* the actual base sequences of a fragment of DNA, but we have developed a model of DNA structure from experimentation. *If* that model is correct, then a technique such as dideoxy sequencing of DNA should work. As you learn about an experimental method, identify the models being used. Consider how the results obtained with the method would be altered if the model were incorrect.

QUESTIONS FOR PRACTICE

Multiple Choice Questions

For questions 1–9, match each experimental goal with techniques useful to attain it. Multiple techniques may be used in achieving one goal. Some techniques will be used more than once.

Experimental Goal	Technique
1. Determine if a particular mRNA is expressed at a specific developmental stage.	a. restriction mapping
2. Separate mRNA from a mixture of mRNA, rRNA, and tRNA prior to construction of cDNA.	b. partial digestion with a restriction enzyme that recognizes a 4-bp site
3. Isolate a particular cDNA clone in the cDNA library using an antibody.	c. polymerase chain reaction
4. Determine gene homology.	d. using expression vector to produce a particular protein
5. Determine the organization of introns in a gene.	e. northern blot analysis
6. Produce relatively large genomic DNA fragments for cloning in a library.	f. southern blot analysis
7. Identify plasmids with inserted DNA sequences.	g. blue/white selection
8. Amplify a specific genomic DNA sequence.	h. using oligo(dT) to isolate RNA with poly(A) tails
9. Determine the size of a fragment of DNA inserted into a plasmid vector.	i. plaque lifts
	j. agarose gel electrophoresis

10. Why are antibiotic resistance markers important components of plasmid cloning vectors?
    a. The plasmid must have resistance to accept DNA inserts.
    b. They allow identification of bacteria that have taken up a plasmid.
    c. They ensure the presence of both the *ori* and *cos* sites.
    d. They ensure that the plasmid can be cut by a restriction enzyme.
11. When a PCR reaction is performed using genomic DNA as a template, a 1.5-kb product is amplified. When the same reaction is set up using cDNA as a template, a 0.8-kb product is amplified. A likely explanation for the different-sized products is that
    a. primers always bind to different sequences in different templates
    b. there is a mutation in the genomic DNA
    c. there is an intron in the gene
    d. the cDNA is degraded
12. The restriction enzyme *Bam*HI cleaves a phosphodiester bond between two GC base pairs at a six-base-pair site. The 5′→3′ sequence of one of the sites is GGATCC. Which statement is true?
    a. *Bam*HI leaves a 5′ overhang.
    b. *Bam*HI leaves a 3′ overhang.
    c. *Bam*HI leaves a blunt end.
    d. DNA cleaved by *Bam*HI cannot be religated.

Answers: 1e, j; 2h; 3d; 4a, f, j; 5a, c, f, j; 6b, j; 7g (possibly also a, k); 8c; 9a, j; 10b, 11c, 12a

Thought Questions

1. What kinds of organisms naturally make restriction enzymes? Of what use are they to the organism where they are naturally made?
2. Address the following questions to explore the kinds of results that could be obtained on a Southern blot made with genomic DNA.
   a. DNA is digested with a restriction enzyme that cuts at a defined 6-bp sequence having 2 GC base pairs and 4 AT base pairs. The genome is 40% GC. What is the *average* fragment size?
   b. The cleaved DNA is separated by size using agarose gel electrophoresis. When the gel is stained, a smear of DNA is seen. Why is a smear and not one or a few bands of DNA seen?
   c. The DNA is transferred in a Southern blot to a membrane. When sequences transferred to the membrane are hybridized to a unique-sequence probe made from a fragment that is 4 kb in length, bands that are 0.5, 3, and 9 kb are seen. How do you interpret this result?
   d. The same blot is hybridized with a probe that has homology to moderately repetitive DNA. The probe has been made from a fragment that is 1.2 kb in size. The blot has about 40 bands, ranging in size from 3 to 17 kb. How do you interpret this result?
3. Construct a restriction map of a 10-kb DNA fragment using the following data:

Enzymes Used	**Sizes of Fragments (in kb)**
*Eco*RI	1, 4, 5
*Bam*HI	4, 6
*Hin*dIII	0.8, 1.5, 7.7
*Eco*RI and *Bam*HI	1, 2, 3, 4
*Eco*RI and *Hin*dIII	0.5, 0.8, 1, 3.2, 4.5
*Bam*HI and *Hin*dIII	0.8, 1.5, 2.5, 5.2
*Bam*HI, *Eco*RI, and *Hin*dIII	0.5, 0.8, 1, 2, 2.5, 3.2

4. Suppose you have cloned 11 kb of eukaryotic genomic DNA that includes a single gene. After analyzing the sequence of about 500 bp at each end of the clone, you believe you have evidence that these sequences contain the 5′ UTR and 3′ UTR sequences of the gene. What might this evidence be? How can you use the information you have in hand, together with Southern and northern blot analyses and PCR to (a) identify the length of the primary transcript of the gene, (b) identify the length of the mature mRNA of the gene, and (c) obtain a cDNA copy of the mature mRNA of the gene?

Thought Questions for Media

After reviewing the media at the *Essential iGenetics* website, try to answer these questions.

1. In the animation depicting restriction mapping, how is agarose gel electrophoresis able to separate fragments of cleaved DNA?
2. The rotating disk method is effective for orienting restriction fragment sites in a circular molecule relative to each other. Explain how it works.
3. Sometimes, when the rotating disk method is being used to locate restriction enzyme sites relative to each other, overlaying maps doesn't allow sites to be positioned. In some cases, an alternative mirror image map must be used. Why?

4. How could you modify the rotating disk method to map linear DNA fragments?
5. Automated DNA sequencers have made sequencing DNA both cheaper and easier. Why is it still important to know how to construct a restriction map?
6. The calibration curve shown in the animation is not linear over the entire length of the gel. However, it is approximately linear between about 1,000 and 10,000 base pairs. Why might it be linear over only part of the length of the gel?
7. Each of two enzymes, enzyme I and enzyme II, cuts a plasmid giving fragments of 2,506 and 5,160 bp. What is the best means to *quickly* and unambiguously determine how many different sites these enzymes recognize and their positions relative to each other?

i

1. What is the rationale behind deleting the *STA1, STA2,* and *STA2K* genes?
2. Why is it important to *entirely* delete each of the *STA1, STA2,* and *STA2K* genes?
3. Why are there differences between the beers made with the three recombinant yeast strains, since each has a deletion of a gene responsible for glucoamylase production?
4. Why is the *CUP1* gene needed in these experiments? How would you demonstrate that it had no effect on fermentation characteristics or beer quality?
5. Refer back to the frame that asks you to insert the *STA1* gene into the YEpD vector. Suppose there were two *Sna*BI sites flanking the *STA1* gene. Even though there is a unique *Sna*BI site in the YEpD vector, why wouldn't you use it instead of the *Bgl*II site?

SOLUTIONS TO TEXT PROBLEMS

13.1 There are many varieties of cloning vectors. One type of cloning vector used to clone recombinant DNA in *E. coli* is plasmids. What features do these vectors have that make them useful for constructing and cloning recombinant DNA molecules?

Answer: Plasmids need three essential features to be utilized as cloning vectors:

1. A bacterial *ori,* or origin of replication sequence, to allow it to replicate in *E. coli*
2. A dominant selectable marker, such as antibiotic resistance, to allow selection of cells harboring the plasmid
3. At least one unique restriction enzyme cleavage site, so that DNA sequences cut with that enzyme can be spliced into the plasmid

Modern plasmid cloning vectors have been engineered to possess additional features that facilitate easier use as cloning vectors:

1. They are present in a high copy number, which facilitates purification of plasmid DNA.
2. They contain many unique restriction sites in a *polylinker* or *multiple cloning site,* to facilitate cloning fragments of DNA obtained after cleavage with a variety of different restriction enzymes.
3. The polylinker is inserted near the 5′ end of the *lac*Z gene, which encodes β-galactosidase. Cells harboring a plasmid with an intact *lac*Z gene, when grown on media with the β-galactosidase substrate X-gal, form blue colonies. Cells harboring a plasmid whose *lac*Z gene has been interrupted by a cloned segment of DNA will

not express functional β-galactosidase and will be white. Thus, one can tell if a bacterial colony harbors a plasmid with DNA insertion by the color of the colony.

4. They contain phage promoters flanking each side of the polylinker. These promoters are used to make in vitro RNA copies of the cloned DNA. These promoters can then be used for a variety of purposes, including making radioactively labeled RNA probes.

13.2 The ability to clone DNA fragments gives molecular biologists the power to understand the structure and function of our genes and to use fragments of DNA to produce proteins. What are the basic elements of research in recombinant DNA technology?

Answer: Recombinant DNA technology involves the isolation, cloning, manipulation, and expression of DNA sequences. Fundamental to it is the ability to cleave DNA using restriction enzymes, insert it into cloning vectors (e.g., plasmids) using an enzyme such as DNA ligase, and introduce (transform) these vectors into host cells, where they can be propagated and possibly express DNA sequences (i.e., genes) of interest. DNA, once cloned, can be repurified and modified, altering parts of the DNA sequence. Since it is essential to be able to "identify" cloned sequences of interest, the "basic elements of research" may also include the ability to screen genomic and cDNA libraries in different ways (expression screens, oligonucleotide probe screens, heterologous probe screens, etc.) that allow the isolation of cloned sequences of interest. The ability to sequence DNA and amplify specific regions using PCR might also be considered basic elements of research, as these methods provide information about DNA molecules that is essential for research.

13.3 Much effort has been spent on developing cloning vectors that replicate in organisms other than *E. coli.*

a. Describe several different reasons one might want to clone DNA in an organism other than *E. coli.*
b. What is a shuttle vector, and why is it used?
c. Describe the salient features of a vector that could be used for cloning DNA in yeast.

Answer:

a. Vectors have been developed for transforming yeast as well as plant and animal cells. These are useful for studying cloned eukaryotic genes in a eukaryotic environment, commercial production of eukaryotic gene products (e.g., drugs, antibodies), developing gene therapy, engineering crop plants, and developing transgenic animals.
b. Shuttle vectors are cloning vectors that can replicate in two or more host organisms. They are used to introduce DNA into organisms other than *E. coli.* A typical shuttle vector has the ability to replicate in *E. coli,* where it is usually easier and faster to do initial cloning and engineering steps. Once the appropriate recombinant DNA molecule is constructed in *E. coli,* it can be transferred into another organism without further subcloning.
c. A yeast shuttle vector should contain dominant selectable markers for both yeast (e.g., *URA3,* which provides for uracil-independent growth when the vector is transformed into a *ura3* yeast) and *E. coli* (e.g., ampicillin resistance); a large number of unique restriction enzyme sites suitable for cloning foreign DNA; and if it is not integrated into the yeast chromosome, sequences that allow it to replicate autonomously, as a plasmid, in both yeast cells and bacteria.

13.4 The human genome contains about 3×10^9 bp of DNA. How many 40-kb pieces would you have to clone into a library if you wanted to be 90 percent certain of including a particular sequence?

Answer: There are two ways to solve this problem.

(1) Deduce the probability relationships from first principles.

$$P \text{ (identifying sequence in \underline{one} 40-kb piece} = 40{,}000 \text{ bp}/3 \times 10^9 \text{ bp}$$
$$= 13.333 \times 10^{-6}$$

$$P \text{ (\underline{not} finding sequence in \underline{one} 40-kb piece)} = 1 - (13.333 \times 10^{-6})$$

$$P \text{ (\underline{not} finding sequence in \underline{n} clones)} = [1 - (13.333 \times 10^{-6})]^n$$

If one wants to be 90 percent certain of finding the sequence, one is willing to allow a 10% chance that the sequence will not be found. Thus,

$$0.10 = [1 - (13.333 \times 10^{-6})]^n$$
$$n = \ln(0.10)/\ln[1 - (13.333 \times 10^{-6})]$$
$$n = 172{,}693 \text{ pieces}$$

(2) Recall from the text presentation that the probability of having any sequence represented in a genomic library is given by $N = \ln(1 - P)/\ln(1 - f)$, where N = number of recombinant DNA molecules that must be screened, P = probability of finding at least one positive clone, and f = fractional proportion of the genome in a single recombinant DNA molecule. Here, $P = 0.90$ and $f = 40{,}000/(3 \times 10^9)$, so $N = 172{,}693$.

13.5 What is a cDNA library, and from what cellular genetic material is it derived?

Answer: A complementary DNA, or cDNA, is a DNA copy of an mRNA. A cDNA library is a collection of clones containing the different cDNAs synthesized from the entire population of mRNA of a particular (usually eukaryotic) tissue or cell. It is constructed as follows: Initially, the mRNAs expressed in a tissue or cell are isolated. The mRNA molecules are purified from the other RNAs present in cells by passing the total cellular RNA over an oligo-dT column, which binds the poly(A) tails of the mRNAs. After the other RNAs are washed through the column, the poly(A)$^+$ mRNAs are eluted from the column. Then, cDNAs are synthesized by annealing an oligo(dT) primer to the poly(A) tail of each mRNA and using the enzyme reverse transcriptase to synthesize a single-stranded DNA copy of each mRNA strand. This results in a collection of DNA-mRNA hybrids. RNase H is added to partially degrade the mRNA strands, leaving single-stranded complementary DNAs with short mRNA fragments attached. DNA polymerase I is added to synthesize a second DNA strand using the short mRNAs as primers. DNA ligase is used to ligate the DNA fragments of the second strand together. Finally, linkers or adapters are ligated onto the ends of the cDNAs (if linkers are used, the linkers are cleaved to generate sticky ends), and each member of the resulting collection of double-stranded DNAs is ligated into a restriction site in a cloning vector and propagated. Since there is a population of mRNA molecules in a particular tissue or cell, this procedure produces a population of clones, or a library, containing different cDNA inserts, that reflects the population of mRNA present in a particular tissue or cell. See text Figures 13.6 and 13.7.

If a cDNA library is constructed using mRNA isolated from a particular tissue, the cDNA inserts represent partial copies of genes transcribed in that tissue. Thus, each clone can be used to identify a gene expressed in that tissue. A cDNA library can be screened for cloned copies of particular mRNA transcripts by using the same methods that are used to screen genomic libraries.

13.6 A colleague has sent you a 4.5-kb DNA fragment excised from a plasmid cloning vector with the enzymes *Pst*I and *Bgl*II (see Table 13.1 for a description of these enzymes and the sites they recognize). Your colleague tells you that within the fragment there is an *Eco*RI site that lies 0.49 kb from the *Pst*I site. Suppose you want to clone this DNA fragment into the plasmid vector pUC19 (described in Figure 13.4).

a. How would you proceed?
b. How would you verify that you have cloned the correct fragment and determine the orientation of the fragment within the pUC19 cloning vector?

Answer:

a. The *Bgl*II enzyme leaves a 5′-GATC overhang, while the *Pst*I enzyme leaves a 3′-ACGT overhang. If the polylinker of the pUC19 vector could be cleaved to leave two identical overhangs, the 4.5-kb fragment could be directionally cloned into the vector. Examination of the restriction enzyme sites available in the polylinker of pUC19 reveals a *Pst*I site, but no *Bgl*II sites. There are two approaches to obtain the required 5′-GATC overhang. Examine the polylinker further and compare the sites in the polylinker to the enzymes described in Table 7.1 to determine whether cleaving any of these sites leaves the same kind of overhang as *Bgl*II. This analysis identifies a *Bam*HI site that, if cut, would leave a 5′-GATC overhang, just like that of *Bgl*II. Thus, cleaving the vector with *Bgl*II would produce the appropriate sticky end. To directionally clone the insert, cleave the pUC19 vector with *Pst*I and *Bam*HI, allow the fragment to anneal to these sticky ends, and use DNA ligase to seal the gap in the phosphodiester backbones.

   Another approach to obtain appropriate sticky ends for directional cloning is to cut the polylinker with the enzyme *Sma*I (which leaves blunt ends), ligate a linker containing a *Bgl*II site onto the blunt end, and cleave the modified vector with *Bgl*II and *Pst*I.

b. Transform the ligated DNA into a host bacterial cell, and plate the cells on bacterial medium containing ampicillin and a substrate (X-gal) for β-galactosidase that turns blue when cleaved. This selects for bacterial colonies that harbor pUC19 plasmids and allows for a blue/white selection to identify colonies that have plasmids with inserts. Pick white colonies (which have an interrupted *lac*Z gene and so do not cleave the substrate) and isolate plasmid DNA from them. Cleave the DNA with restriction enzymes and analyze the products using agarose gel electrophoresis to verify that the appropriate-sized fragments are recovered. Restriction with *Eco*RI should give two fragments, one that is 2.686 + 0.49 = 3.176 kb (vector plus the 0.49-kb *Eco*RI fragment of the insert) and one that is 4.5 – 0.49 = 4.01 kb (the insert minus the 0.49-kb fragment of the insert). A set of double digests (*Eco*RI + *Pst*I; *Eco*RI + *Bam*HI) will also be informative.

13.7 Suppose you wanted to produce human insulin (a peptide hormone) by cloning. Assume that you could do this by inserting the human insulin gene into a bacterial host, where, given the appropriate conditions, the human gene would be transcribed and then translated into human insulin. Which would be better to use as your source of the gene: human genomic insulin DNA or a cDNA copy of this gene? Explain your choice.

Answer: It would be preferable to use cDNA. Human genomic DNA contains introns, while cDNA synthesized from cytoplasmic poly(A)$^+$ mRNA does not. Prokaryotes do not process eukaryotic precursor mRNAs having intron sequences, so genomic clones will not give appropriate translation products. Since cDNA is a complementary copy of a functional mRNA molecule, the mRNA transcript will be functional, and when translated, human (pro-)insulin will be synthesized.

13.8 You are given a genomic library of yeast prepared in a bacterial plasmid vector. You are also given a cloned cDNA for human actin, a protein that is conserved in protein sequences among eukaryotes. Outline how you would use these resources to attempt to identify the yeast actin gene.

Answer: Radioactively (or nonradioactively) label the human actin cDNA, and use it as a heterologous probe to screen the yeast genomic library. To radioactively label the human actin cDNA using a random primer method, denature the double-stranded cDNA by boiling, allow "random" short oligonucleotides (primers) to anneal to a cDNA strand, and synthesize DNA complementary to the cDNA using one or more labeled dNTPs and the Klenow fragment of DNA polymerase I. Once a probe has been made, clones harboring plasmids with inserts encoding the yeast actin gene can be identified as follows: Plate the library onto bacterial media in Petri plates at a density that allows separate colonies to be identified. Overlay the bacterial colonies with a positively charged membrane that can bind DNA, and then lift it off so that some of each bacterial colony is attached to the membrane. Lyse the bacterial cells bound to the membrane in situ using an alkaline solution, so that the plasmid DNA harbored by the cells binds to the membrane in single-stranded form. Allow the probe to hybridize with sequences on the membrane. Detect the location of the bound probe (e.g., use autoradiography if the probe is radioactively labeled), and then align the membrane with the original Petri plate from which colonies were lifted. Identify the clone having the hybridizing cDNA insert. This clone will have a yeast genomic DNA insert that has sequence homology to the human actin gene probe and most likely is a yeast actin gene. This can be assessed by obtaining and analyzing its DNA sequence.

13.9 A cDNA library is made with mRNA isolated from liver tissue. When a cloned cDNA from that library is digested with the enzymes *Eco*RI (E), *Hin*dIII (H) and *Bam*HI (B), the restriction map shown in the figure at the top of the next page, part (a), is obtained. When this cDNA is used to screen a cDNA library made with mRNA from brain tissue, three identical cDNAs with the restriction map shown in part (b) are obtained. When either cDNA is used to synthesize a uniformly labeled ^{32}P-labeled probe and the probe is allowed to hybridize to a Southern blot prepared from genomic DNA digested singly with the enzymes *Eco*RI, *Hin*dIII, and *Bam*HI, an autoradiograph shows the pattern of bands seen in part (c). When either cDNA is used to synthesize a uniformly labeled ^{32}P-labeled probe and used to probe a northern blot prepared with poly(A)+ RNA isolated from liver and brain tissues, the pattern of bands in part (d) is seen. Fully analyze these data and then answer the following questions.

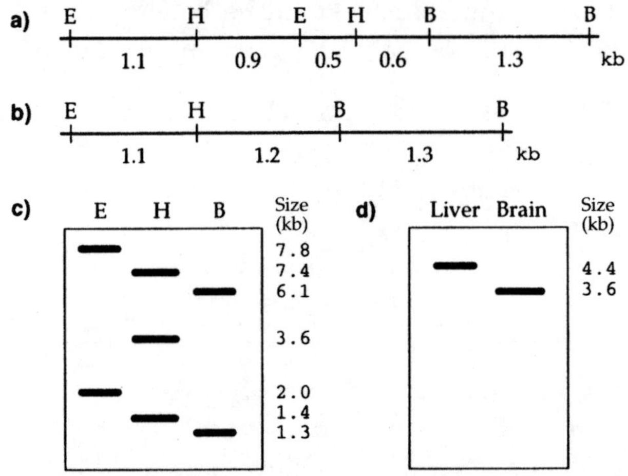

a. Do these cDNAs derive from the same gene?
b. Why are different-sized bands seen on the northern blot?
c. Why do the cDNAs have different restriction maps?
d. Why are some of the bands seen on the whole-genome Southern blot different sizes than some of the restriction fragments in the cDNAs?

Answer:

a. Since both cDNAs hybridize to the same bands on a genomic Southern blot, they are copies of mRNAs transcribed from the same sequences. Therefore, it is likely that they are from the same gene.
b. Different bands on the northern blot indicate that the primary mRNA for this gene may be processed differently in brain and liver tissue. For example, it is possible that the 0.8-kb size difference between the two bands reflects a 0.8-kb intron that is spliced out in brain tissue but is not spliced out in liver tissue.
c. The two cDNAs are copies of mRNAs found in two different tissues. The northern blot indicates that there are some differences between the mRNAs in the different tissues. Thus, it is not surprising that the restriction maps are not identical. Note that the ends of the restriction maps are identical (the same *Eco*RI-*Hin*dIII and *Bam*HI fragments), while the internal regions are not (the brain cDNA lacks the 0.5-kb *Eco*RI-*Hin*dIII fragment and some of each adjoining fragment).
d. The genomic Southern blot gives an indication of the gene organization at the DNA level, while the cDNA maps give an indication of the structure of the mRNA transcript(s). When the cDNA is used to probe genomic DNA sequences, it will hybridize to any sequences that are transcribed. Since restriction sites in the genome do not delineate where the transcribed regions are, the probe will hybridize to genomic DNA fragments that are only partly transcribed. That is, the probe will hybridize to transcribed sequences that are "connected to" nontranscribed sequences. Thus, the large (7.8-, 7.4-, 6.1-, and 3.6-kb) bands reflect the parts of the cDNA that hybridize to genomic DNA fragments that are only partly transcribed. Since they are the same fragment sizes that appear in the liver cDNA, the smaller fragments (2.0, 1.4, and 1.3 kb) represent fragments that are entirely transcribed. Based on these data, a possible gene organization is illustrated at the top of the next page:

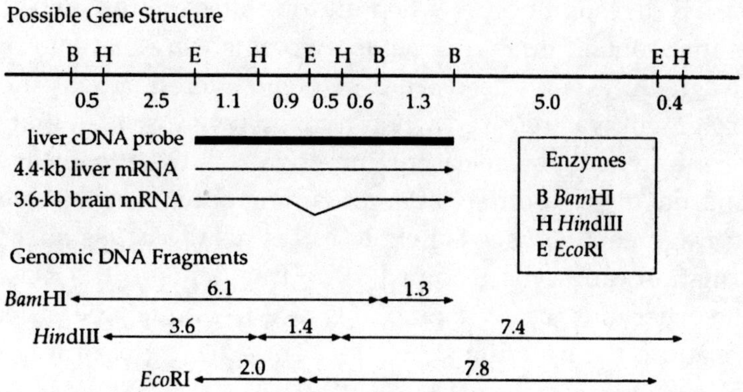

13.10 Restriction endonucleases are used to construct restriction maps of linear or circular pieces of DNA. The DNA usually is produced in large amounts by recombinant DNA techniques. Generating restriction maps is like putting the pieces of a jigsaw puzzle together. Suppose we have a circular piece of double-stranded DNA that is 5,000 bp long. If this DNA is digested completely with restriction enzyme I, four DNA fragments are generated: Fragment *a* is 2,000 bp long; *b* is 1,400 bp long; *c* is 900 bp long; and *d* is 700 bp long. If, instead, the DNA is incubated with the enzyme for a short time, the result is incomplete digestion of the DNA: Not every restriction enzyme site in every DNA molecule will be cut by the enzyme, and all possible combinations of adjacent fragments can be produced. From an incomplete digestion experiment of this type, fragments of DNA are produced from the circular piece of DNA that contains the following combinations of the above fragments: *a-d-b*, *d-a-c*, *c-b-d*, *a-c*, *d-a*, *d-b*, and *b-c*. Lastly, after digesting the original circular DNA to completion with restriction enzyme I, the DNA fragments are treated with restriction enzyme II under conditions conducive to complete digestion. The resulting fragments are 1,400, 1,200, 900, 800, 400, and 300 bp. Analyze all the data to locate the restriction enzyme sites as accurately as possible.

Answer: Sort out the results for the first enzyme by tabulating the results:

Complete Digestion		Partial Digestion	
Fragment	**Size (bp)**	**Fragment**	**Size (bp)**
a	2,000	*d-a-c*	3,600
b	1,400	*a-d-b*	3,100
c	900	*c-b-d*	3,000
d	700	*a-c*	2,900
Total plasmid	5,000	*d-a*	2,700
		b-c	2,300
		d-b	2,100

Consider the following: If an enzyme cuts a circular molecule once, it will produce one fragment. If an enzyme cuts a circular molecule twice, it will produce two fragments. (Diagram these situations to convince yourself of this.) Since four fragments are produced when enzyme I completely cleaves the plasmid, enzyme I must cut the plasmid at four sites. A partial digestion occurs when all four sites are not cut. For the partial digestion fragments that contain three fragments, two cuts were made in neighboring sites. Thus, the *d-a-c* fragment was released when two cuts were made at sites flanking fragment *b*, the *a-d-b* fragment was released when two cuts were made at sites flanking fragment *c*, and the *c-b-d* fragment was released when two cuts were made at sites flanking fragment *a*. For the partial digestion fragments containing only two fragments, cuts were made that flank both fragments. Thus, *a* is next to *c*, *d* is next to *a*, *b* is next to *c*, and *d* is next to *b*. This information can be used to order the fragments in the plasmid. Since *a* is next to *c* and *d*, the order must be *c-a-d*. Since *b* is next to *c* and *d*, the order must be *d-b-c*. Since *d* is next to *a* and *b*, the order must be *a-d-b*. Thus, the order of fragments in the plasmid is *c-a-d-b*.

When the plasmid is cleaved with both enzyme I and enzyme II, six fragments are produced. This indicates that the two enzymes together recognize six sites. Since enzyme I cleaves at four sites, enzyme II must cleave at two sites. Since the 1,400- and 900-bp fragments produced when enzyme I cleaves the plasmid remain intact in the double digestion, and the 2,000- and 700-bp fragments do not remain intact in the double digestion, enzyme II must cleave at sites within the 2,000- and 700-bp fragments (*a* and *d*). Given the 1,200- and 800-bp fragments produced with the double digestion, enzyme II must cleave *a* 800 bp from an enzyme I site. Given the 400- and 300-bp fragments produced with the double digestion, enzyme II must cleave *d* 300 bp from an enzyme I site. This gives the following map:

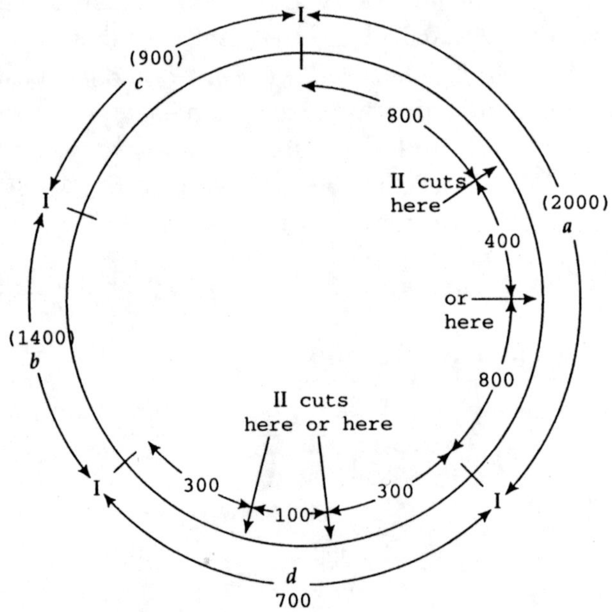

The ambiguities in enzyme II site positions could be resolved if one knew what size fragments were produced when the plasmid was digested to completion with enzyme II alone.

13.11 A piece of DNA 5,000 bp long is digested with restriction enzymes A and B, singly and together. The DNA fragments produced are separated by DNA electrophoresis and their sizes are calculated, with the following results:

Digestion with		
A	B	A + B
2,100 bp	2,500 bp	1,900 bp
1,400 bp	1,300 bp	1,000 bp
1,000 bp	1,200 bp	800 bp
500 bp		600 bp
		500 bp
		200 bp

Each A fragment is extracted from the gel and digested with enzyme B, and each B fragment is extracted from the gel and digested with enzyme A. The sizes of the resulting DNA fragments are determined by gel electrophoresis, with the following results:

A Fragment		Fragments Produced by Digestion with B	B Fragment		Fragments Produced by Digestion with A
2,100 bp	→	1,900, 200 bp	2,500 bp	→	1,900, 600 bp
1,400 pb	→	800,600 bp	1,300 bp	→	800, 500 bp
1,000 bp	→	1,000 bp	1,200 bp	→	1,000, 200 pb
500 bp	→	500 bp			

Construct a restriction map of the 5,000-bp DNA fragment.

Answer: Construct a map stepwise, considering the relationship between the fragments produced by double digestion and the fragments produced by single-enzyme digestion. Start with the larger fragments. The 1,900-bp fragment produced by digestion with both A and B is part of the 2,100-bp fragment produced by digestion with A and the 2,500-bp fragment produced by digestion with B. Thus, the 2,500-bp and 2,100-bp fragments overlap by 1,900 bp, leaving a 200-bp A-B fragment on one side and a 600-bp A-B fragment on the other. One has:

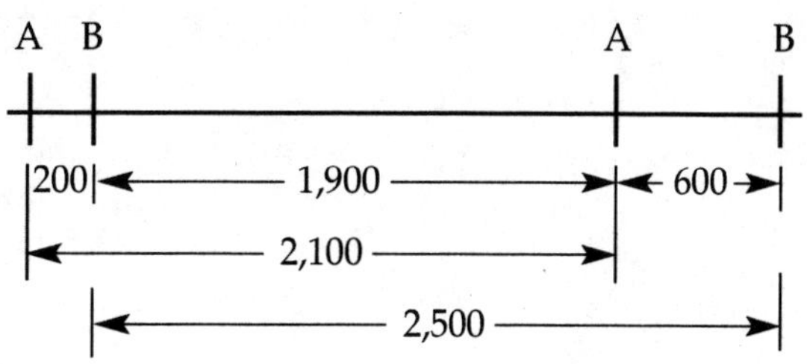

The map is extended in a stepwise fashion until all fragments are incorporated into the map. The restriction map is:

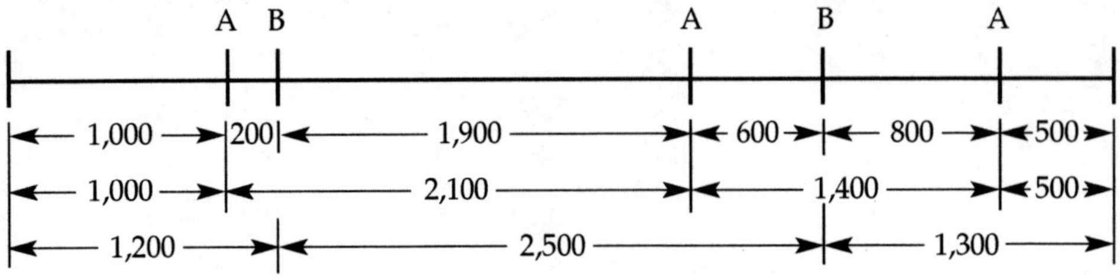

13.12 Draw the banding pattern you would expect to see on a DNA-sequencing gel if you annealed the primer 5′-CTAGG-3′ to the following single-stranded DNA fragment and carried out a dideoxy sequencing experiment. Assume that the dNTP precursors were all labeled.

3'-GATCCAAGTCTACGTATAGGCC-5′

Answer: The primer will anneal to the fragment and be extended at its 3′ end in four separate reactions, each with small amounts of a different dideoxynucleotide. In each reaction, some chains will be prematurely terminated when the dideoxynucleotide is incorporated. By using labeled dNTP precursors, all extension products will be labeled and be observed as distinct, labeled bands after separation on a denaturing polyacrylamide gel and signal detection.

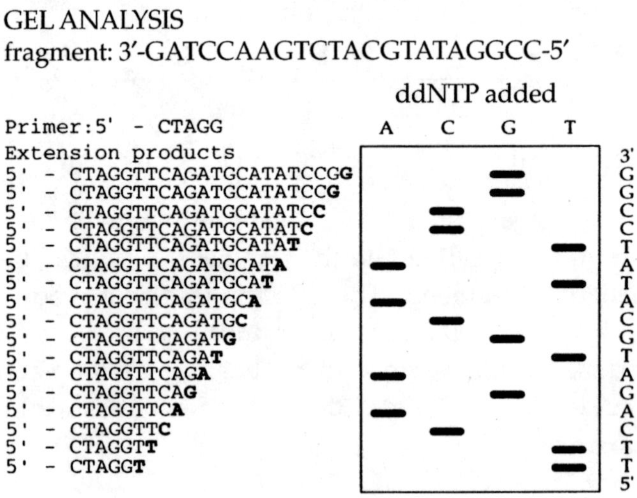

13.13 PCR fragments can be separated by gel electrophoresis, and the size of a fragment can be estimated. Explain the process involved in determining the size of the PCR fragment in question.

Answer: In gel electrophoresis, DNA fragments are separated by size. Since DNA is negatively charged due to its phosphates, it will migrate toward the positive pole in an electric field. Since, in an agarose gel, smaller DNA fragments will move more readily through the pores in the gel, smaller DNA fragments move through the gel more rapidly than larger DNA fragments. [Although larger DNA fragments have more phosphate groups and hence more *total* negative charge, they have the same amount of charge density (negative charge per unit mass) as smaller DNA fragments. Hence, the DNA is separated by size in this gel, not by charge.] To determine the size of a PCR fragment, it should be loaded into one well of an agarose gel, and a size standard, or marker, DNA sample should be loaded into another well of the gel. A current should be applied to induce the DNA samples to migrate through the gel. After electrophoresis, the gel should be stained with a dye that binds DNA, such as ethidium bromide. Ethidium bromide will complex with DNA and fluoresces under ultraviolet light. The fluorescent image should be photographed, and the distance each DNA band has migrated from the loading should be measured. Since the molecular sizes of the DNA fragments in the marker lane are known, a calibration curve can be drawn to relate the DNA size (in log kb) to the migration distance (in mm). Then the migration distance for the DNA band of the PCR product can be measured. By comparing it to the calibration curve, its size can be determined. See text Figure 13.9.

14

APPLICATIONS OF RECOMBINANT DNA TECHNOLOGY

CHAPTER OUTLINE

ANALYSIS OF BIOLOGICAL PROCESSES

Glucose Repression of Transcription of the Yeast *GAL1* Gene

DNA MOLECULAR TESTING FOR HUMAN GENETIC DISEASE MUTATIONS

Concept of DNA Molecular Testing

Purposes of Human Genetic Testing

Examples of DNA Molecular Testing

ISOLATION OF HUMAN GENES

Cloning the Cystic Fibrosis Gene

DNA TYPING

DNA Typing in a Paternity Case

Other Applications of DNA Typing

GENE THERAPY

COMMERCIAL PRODUCTS

GENETIC ENGINEERING OF PLANTS

REVIEW OF KEY TERMS, SYMBOLS, AND CONCEPTS

In your own words, write a brief, precise definition of each term in the groups on the following page. Check your definitions using the text. Then develop a concept map using the terms in each list.

THINKING ANALYTICALLY

Much of the excitement in modern genetics comes as genetic principles and recombinant DNA technology are combined and applied to address fundamental biological questions as well as practical human problems. The material in this chapter introduces many examples—methods used to investigate biological processes, applications of technology for improving human health

1	2	3	4
recombinant DNA technology transcriptional regulation regulatory gene northern blotting PCR, RT-PCR glucose repression	genetic vs. diagnostic testing DNA molecular testing newborn, disease screening prenatal diagnosis carrier detection DNA typing, fingerprinting, profiling, marker polymorphism, mutation inclusion, exclusion result RFLP STR, microsatellite VNTR, minisatellite monolocus, multilocus probe allele-specific oligonucleotide (ASO) hybridization genetically modified organism	positional cloning genetic linkage in situ hybridization chromosome walking chromosome jumping zoo blot, Noah's ark blot pseudogene candidate gene	gene therapy somatic gene therapy transgene trangenic cell, organism biotechnology DNA typing

and agriculture, determination of paternity, and applications in forensics. As you read through the examples, look for how our understanding of genetic principles and recombinant DNA technology allows for inventive, pragmatic application. Analyze the example by asking:

- What questions are being asked?
- Why are they important? How do they relate to a fundamental biological question or practical problem?
- What methods are being used to address the questions?
- What genetic principles or recombinant DNA applications underlie a manipulation or investigation? (What did we need to know before being able to do this?)
- Are multiple investigatory approaches being combined?
- How is the method effective in answering the question?

After you have read through the examples, go back and reflect more generally on the types of questions and hypotheses that can be investigated and whether there are limitations to the method or application.

QUESTIONS FOR PRACTICE

Multiple Choice Questions

1. Positional cloning
   a. requires knowledge of the gene product before the gene can be cloned
   b. generates a transgenic organism that expresses a gene only in certain tissues
   c. isolates a disease gene based on its approximate chromosomal location
   d. is when a cDNA has been cloned into a specific orientation in an expression vector
2. Chromosome walking is
   a. used to obtain a set of overlapping clones from a genomic library
   b. used to obtain a set of overlapping clones from a cDNA library
   c. used to jump between chromosomal locations without cloning intervening DNA
   d. impossible in eukaryotes because of the amount of interspersed repetitive DNA
3. During the positional cloning of a disease gene, four candidate genes are identified. What would provide the most convincing evidence that one of them is responsible for the disease?
   a. A zoo blot shows that one gene is also found in other organisms.
   b. Polymorphisms are present in one of the genes in affected individuals.
   c. One of them is expressed in the tissue affected by the disease.
   d. Mutational changes are present in one of the genes in affected individuals.
4. What is the difference between an STR and a VNTR?
   a. An STR is the same as a microsatellite, while a VNTR is the same as a minisatellite.
   b. Both are tandem repeats: STRs are 2 to 4 bp long; VNTRs are 5 to 10s of base pairs long.
   c. Both are tandem repeats: STRs are 5 to 10s of bp long; VNTRs are 2 to 4 bp long.
   d. STRs are simple tandem repeated sequences found only in one location, while VNTRs are always dispersed in different regions of the genome.
5. What is the difference between a monolocus probe and a multilocus probe?
   a. A monolocus probe detects alleles at one gene, while a multilocus probe detects alleles at different genes.
   b. A monolocus probe detects one of two alleles at one gene, while a multilocus probe detects both alleles at one gene.
   c. A monolocus probe detects STR or VNTR sequences at one locus (site) in the genome, while a multilocus probe detects STR or VNTR sequences at a number of loci (sites) in the genome.
   d. A monolocus probe detects STR sequences, while a multilocus probe detects VNTR sequences.
6. An ASO hybridization is a method
   a. to assess genotypes: *a*llele-*s*pecific *o*ligonucleotide hybridization. PCR products differing by one base are distinguished by hybridization with labeled oligonucleotides.
   b. used to screen jumping libraries: "*a*ll *s*imilar *o*bjects" hybridization. A probe is used to identify clones containing similar sequences.
   c. to distinguish between two similar transgenic animals: *a*lmost the *s*ame *o*rganisms. Blood taken from each is typed for a set of DNA markers, and the results are compared.
   d. for testing for the presence of GMOs in foods: *a*griculturally *s*ensitive *o*ligonucleotide hybridization. Since transgenic plants are modified using expression vectors, an oligonucleotide probe can be used to test if foods contain it and hence contain GM plant material.

7. Which one is *not* an example of a product produced by a biotechnology company?
   a. human growth hormone
   b. DNase
   c. bacteria that can accelerate the degradation of oil pollutants
   d. aspirin
8. Suppose DNA typing is used in a paternity case. How do exclusion results differ from inclusion results?
   a. Exclusion results are easier to prove; one just needs to show that the male in question has no alleles in common with the baby.
   b. Inclusion results require positive identity to be established and usually requires testing for alleles at many genes.
   c. Inclusion results require calculation of the relative odds that an allele came from the accused or from another person and requires knowing the frequencies of VNTR and STR alleles in many ethnic groups.
   d. all of the above

Answers: 1c, 2a, 3d, 4b, 5c, 6a, 7d, 8d

Thought Questions

1. When cells are subjected to a heat shock (a transient high-temperature pulse), they become stressed and stop transcribing most genes. They start transcribing others that help them deal with the stress of the heat shock. How would you assess if transcription of a gene you have cloned is repressed or activated by heat shock?
2. After developing a GMO, some companies secure patents on it. Suppose a company sells seed for a GMO to a farmer under the condition that the farmer may plant the seed but may not retain seed stock from any crop that is harvested. The seed is expensive, and the farmer only buys seed from the company once. How would the company check if the farmer has used seed harvested from the first crop?
3. Suppose the farmer described in question 2 is quite poor and also no longer retains the seed stock he used prior to planting seed for the GMO. What ethical issues arise?
4. While many mutations causing disease result in proteins that are nonfunctional or that have altered function (such as in sickle-cell anemia), some are caused by mutant alleles that fail to make a protein product. Therapeutic approaches will differ in this situation. Therefore, when a disease gene is cloned, it is important to address if the disease phenotype is caused by an abnormal protein or by the absence of a protein. Suppose you have cloned a gene for a disease and notice that a restriction site for *Eco*RI has been altered in a transcribed region of the gene. How would you experimentally address this issue if you are given a tissue biopsy from a patient with the disease?
5. In U.S. courts, DNA typing is not yet a generally accepted method for proving parenthood or guilt. However, it is often accepted as providing evidence that excludes an accused individual. What is the basis for the difference in the level of acceptance of DNA testing? Do you see this changing over time?
6. Some indigenous peoples have taken positions against the use of GMOs based on a cultural viewpoint that they alter the spiritual integrity of living organisms. Some Maori (an indigenous people in New Zealand), for example, have asked that transgenic sheep that produce milk with pharmaceutically important compounds not be raised on their land and that genetically modified trees harvested for lumber not be grown on their land. How do you

respond to these types of concerns, and should these concerns be more generally raised? (Recall the text citing that "between 50 and 75 percent of produced and processed foods in a supermarket may be genetically modified or contain GMOs.")

7. Even though CF is the most common lethal genetic disease in the United States today, it is not tested for routinely in newborns. Why might this be the case?
8. What is the difference between a genetic test and a diagnostic test if each is also a DNA molecular test? How do the ethical and legal issues surrounding these two tests differ?
9. Somatic gene therapy holds great promise for treating genetic disease. What is the underlying basis for somatic gene therapy? What ethical and legal issues are associated with its implementation?

Thought Questions for Media

After reviewing the media at the *Essential iGenetics* website, try to answer these questions.

1. For one primer pair, how many different PCR products can be obtained in one individual? How many different alleles can be detected with one primer pair?
2. For five primer pairs, how many different PCR products can be obtained in one individual? How many different alleles can be detected with five primer pairs?
3. Are alleles detected as STRs dominant or recessive or neither? Why?
4. For one primer pair, how many different patterns of PCR products can be obtained in two individuals? For five primer pairs, how many different patterns of PCR products can be obtained in two individuals?
5. Why is it important in forensics to perform DNA typing on loci that are highly polymorphic?
6. How is the profile probability used to determine the likelihood that two matching samples come from the same individual?

SOLUTIONS TO TEXT PROBLEMS

14.1 Imagine that you have been able to clone the structural gene for an enzyme in a catecholamine biosynthetic pathway from the adrenal gland of rats. How could you use this cloned DNA as a probe to determine whether this same gene functions in the rat brain?

Answer: If the same gene functions in the brain, the gene for the enzyme must be transcribed into a precursor mRNA, processed to a mature mRNA, and then translated to produce the functional enzyme. Thus, transcripts for the gene should be found in the brain. To address this issue, label the cloned DNA, and use it to probe a northern blot having mRNA isolated from brain tissue. If the mRNA is rare, it may be prudent to use mRNA isolated from a specific region of the brain (e.g., the hypothalamus). An alternative, quite sensitive approach would be to sequence the cloned DNA, analyze the sequence to identify the coding region, and then design PCR primers that could be used to amplify cDNA made from mRNA isolated from various brain regions. To perform this RT-PCR (reverse transcriptase PCR), one would isolate mRNA, reverse transcribe it into cDNA, and then perform PCR. Obtaining an RT-PCR product in such an investigation would provide evidence that the gene is transcribed in the brain. In this alternative method, it would be important to be sure that no genomic DNA was present in the PCR amplification mixture, as the gene for the enzyme would be found in genomic DNA in both tissues.

14.2 Imagine that you find a RFLP in the rat genomic region homologous to your cloned catecholamine synthetic gene from question 14.1 and that in a population of rats displaying this polymorphism, there is also a behavioral variation. You find that some of the rats are normally calm and placid, but others are hyperactive, nervous, and easily startled. Your hypothesis is that the behavioral difference is caused by variations in your gene. How could you use your cloned sequence to test this hypothesis?

Answer: Initially, perform a genomic Southern analysis to see if the RFLP is limited to animals with or without the behavioral phenotype. Isolate DNA from the affected and normal rats, cleave it with the restriction enzyme that was used to identify the RFLP, separate the DNA by size on an agarose gel, blot the gel in the manner of Southern to a membrane, make a probe from DNA in the cloned catecholamine synthetic gene region, hybridize the probe to the Southern blot, and detect the location of the probe signal. If the data show that normally behaving rats and abnormally behaving rats share identical RFLPs, the hypothesis is not supported. If the data show that the presence of the RFLP correlates with the behavior of the rats (e.g., normally behaving rats never have the RFLP and abnormally behaving rats always have the RFLP), the hypothesis is supported. However, these data do not demonstrate that the behavioral difference is *caused* by the DNA difference.

To gather additional evidence that the DNA difference *causes* a variation in the gene that causes the behavioral difference, try to demonstrate that there is a correlation between the presence of the RFLP, an altered enzyme activity of the catecholamine synthetic gene, and the behavior. Determine if animals with the RFLP have different levels of enzymatic activity (for catecholamine synthesis) or catecholamine production than animals without the RFLP. With the earlier results, this provides a correlation between the presence of the RFLP, the behavioral trait, and the function of the gene.

Note that this still does not demonstrate that the RFLP itself is the mutation affecting the gene. There are two possibilities. First, the RFLP affects the catecholamine synthetic gene by affecting either its transcription, its transcript processing, or its protein product. Second, the RFLP is close to another mutation that affects the gene. It is therefore important to determine the location of the RFLP relative to the location of the gene. If the RFLP lies within the coding region of the gene, it may result in a different, perhaps nonfunctional protein being produced. If it lies within a regulatory region (e.g., within the promoter or in an enhancer or silencer), it could affect the amount of transcription of the gene and so affect the amount of enzyme that is produced. If it lies within an intron, it may affect mRNA processing and so affect the amount of mature mRNA and gene product produced. To address these issues, sequence the region containing the RFLP, and use this information, together with restriction mapping, to identify where the RFLP is relative to the transcribed region of the gene. If the RFLP lies in a region that might reasonably affect mRNA processing or the level of transcription of the gene, experimentally assess if this is the case by using the methods described in the answer to question 14.1. That is, determine if the presence of the RFLP correlates with the level of mature mRNA produced from the gene. Alternatively, check to see if the RFLP causes a missense or nonsense mutation in the coding region of the gene, producing an altered protein.

If this analysis does not demonstrate that the RFLP itself causes a defect in transcription, transcript processing, or the protein product, it remains possible that another, nearby mutation is the cause of the gene alteration or that the RFLP is the cause, but it affects the gene in a manner that is not easily recognized (e.g., it lies in an enhancer far from the gene that is not easily identified). To investigate this possibility, cross the affected and unaffected rats to control strains, and follow the segregation of the RFLP and the behavioral phenotype. If the two traits cosegregate, that is, if the RFLP is always present with one of the behavioral traits, this provides evidence that the two are linked. If the RFLP and the behavioral difference can be separated by recombination, the RFLP itself cannot be the molecular basis of the behavioral phenotype, but a closely linked mutation might be. The frequency of recombination will provide an estimate of how close the DNA region with the RFLP is to the site of the mutation causing the behavioral phenotype. Sequence analysis of the region may lead to the identification of the causal DNA alteration.

14.3 DNA was prepared from small samples of white blood cells from a large number of people. Ten different patterns were seen when these DNAs were all digested with *Eco*RI, then subjected to electrophoresis and Southern blotting. Finally, the blot was probed with a radioactively labeled cloned human sequence. The following figure shows the ten DNA patterns taken from ten people.

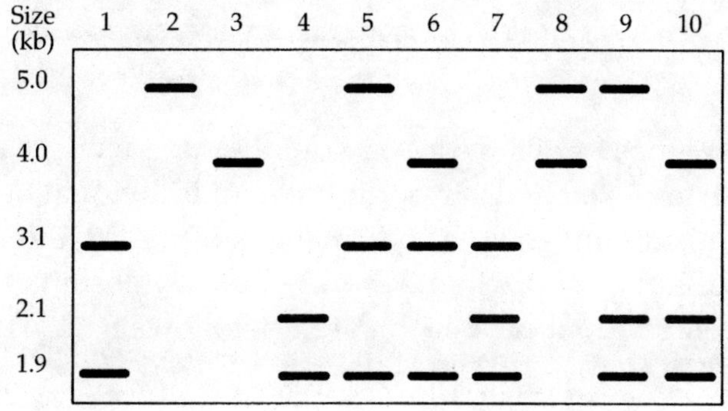

a. Explain the hybridization patterns seen in the ten people in terms of variation in *Eco*RI sites.

b. If the individuals whose DNA samples are in lanes 1 and 6 on the blot were to produce offspring together, what bands would you expect to see in DNA samples from these offspring?

Answer:

a. The probe hybridizes to the same genomic region in each of the ten individuals. Different patterns of hybridizing fragments are seen because of polymorphism of the *Eco*RI sites in the region. If a site is present in one individual but absent in another, different patterns of hybridizing fragments will be seen. This provides evidence of restriction fragment length polymorphism. To distinguish between sites that are invariant and those that are polymorphic, analyze the pattern of bands that appear. Notice that the sizes of the hybridizing bands in individual 1 add up to 5 kb, the size of the band in individual 2 and the largest hybridizing band. This suggests that there is a polymorphic site within a 5-kb region. This is indicated in the diagram below, where the asterisk over site *b* depicts a polymorphic *Eco*RI site:

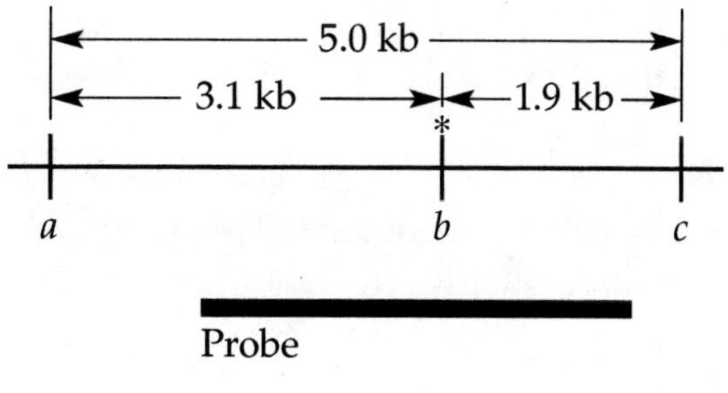

Notice also that the size of the band in individual 3 equals the sum of the sizes of the bands in individual 4. Thus, there is an additional polymorphic site in this 5-kb region. Since the 1.9-kb band is retained in individual 4, the additional site must lie within the 3.1-kb fragment. This site, denoted *x*, is incorporated into the diagram below. Notice that since the 1.0-kb fragment flanked by sites *a* and *x* is not seen on the Southern blot, the probe does not extend into this region.

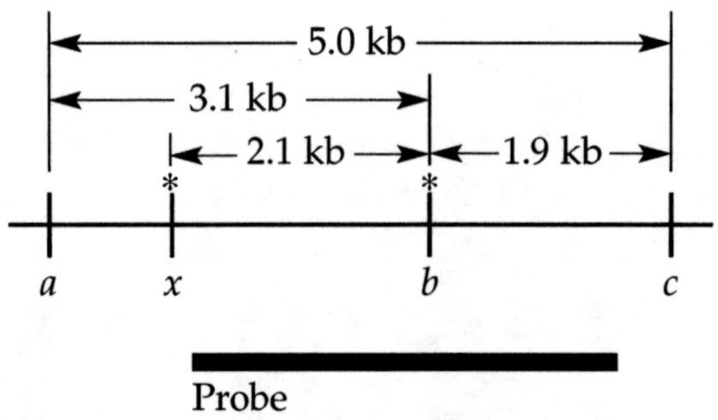

Depending on whether *x* and/or *b* are present, one will see either 5-kb, 3.1-, and 1.9-kb, 2.1- and 1.9-kb, or 4-kb bands. In addition, if an individual has chromosomes with different polymorphisms, one can see combinations of these bands. Thus, individual 5 has one chromosome that lacks sites *x* and *b* and one chromosome that has site *b*. The chromosomes in each individual can be tabulated as follows:

Individual	Sites on Each Homologue	Homozygote or Heterozygote?
1	*a, b, c*	homozygote
2	*a, c*	homozygote
3	*x, c*	homozygote
4	*x, b, c*	homozygote
5	*a, c/a, b, c*	heterozygote
6	*x, c/a, b, c*	heterozygote
7	*a, b, c/x, b, c*	heterozygote
8	*a, c/x, c*	heterozygote
9	*a, c/x, b, c*	heterozygote
10	*x, c/x, b, c*	heterozygote

b. Since individual 1 is homozygous, chromosomes with sites at *a*, *b*, and *c* will be present in all of the offspring, giving bands at 3.1 and 1.9 kb. Individual 6 will contribute chromosomes of two kinds, one with sites at *x* and *c* and one with sites at *a*, *b*, and *c*. Thus, if this analysis is performed on their offspring, two equally frequent patterns will be observed: a pattern of bands at 3.1 and 1.9 kb and a pattern of bands at 4, 3.1, and 1.9 kb. This is just like the patterns seen in the parents.

14.4 The maps of the sites for restriction enzyme R in the wild-type and mutated cystic fibrosis genes are shown schematically in the following figure.

Samples of DNA obtained from a fetus (F) and her parents (M and P) were analyzed by gel electrophoresis followed by the Southern blot technique and hybridization with the radioactively labeled probe designated "CF probe" in the previous figure. The autoradiographic results are shown in the following figure.

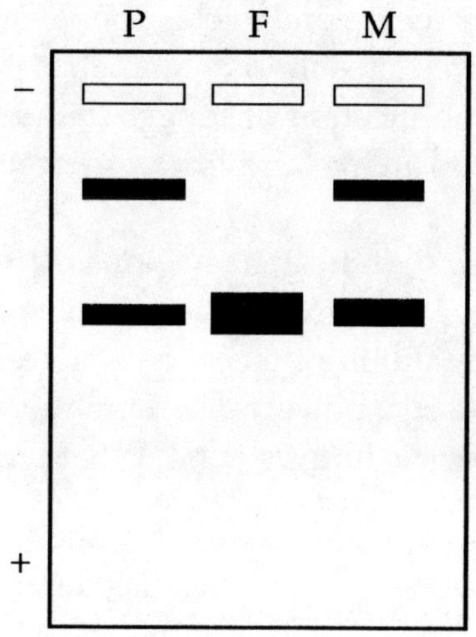

Given that cystic fibrosis is a recessive mutation, will the fetus be affected? Explain.

Answer: Chromosomes bearing CF mutations have a shorter restriction fragment than chromosomes bearing wild-type alleles. Both parent lanes (M and P) have two bands, indicating that each parent has a normal and a mutant chromosome. The parents are therefore heterozygous for the CF trait. The fetus lane (F) shows only one (lower-molecular-weight) band. The size of the band indicates that the fetus has only mutant chromosomes. The intensity of the band is about twice that of the same-sized band in the parent lanes. This is because the diploid genome of the fetus has two copies of the fragment, while the diploid genome of each parent only has one. Since the fetus is homozygous for the CF trait, it will have CF. (Do not confuse the black bands with the open boxes, which represent the origin, the location where the DNA was loaded before electrophoresis.)

14.5 Polymerase chain reaction (PCR) was introduced as a rapid and efficient way to amplify specific fragments of DNA. What is required to amplify a specific region (the target region) of DNA in a genome?

Answer: To amplify a specific region, one needs to know the sequences flanking the target region so that primers able to amplify the target region can be designed. Once primers are synthesized, the polymerase chain reaction can be assembled. It contains a DNA template (in this case, genomic DNA), the pair of primers that flank the DNA segment targeted for amplification, a heat-resistant DNA polymerase (*Taq*), the four dNTPs (dATP, dTTP, dGTP, and dCTP), and an appropriate buffer. See Figure 13.14.

14.6 What are the steps in one cycle of the polymerase chain reaction?

Answer: In the polymerase chain reaction, each cycle consists of denaturation, annealing, and extension. These steps occur at different temperatures and are repeated to amplify the desired fragment. In one cycle, the reaction mixture is first heated to 94°C to denature the DNA template. Then it is cooled to a temperature that allows the primers to anneal to specific sites on the DNA template. Then the reaction is heated to 72°C, where DNA polymerase extends the primers at their 3′ ends. As shown in Figure 13.14, DNA fragments are produced in a unit length equal to the length of DNA between the two primer sites after the first cycle. After the second cycle, double-stranded fragments of unit length are produced. Subsequent cycles result in a geometric increase in the number of unit-length molecules, but linear amplification of longer-than-unit-length molecules. Consequently, the PCR reaction produces a large amplification of unit-length molecules.

14.7 Geneticists were trying to find the gene responsible for cystic fibrosis. They did not know the identity of its protein product. However, they did know that something about salt transport was disrupted, although they knew nothing about the nature of the defective protein. The geneticists used a powerful combination of classic and molecular approaches and in 1989 located the cystic fibrosis gene. What were the steps they used?

Answer: Cloning of the gene responsible for cystic fibrosis involved sequential homing in on the gene, using both genetic and molecular data.

1. *Linkage analysis.* First, RFLP markers linked to the CF gene were identified. As the patterns of inheritance of CF were tracked in families, DNA from family members was analyzed using Southern blot analysis to identify RFLP markers that showed genetic linkage to the CF locus.
2. *Chromosomal localization.* The RFLP marker was used as a probe in in situ hybridization to identify the chromosome on which the CF gene was located—chromosome 7. Additional nearby RFLPs were then identified. These were used in genetic linkage analysis to identify flanking markers that defined which region of chromosome 7 harbored the CF gene.
3. *Cloning genomic DNA.* A combination of chromosomal walking and jumping was used to identify a large number of overlapping genomic DNA clones spanning about 500 kb in the CF region.
4. *Identifying CF candidate gene.* To identify the CF gene among these DNA sequences, subcloned segments of the region were used as probes on zoo blots to detect if any were conserved among other organisms. Five clones with conserved gene sequences were identified. Linkage analysis excluded two from being responsible for CF. A third was excluded because it was a pseudogene. A fourth was excluded because it was not expressed in tissues affected in CF patients. The fifth subclone was used to synthesize a probe that was used to screen a cDNA library made from mRNA expressed in cultured sweat gland tissue cells. These cells were chosen since they might express genes involved in salt transport. The probe identified a single 6.5-kb band on a northern blot and allowed the identification of a candidate cDNA clone. The cDNA clone was used to analyze the genomic clones in more detail.

5. *Confirming that candidate gene is responsible for CF.* The DNA sequences of the candidate CF gene were compared in normal and affected individuals. After a mutational change (a 3-bp deletion) was found in the gene from one CF patient, DNA changes were identified in the gene in additional CF patients.
6. *Characterizing the protein produced by the CF gene.* Computerized analyses of the predicted CF protein provided a basis to propose a structure for the CF protein. In turn, this was used to develop testable hypotheses about its function.

14.8 What is DNA fingerprinting? How could this method be used to establish parentage? How is it used in forensic science laboratories?

Answer: DNA fingerprinting is the characterization of an individual in terms of the set of DNA markers the person has. For DNA fingerprinting, DNA markers are usually chosen so that they are highly polymorphic in a population, and they may include RFLPs, simple tandem repeat polymorphisms of 2 to 4 bp in length (STRs, microsatellites), and variable numbers of tandem repeat polymorphisms of 5 to 10s of bp in length (VNTRs, minisatellites). They can be derived either from just one locus or more than one locus (monolocus or multilocus probes). Since all individuals except for identical (monozygotic) twins have different genomes, individuals of a species differ in terms of their DNA fingerprints.

Each parent donates one allele at a DNA marker to their offspring. Therefore, the set of DNA markers a child has come from his/her parents. If two parents and a child are tested, all of the DNA markers present in the child must also be present in the set of markers in the two parents. However, the parents can have additional markers, since they only donate one of their two alleles at each marker. If a child has no alleles in common with a suspected parent, the suspected individual can be excluded from consideration as the child's parent. If alleles are shared, the likelihood that the set of alleles came from the suspected parent must be calculated and compared with the likelihood that the set of alleles came from another person. If the alleles present in the child and one known parent are known, then one can infer at least some of the alleles that were contributed by the other parent. Usually, multiple DNA markers are used to minimize possible inaccuracy and provide for a better statistical evaluation of the results, showing that the results obtained are not based on sampling variation. The calculation of the relative odds that a set of alleles derives from a suspected parent or from another person depends on knowing the frequencies of the marker alleles in the child's ethnic population. These types of data should be used together with other, non-DNA-based evidence to support a claim of parentage.

In forensic science laboratories, DNA fingerprinting is used to match (or exclude) the DNA fingerprint of a suspected individual with the DNA fingerprint provided by physical evidence (blood, skin, hair, semen, saliva, etc.) gathered at the scene of a crime. As in the analysis for parentage, DNA fingerprinting is very useful for excluding suspects who do not share alleles with that found in the physical evidence. Inclusion results, results where an individual is positively identified as being responsible for a crime, are more difficult to obtain. As for parentage, a calculation of the relative odds that a set of alleles derive from a suspect or from another person must be made, and this requires good population statistics. In addition, there must be evidence that there were no errors in collecting or processing samples. Thus, DNA evidence is often more useful for excluding suspects than for proving their guilt, and when used to support guilt, must usually be supported by additional, non-DNA-based evidence.

14.9 One application of DNA fingerprinting technology has been to identify stolen children and return them to their parents. Bobby Larson was taken from a supermarket parking lot in New Jersey in 1978, when he was 4 years old. In 1990, a 16-year-old boy called Ronald Scott was found in California, living with a couple named Susan and James Scott, who claimed to be his parents. Authorities suspected that Susan and James might be the kidnappers and that Ronald Scott might be Bobby Larson. DNA samples were obtained from Mr. and Mrs. Larson and from Ronald, Susan, and James Scott. Then DNA fingerprinting was done, using a polymorphic probe for a particular VNTR family, with the results shown in the figure below. From the information in the figure, what can you say about the parentage of Ronald Scott? Explain.

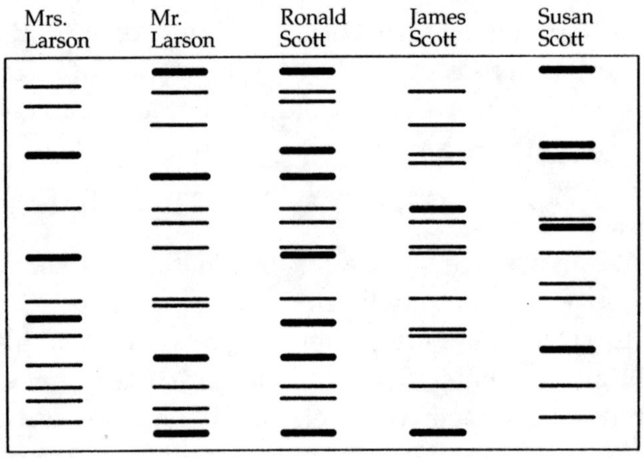

Answer: James and Susan Scott are not the parents of "Ronald Scott." There are several bands in the fingerprint of the boy that are not present in either James or Susan Scott and thus could not have been inherited from either of them (e.g., bands a and b in the answer figure). In contrast, whenever the boy's DNA exhibits a band that is missing from one member of the Larson couple, the other member of the Larson couple has that band (e.g., bands c and d). Thus, there is no band in the boy's DNA that he could not have inherited from one or the other of the Larsons. These data thus support an argument that the boy is, in fact, Bobby Larson. These data should be used together with other, non-DNA-based evidence to support the claim that the boy is Bobby Larson.

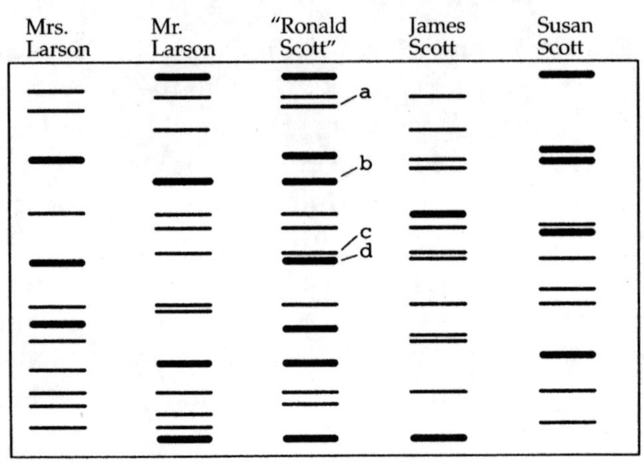

14.10 In 1990, the first human gene therapy experiment on a patient with adenosine deaminase deficiency was done. Patients who are homozygous for a defective gene for this enzyme have defective immune systems and risk death from diseases as simple as a common cold. Which cells were involved, and how were they engineered?

Answer: In this gene therapy experiment, T cells were isolated from the patient and a viral vector was used to introduce the normal adenosine deaminase gene. The cells were then reintroduced into the patient.

14.11 You have inserted human insulin cDNA in the cloning vector pUC19 and transformed the clone into *E. coli*, but insulin was not expressed. Propose some hypotheses to explain why not.

Answer: If genomic DNA had been used, there could be concerns that an intron in the genomic DNA was not removed, since *E. coli* does not process RNAs as eukaryotic cells do. However, if the cDNA is a copy of a mature mRNA, this should not be a concern. There are other potential concerns, however. First, depending on the nature of the sequence inserted, a fusion protein with β-galactosidase may have been produced, and not just human insulin. In pUC19, the polylinker is within the β-galactosidase gene. Sequences inserted into the polylinker, if inserted in the correct reading frame (the same one as used for β-galactosidase), will be translated into a β-galactosidase fusion protein. If this is acceptable, it would be important to ensure that only the ORF (the open reading frame) of the insulin gene is properly inserted into the polylinker of the pUC19 vector. For the insulin ORF to be properly inserted, it must be inserted in the correct reading frame so that premature termination of translation does not occur and the correct polypeptide is produced. One could not use a complete copy of the human mRNA transcript for insulin. If transcribed, it would have features of eukaryotic transcripts but not features required for prokaryotic translation. Indeed, some of its 5′ UTR and 3′ UTR sequences may interfere with prokaryotic transcription and translation. For example, it will lack a Shine-Dalgarno sequence to specify where translation should initiate and identify the first AUG codon. In the pUC19 vector, a Shine-Dalgarno sequence is supplied after the promoter for the *lac*Z gene, since without an insert in the polylinker, β-galactosidase is produced. However, the cDNA may have 5′ UTR sequences that interfere with translation initiation in prokaryotes or that contain stop codons, terminating translation of the β-galacosidase fusion protein. Second, the cDNA may encode a protein that is posttranslationally processed to become human insulin. The protein produced in *E. coli* may not be processed.

14.12 The ability to place cloned genes into plants raises the possibility of engineering new, better strains of crops such as wheat, maize, and squash. It is possible to identify useful genes, isolate them by cloning, and insert them directly into a plant host. Usually, these genes bring out desired traits that allow the crops in question to flourish. Why is there such concern by consumers about this process? Do you feel that the concern is justified? Defend your answer.

Answer: There is clearly a huge potential for success in crop modification. In addition to the examples presented in the text, visit the websites listed below, two of many sites that describe the potential value of a modified, vitamin-A-enhanced rice (golden rice). Such a product could, in principle, help millions of people suffering from vitamin A deficiency. Vitamin A deficiency is associated with blindness and increased disease susceptibility.

http://www.disasterrelief.org/Disasters/000808gmofoods/

http://www.monsantoindia.com/news/archives/August2000/040800monsanto.html

There are multiple, different reasons consumers have concerns about this and other GMOs. First, some individuals find that this type of genetic alteration is in conflict with their culturally held views of the spiritual integrity of living organisms (see Thought Question 7). Second, some individuals feel that the dangers associated with genetically modified organisms are unknown or not well enough understood. Several different types of concerns exist. For example, a gene placed in a particular crop species might be transferred laterally, perhaps via some infectious agent, to another species, where it is undesirable; the gene may confer properties that are deleterious to the environment, which only appear years after introduction and which would not be apparent in controlled testing environments; the altered strain may deleteriously alter existing ecological balances; the altered gene may negatively impact on human health in unknown ways; there may be significant concerns about food safety. Third, there are concerns about the maintenance of genetic diversity in crops. What will be the effect on seed stocks if only a few highly desirable strains of some crops are grown? Will this result in increased susceptibility to potential pathogens in the future? Fourth, there are economic concerns: How ethically sound are the business practices of companies involved in these efforts? Should science be financed by corporations? Are these developments in the best interests of underdeveloped nations? Finally, it is a new and largely unknown area. There is much that is unknown, and regulatory agencies are themselves entering a new territory and are being offered conflicting "expert" opinions.

15

GENOME ANALYSIS

CHAPTER OUTLINE

REVIEW OF KEY TERMS, SYMBOLS, AND CONCEPTS

In your own words, write a brief, precise definition of each term in the groups on the following page. Check your definitions using the text. Then develop a concept map using the terms in each list.

THINKING ANALYTICALLY

Analysis of genomes is complex. Not only are there multiple approaches to mapping and sequencing a genome, but also a huge amount of information has been gathered through sequence analysis, and an entirely new area—bioinformatics—has emerged to analyze and apply this information. As in previous chapters, organizing information into a contextual framework will help you understand the logic behind the different approaches and retain information obtained from sequence analyses. It will help you to organize the information into separate units.

After reading through the chapter, sketch out answers to questions addressing its main topics, such as: How is a genome mapped? What different types of markers are used? Why is mapping helpful to sequencing a genome? What is a direct shotgun approach? In what different ways are sequences analyzed? What similarities and differences have been seen in the genomes of organisms sequenced to date?

1	2
structural genomics	genomics
physical vs. genetic mapping	structural genomics
Human Genome Project (HGP)	functional genomics
shotgun approach, direct shotgun approach	comparative genomics
sequencing, assembly, finishing	genome annotation
map: genetic, physical, cytogenetic, restriction, clone contig	bioinformatics
polymorphism, DNA marker, genetic marker	database
RFLP, VNTR, STR, SNP	ORF
anonymous probe	homologous gene, homologue
minisatellite, microsatellite	homology
oligonucleotide hybridization analysis	homology search
DNA microarray, DNA chip, GeneChip®	BLAST
oligonucleotide array, probe array	*FUN* gene
chromosomal band; p, q chromosome arms	orphan family, single orphan
dideoxy sequencing	protein domain
ORF, ECE	gene knockout
NIH, DOE	loss-of-function mutation
	transcriptome
	proteome
	gene expression profile
	transcriptional fingerprint
	probe array
	proteomics
	Human Genomic Project (HGP)
	ELSI program
	Bacteria, Archaea, Eukarya

Once you can answer questions like these that relate to structural genomics, identify how sequence information is used in answering questions related to functional and comparative genomics. What are the aims of functional genomics? What methods are used in functional genomics? What new technologies have been developed that allow simultaneous analysis of many genes at once? What are the practical applications of these technologies? What are the aims of comparative genomics? What human problems can it help solve? What biological questions can it address? What tools are used in comparative genomics?

Finally, step back and consider that modern genetics increasingly interfaces with many different aspects of human health and human society. What ethical, legal, and social issues arise with the advent of genomics and its related technologies? Ask yourself if all of these are unquestionably for the good of society or whether some need to be carefully monitored and/or regulated? Step back and think outside the box wherein scientists only pursue information and research for its own sake, justified by the potential for positive benefit to society. What dangers are associated with genomics? How, as a society, can be ensure that the "good" of genomics is utilized while minimizing its dangerous aspects?

QUESTIONS FOR PRACTICE

Multiple Choice Questions

1. Which of the following is *not* a goal of the Human Genome Project?
   a. Identify all the genes in human DNA.
   b. Store DNA sequence and gene-related information in public databases.
   c. Transfer technologies related to genomics to the private sector.
   d. Identify the gene(s) associated with each human disease.
2. Genetic crosses are used for all of the following *except*
   a. to establish the locations of genetic markers on chromosomes
   b. to determine the genetic distance between markers
   c. to determine the number of base pairs between markers
   d. to determine the recombination frequency between markers
3. Which DNA marker is most abundant in eukaryotic genomes?
   a. STR
   b. VNTR
   c. SNP
   d. EST
4. DNA microarrays are
   a. aligned sets of printed DNA sequences, ordered according to their map positions
   b. ordered grids of DNA molecules of known sequence fixed at known positions on a solid substrate
   c. aligned sets of printed STRs (microsatellites) in a database
   d. ordered grids of homologous genes from different organisms
5. Which of the following can be done using computerized algorithms for prokaryotic, but not eukaryotic, genomes?
   a. Identify the locations of all protein-coding regions.
   b. Identify the locations of known repeated sequences.
   c. Identify the locations of known transposable elements.
   d. Identify the locations of all rRNA genes.
6. Which of the following is *not* true about archaeon genomes?
   a. Some have large main circular chromosomes as well as multiple circular extrachromosomal elements.
   b. The genes involved in energy production, cell division, and metabolism are similar to their counterparts in bacteria.
   c. The genes involved in DNA replication, transcription, and translation are similar to their counterparts in eukaryotes.
   d. They have a higher GC content compared with either bacteria or eukaryotes.
7. Why could Celera Genomics obtain only the sequence of about 97 percent of the genomes of *Drosophila* and humans?
   a. In both organisms, simple telomeric sequences are very difficult to sequence, and these make up about 3 percent of the genome.
   b. In both organisms, long ribosomal DNA repeats comprise about 3 percent of the genome. Since these are identical, it is impossible to sequence each individual repeat.

c. In both organisms, approximately 3 percent of the sequences were in error.
d. In both organisms, heterochromatic regions mostly near the centromeres are unclonable, making sequences in these regions impossible to obtain.

8. Arguments can be made for using simpler organisms like *Drosophila* to study complex diseases like cancer. Which one reason below is the most compelling?
   a. *Drosophila* flies do not get cancer, and it would be useful to understand why.
   b. It has been impossible to find genes associated with cancer in humans.
   c. *Drosophila* flies have homologues for about 61 percent (177 of 289) of human disease genes, making them good models to study the functions of genes involved in human disease.
   d. *Drosophila* flies are usually found near human populations and may cause cancer.
9. What is BLAST?
   a. the Basic Local Alignment Search Tool, used for homology searches when a DNA or amino acid sequence is compared to other sequences in a database
   b. the Big Linear Accelerator Sequencing Tool, a device funded by the Department of Energy that is able to read 10,000 bp of DNA from one sequencing reaction
   c. the Best Legal Analysis of Sequence Technology, a document from the American Bar Association that legally analyzes situations faced by biotechnology companies that introduce genetically modified organisms into the environment
   d. the Beer Legal Age Screening Tool, a DNA chip used to detect the length of simple telomeric repeats (Since these shorten with each cell division in somatic cells, this can be used to quickly determine whether an individual is over the legal drinking age.)
10. What is the ELSI program?
    a. the world's largest bioethics program
    b. a program funding studies of the ethical, legal, and social issues related to the availability of genetic information
    c. a program that is likely to impact legislation concerning the privacy of genetic information, the safe and effective introduction of genetic information in the clinical setting, the fair use of genetic information, and the education of professionals and the public
    d. all of the above

Answers: 1d, 2c, 3c, 4b, 5a, 6d, 7d, 8c, 9a, 10d

Thought Questions

1. What are the arguments for and against the ability of states or the federal government to collect genetic data on their populace?
2. Huntington disease is a neurodegenerative disorder inherited as an autosomal dominant disease and is caused by a single gene mutation. It affects about 30,000 Americans. A child of an affected parent has a 50 percent chance of inheriting the disease allele. While there is intensive research into treatments for Huntington disease, there is at present no cure. Carriers of the disease allele will eventually get the disease. Why might someone who has a first-degree relative with the disease want to be tested for the disease allele? Why might the person not want to be tested? Should there be testing for a genetic disease for which there is no cure?

3. Unlike Huntington disease, many different genes appear to be associated with other neurodegenerative diseases, such as Alzheimer disease. Having mutations at one of these genes increases the risk of getting the disease, but does not predict with certainty that an individual will get the disease. Other differences from Huntington disease are that there are disease forms that are apparently not caused by gene mutations, and prevention and treatment options have been proposed. Alzheimer disease affects over 4 million Americans. Would you modify your answers to the questions posed in Thought Question 2? Under what circumstances, if any, would genetic testing for genes related to Alzheimer disease be warranted?
4. How are the challenges faced by proteomics different from the challenges faced by genomics?
5. DNA microarrays can be used to distinguish between two similar cancers affecting the same tissue. How is this done, and why might this be therapeutically helpful?
6. What impact does the completion of a high-resolution map and detailed sequence of the human genome have on positional cloning of disease genes?
7. What is meant when two genes are referred to as homologues? For a gene encoding a protein product, is it better to search for homologues based on DNA sequence or amino acid sequence? Why?
8. Explore the NCBI website for BLAST at *http://www.ncbi.nlm.nih.gov/BLAST/*. What different forms of BLAST have been developed, and what are they used for? Why is it important statistically to evaluate matches between an obtained sequence and sequences in a database as similar sequences are identified?
9. After the human genome sequence was completed, searches were performed to identify possible genes. About 35,000 were found. What different criteria might be used to quickly provide evidence that a DNA sequence is part of a functional gene? (Hint: Think about the value of homology searches.)
10. Within different ethnic groups, some alleles are more common than others. Consequently, for diseases influenced by genetic factors, different ethnic groups vary in their frequency of disease alleles. What challenges does this present for genetic testing and for preventive medicine based on the identification of genetically at-risk populations?

Thought Questions for Media

After reviewing the media at the *Essential iGenetics* website, try to answer these questions.

1. After shearing the DNA to fragments that are approximately 2 kb in size, why are the DNA fragments treated with S1 nuclease?
2. After the 2-kb DNA fragments are cloned into plasmid vectors, they are transformed into bacteria, the plasmids isolated, and their inserts sequenced. Since only about 500 bp of insert DNA can be obtained from each end of an insert, how is the remaining 1 kb of DNA obtained?
3. How are overlapping clones identified to construct a clone contig?
4. Why might there be gaps in an assembled sequence?
5. What are the differences and similarities between a Southern blot and a colony blot?
6. The animation illustrated how microarrays can be designed to identify mutations in a gene as well as to analyze differences in patterns of gene expression.
    a. What differences are there in the design of these two types of microarrays?
    b. What differences are there in labeling the target DNA samples for these two types of analyses?

1. Why is it important to expose the microarray to two differently labeled probes simultaneously, instead of using two microarrays with a single probe on each?
2. What are the sources of the tissues for the mRNA samples used to develop the probes for the microarrays? How important is it for the tumor tissue sample to be composed only of tumor cells and not be a mixture of tumor and normal cells?
3. Is the selection of drugs for Maeve's therapy based on a genotype? Tumor cells usually harbor multiple somatic mutations and often also contain chromosomal rearrangements. If the drug selection is based on a genotype, is it based on her genotype or the genotype of the tumor? If not, what is it based on?
4. Specific genes are placed on microarrays. What information was needed to design the microarray used in this investigation?
5. How could you use microarrays to identify additional genes that might be helpful in identifying more effective therapies for breast cancer?

SOLUTIONS TO TEXT PROBLEMS

15.1 RFLP, VNTR, STR, and SNP are abbreviations for DNA markers used to map eukaryotic genomes. Explain what each abbreviation stands for and what each marker is.

Answer: RFLPs are restriction fragment length polymorphisms, and they reflect the presence or absence of a restriction enzyme site within DNA. A site present in one chromosome may be absent in another due to a change at a single base (in which case it would also be a SNP), although chromosomal aberrations can also result in a RFLP (e.g., a deletion removing the site, a chromosomal rearrangement repositioning the site). If a RFLP is assessed in a genomic Southern blot, a probe spanning the site will detect different patterns of hybridization. Probes for RFLPs are not obtained in any directed way, and so they are relatively uncommon markers. However, since SNPs (see below) often affect a restriction site, and SNPs are quite common, in principle a great many RFLPs exist but have not been detected or characterized. Although RFLPs were the first type of DNA markers studied, it is currently easier to detect a SNP. Consequently, RFLPs are not extensively used in current mapping studies.

VNTRs and STRs are tandemly repeated sequences, with VNTRs (variable number of tandem repeats, or minisatellites) being tandem repeats of 5 to a few 10s of base pairs and STRs (simple tandem repeats, or microsatellites) being repeats of 1 to 4 bp. Since STRs typically have 10 to 20 copes of the 1-to-4 bp repeat, they can be readily detected using PCR (which works well for amplifying sequences less than 300 bp long). Some VNTRs can also be detected by PCR using primers that flank the repeating sequence, but more often, they have many copies of a relatively long sequence, making them less than ideal for PCR detection. In this case, they are detected by restriction enzyme digestion (using enzymes that flank the repeat), Southern blotting, and probing with a labeled DNA fragment containing several copies of the repeating sequence. Only about 100 VNTRs are known in the human genome. In contrast, STRs are evenly distributed throughout the genome and are found at a higher frequency. There are about 100,000 loci at an average frequency of 1 every 30,000 bp. This, plus their easier detection using PCR, makes them more useful than VNTRs in mapping studies.

SNPs are single nucleotide polymorphisms found in at least 1 percent of the human population. They can be typed by oligonucleotide hybridization analysis: A short labeled oligonucleotide probe is hybridized under high stringency to target DNAs. Under these conditions, only a perfect match is allowed, and so target DNAs containing the same allele as the oligonucleotide are identified. Simultaneous typing of hundreds of SNPs can be done using DNA microarrays. SNPs are widely and uniformly distributed throughout the human genome: One SNP can be found in every 500 to 1,000 bp of DNA, and about 3 million SNPs can be found in the entire human genome. SNPs are therefore more frequent than the other types of DNA markers.

15.2 The frequency of individuals in a population with two different alleles at a DNA marker is called the marker's heterozygosity. Why would an STR DNA marker with nine known alleles and a heterozygosity of 0.79 be more useful for mapping studies than a nearby STR having three alleles and a heterozygosity of 0.20?

Answer: If an individual is homozygous for an allele at an STR, all of their gametes have the same STR allele. The STR cannot be used as a marker to distinguish the recombinant- and parental-type gametes of the individual. In a population, individuals are more likely to be heterozygous for STRs with more alleles and higher levels of heterozygosity. The recombinant- and parental-type offspring may be able to be distinguished in individuals heterozygous for an STR, making crosses informative for mapping studies.

15.3 Abbreviations used in genomics typically facilitate the quick and easy representation of longer tongue-twisting terms. Explore the nuances associated with some abbreviations by answering the following questions: Could a RFLP, VNTR or STR be identified as a SNP? Why or why not?

Answer: A SNP is a single nucleotide polymorphism. Since a single base change can alter the site recognized by a restriction endonuclease, a SNP can also be a RFLP. Since STRs and VNTRs are based on variable number of tandemly repeated sequences (1 to 4 and 5 to 10s of base pairs, respectively), they will not usually be SNPs.

15.4 Genomes can be sequenced using a direct shotgun approach or a mapping approach.

a. What is the difference between these approaches?
b. What limitations have traditionally been associated with the direct shotgun approach, and how have they been overcome?

Answer:

a. In a direct shotgun approach, the genome is broken into partially overlapping fragments, each fragment is cloned and sequenced, and the genome sequence is assembled using computerized algorithms. The Celera group used a modification of this approach to sequence the human genome. They robotically sequenced 300 to 500 bp at the ends of many clones, keeping track of which sequences were from the ends of the same clone. In total, they obtained about five times the amount of sequence of the entire genome. This ensured that they would have sufficient overlap among sequences so that they could assemble a nearly complete (97 percent) sequence.

   In a mapping approach, high-resolution genetic and physical maps of the genome are obtained and integrated. This allows the position of DNA markers to be mapped onto clone contigs and gives an ordered assembly, without any gaps, of partially overlapping clones that comprise the DNA of each chromosome. For the human genome, this was done using BAC clones. The sequence of each BAC clone is obtained using a shotgun approach: The insert of the clone is cut out, sheared mechanically into a partially overlapping set of fragments appropriate for sequencing, cloned into a plasmid, the plasmid inserts sequenced, and the BAC clone sequence assembled using computerized algorithms. The contiguous sequence of each chromosome is obtained by assembling the sequences of the BAC contig.

b. The direct shotgun approach was used initially only for small genomes of less than 100 kb. This is because it was time consuming and expensive to sequence the huge number of clones that would be required in a larger genome and because it was difficult to accurately assemble the huge number of relatively small 300-to-500-bp sequence reads that would be obtained in larger genomes. It was difficult to assemble sequence reads because of the frequency of repetitive DNA sequences and because of the lack of sophisticated computer algorithms.

Two technical advances and several strategic changes allowed these difficulties to be overcome. The technical advances consisted of the development of sophisticated computer algorithms for assembling sequences and the development of robotic procedures for DNA sequencing. Strategically, the Celera group realized that they could obtain more complete sequence, even for regions that had some repetitive DNA sequences, if they sequenced the ends of many partially overlapping clones from two different libraries, one with 2-kb inserts and one with 10-kb inserts. Sequencing of the ends of the clones with 2-kb inserts and keeping track of pairs of sequence reads from one clone allowed them to assemble contigs covering most of the genome. However, since many repeated sequences are about 5 kb in length, one end of a clone with a 2-kb insert may be unique sequence, while the other end may lie in a repetitive sequence. The repetitive sequence will be found in many clones in the library and come from regions dispersed throughout the genome. Therefore, if one only had sequence data from the 2-kb insert clones, it would be impossible to determine, for a clone with a repetitive sequence at one end, which other clone is the true overlapping one from the genome. Now, if the repeat is about 5 kb in length, it can be found entirely within some of the 10-kb clones, so that some of them will have unique sequences flanking a repeated DNA sequence. Therefore, sequencing the ends of clones from the 10-kb library will circumvent the problem associated with this type of repetitive DNA sequence. Sequence information from the clones with 10-kb inserts also provides independent confirmation of contig-sequence structure.

15.5 How has genomic analysis provided evidence that Archaea is a branch of life distinct from Bacteria and Eukarya?

Answer: Sequencing of the genomes of Archaea has shown that their genes are not uniformly similar to those of either Bacteria or Eukarya. While most of the archaeon genes involved in energy production, cell division, and metabolism are similar to their counterparts in Bacteria, the genes involved in DNA replication, transcription, and translation are similar to their counterparts in Eukarya.

15.6 How do the following compare between *S. cerevisiae* and *C. elegans?*

a. the percentage of genes with introns

b. the number of tRNA genes

c. the number of ORFs
d. the (approximate) percentage of the genome devoted to exons

Answer:

a. In *S. cerevisiae,* there are 6,183 ORFs, 120 to 150 rRNA genes, 37 snRNA genes, and 262 tRNA genes. Of these 6,602 to 6,632 genes, 233 ORFs, and 80 tRNA genes have introns, so about 4.7% have introns. In *C. elegans,* each gene has an average of five introns, so this percentage is close to (or perhaps at) 100%.
b. In *S. cerevisiae,* there are 262 tRNA genes. In *C. elegans,* there are 659 tRNA genes, about 2.5-fold more.
c. In *S. cerevisiae,* there are 6,183 ORFs. In *C. elegans,* there are 19,099 ORFs, about 3.1-fold more.
d. In *S. cerevisiae,* about 70% of the genome is devoted to ORFs, 95.3% of which have no introns, so about 70% of the genome is devoted to exons. In *C. elegans,* only about 27% of the genome is devoted to exons.

15.7 What is bioinformatics, and what is its role in structural, functional, and comparative genomics?

Answer: Bioinformatics is the use of computers and information science to study genetic information and biological structures. It results from the fusion of biology, mathematics, computer science and information science. It has become an important field with the widespread use of databases holding sequence, genetic, functional, structural (e.g., protein structure), and bibliographic information. In structural genomics, it is used to assemble genome (and other) sequences, find genes within databases containing sequence and genetic information, align sequences to determine their degree of matching, predict gene structure, and identify the locations of putative genes. In functional genomics, it is used to predict the function of genes based on homology to DNA or amino acid sequences and to analyze information on gene function (e.g., databases containing information on how the function of a gene can alter the transcriptome or proteome). In comparative genomics, it is used to compare homologous gene sequences and compare genomes in terms of similarities and differences between gene and nongene sequences.

15.8 What is the difference between a gene and an ORF? How might you identify the functions of ORFs whose functions are not yet known?

Answer: At the physical level, a gene is a sequence of DNA that includes both a transcribed DNA sequence that is an RNA gene product (for rRNA, snRNA, and tRNA) or used to direct the synthesis of a protein (for mRNA) and the regulatory sequences that direct its transcription (e.g., the promoter). At a functional level, genes are associated with the functions of their products and by the phenotypes of mutations that alter or eliminate the functions of their products. In contrast, an ORF is a potential open reading frame, the segment of a mRNA (mature mRNA in eukaryotes) that directs the synthesis of a polypeptide by the ribosome. Therefore, genes have both transcribed and untranscribed sequences and, in eukaryotes, introns, which ORFs do not. For some ORFs, we have experimental evidence (cDNA sequence, detected protein products) that they exist. For other ORFs, we have predicted their existence based on genomic sequence information, but lack experimental proof.

To determine the function of ORFs whose functions are not yet known, two broad approaches can be used. First, homology searches can be performed. In these, computer-based comparisons are made between the ORF sequence and all sequences (DNA or protein) in a database. The extent of sequence similarity is used to infer whether the ORF encodes a protein with the same or similar function to that of a gene in the database. Since proteins can have multiple functional domains (e.g., a catalytic domain and a DNA-binding domain), sometimes this can give insight into the partial function of the protein product of the ORF. Second, an experimental approach can be used. In some organisms such as yeast or mice, this may involve "knocking out" the function of the gene by deleting it and analyzing the resulting mutant phenotype. In organisms such as humans, where this experimental approach would be unethical, a homologue in a model organism may be investigated instead. Alternate approaches may involve demonstrating that the ORF encodes a protein and characterizing the protein product and its interactions within cells (e.g., using the yeast two-hybrid system described in Chapter 14).

15.9 Once a genomic region is sequenced, computerized algorithms can be used to scan the sequence to identify potential ORFs.

a. Devise a strategy to identify potential prokaryotic ORFs by listing features assessable by an algorithm checking for ORFs.
b. Why does the presence of introns within transcribed eukaryotic sequences preclude direct application of this strategy to eukaryotic sequences?
c. How might you modify the strategy to overcome the problems posed by the presence of introns in transcribed eukaryotic sequences?

Answer:

a. Since prokaryotic ORFs should reside in transcribed regions, they should follow a bacterial promoter, containing –35 (TTGACA) and –10 (TATAAT) consensus sequences. Within the transcribed region, but before the ORF, there should be a Shine-Dalgarno sequence (UAAGGAGG) used for ribosome binding. Nearby should be an AUG (or GUG, in some systems) initiation codon. This should be followed by a set of in-frame sense codons. The ORF should terminate with a stop (UAG, UAA, UGA) codon.
b. Eukaryotic introns introduce nontranslated sequences into the coding region of a gene. These sequences will be spliced out of the primary mRNA transcript before it is translated. Their presence in eukaryotic DNA must be taken into account, as the reading of introns as coding sequences would introduce amino acids not normally found in eukaryotic proteins, as well as frameshifts and chain termination signals.
c. Eukaryotic introns typically contain a GU at their 5′ ends, an AG at their 3′ ends, and a YNCURAY branch-point sequence 18 to 38 nucleotides upstream of their 3′ ends. To identify eukaryotic ORFs in DNA sequences, scan sequences following a eukaryotic promoter for the presence of possible introns by searching for sets of these three consensus sequences. Then try to translate sequences obtained if potential introns are removed, testing whether a long ORF with good codon usage can be generated. Codon usage refers to the finding that individual species vary in terms of which of the degenerate set of codons they use. For example, there are six codons for the amino acid arginine. A specific organism will use some of these more frequently than others. A more difficult problem arises since alternative mRNA splicing exists at many genes. In these cases, more than one possible ORF may be found in a given DNA sequence. It

is not yet possible to predict how a set of exons at a gene might be alternatively spliced solely from knowledge of the DNA sequence.

15.10 Annotation of genomic sequences makes them much more useful to researchers. What features should be included in an annotation, and in what different ways can they be depicted? For some examples of current annotations in databases, see the following websites:

http://genome-www.stanford.edu/Saccharomyces/maps.html (*S. cerevisiae*)
http://www.flybase.org and http://flybase.bio.indiana.edu:82/annot/ (*Drosophila*)
http://www.tigr.org/tbd/ath1/htmls/ath1.html (*Arabidopsis*)
http://www.ncbi.nlm.nih.gov/genome/guide/ (humans)
http://www.cybergenome.com/ (compilation of links to many organisms)

Answer: An annotation can vary in its level of completeness, depending on what can be inferred based on homology and what experimental data are available. An annotation could include:

1. The location of a transcribed region—graphically depicted with links embedded within the symbols used to depict transcripts
   a. Physical map coordinates
      i. Links to clones, DNA markers in the region
   b. Genetic map coordinates
      i. Links to genes, mapping data in the region
   c. Location of introns, exons, alternative splice sites, alternative promoters, poly(A) addition sites, etc.
      i. Physical map coordinates
      ii. Graphical depictions
2. The types of evidence for a transcribed region
   a. Prediction based on computer algorithms that assess possible sequences of a promoter, conceptual ORF(s), appropriate splicing sites, homology with other genes, etc.
   b. Documentation of splice sites, alternative transcript forms by comparison of cDNA and genomic sequences, etc.
3. The inferred function of the gene product
   a. Links to reports, database entries, publications
   b. Evidence supporting the inferred function
      i. Inferences based on homology
      ii. Experimental evidence
4. Information about gene regulation
   a. Genes regulated in a similar manner
   b. Unique features of the gene and its regulation

15.11 One of the central themes in genetics is that an organism's phenotype results from an interaction between its genotype and the environment. Because some diseases have strong environmental components, researchers have begun to assess how disease phenotypes arise from the interactions of genes with their environments, including the genetic background in which the genes are expressed. (See http://pga.tigr.org/desc.shtml for additional discussion.) How might DNA microarrays be useful in a functional genomic approach to understanding human diseases that have environmental components, such as some cancers?

Answer: One approach is to use model organisms (e.g., transgenic mice) that have been developed as models to study a specific human disease. Expose them and a control population to specific environmental conditions, and then simultaneously assess disease progression and alterations in patterns of gene expression using microarrays. This would provide a means to establish a link between environmental factors and patterns of gene expression that are associated with disease onset or progression.

15.12 How does a cell's transcriptome compare with its proteome?

a. For a specific eukaryotic cell, can you predict which has more total members? Can you predict which has more unique members?
b. Suppose you are interested in characterizing changes in the pattern of gene expression in the mouse nervous system during development. Describe how you would efficiently assess changes in the transcriptome from the time the nervous system forms during embryogenesis to its maturation in the adult.
c. How would your analyses differ if you were studying the proteome?

Answer: The transcriptome is the set of RNAs present in a cell at a particular stage and time, while the proteome is the set of proteins present in the cell at that stage and time.

a. It is likely that the proteome has both more total as well as unique members. It is likely that it has more total members since multiple copies of a protein can be translated from a single mRNA transcript. It is also likely that it has more unique members since a given transcript can in principle give rise to different protein isoforms. Once translated, proteins can be posttranslationally or cotranslationally modified in different ways: phosphorylation, glycosylation, methylation, proteolytic processing, etc. If proteins translated from a single transcript are modified in different ways, multiple protein isoforms are produced.
b. This analysis could be performed using DNA chips. RNA could be isolated from the nervous system of different developmental stages, and the transcriptional profile in each stage could be assessed using DNA chips. To do this type of analysis, RNA from one developmental stage would be labeled with one fluorochrome—say a fluorescent green dye. RNA from a second developmental stage would be labeled with another fluorochrome, say a fluorescent red dye. The labeled probes would be hybridized to the DNA chip, and the relative red:green fluorescence bound to a single site on the chip can be used to infer the relative amounts of gene expression of the gene located at that site on the chip. This could be done for many different developmental stages.
c. For the proteome, one would need to assess changes in the proteins produced over time. Thus, one would isolate proteins from the nervous system of different developmental stages and assess the relative abundance of individual proteins using an analytical method such as quantitative two-dimensional gel electrophoresis.

15.13 Distinguish between structural, functional, and comparative genomics by completing the following exercise. The list describes specific activities and goals associated with genome analysis. Indicate the area associated with the activity or goal by placing a letter (S, structural; F, functional; C, comparative) next to each item. Some items will have more than one letter associated with them.

__________	Aligning DNA sequences within databases to determine the degree of matching
__________	Annotation of sequences within a sequenced genome
__________	Characterizing the transcriptome and proteome present in a cell at a specific developmental stage or in a particular disease state
__________	Comparing the overall arrangements of genes and nongene sequences in different organisms to understand how genomes evolve
__________	Describing the function of all genes in a genome
__________	Determining the functions of human genes by studying their homologues in nonhuman organisms
__________	Developing a comprehensive two-dimensional polyacrylamide gel electrophoresis map of all proteins in a cell
__________	Developing a physical map of a genome
__________	Developing DNA microarrays (DNA chips)
__________	Identifying homologues to human disease genes in organisms suitable for experimentation
__________	Identifying a large collection of simple tandem repeat or microsatellite sequences to use as DNA markers within one organism
__________	Making gene knockouts and observing the phenotypic changes associated with them

Answer:

S, F, C	Aligning DNA sequences within databases to determine the degree of matching
F, C	Annotation of sequences within a sequenced genome
F	Characterizing the transcriptome and proteome present in a cell at a specific developmental stage or in a particular disease state
C	Comparing the overall arrangements of genes and nongene sequences in different organisms to understand how genomes evolve
F	Describing the function of all genes in a genome
F, C	Determining the functions of human genes by studying their homologues in nonhuman organisms
F	Developing a comprehensive two-dimensional polyacrylamide gel electrophoresis map of all proteins in a cell
S	Developing a physical map of a genome
S, F, C	Developing DNA microarrays (DNA chips)

F, C ______ Identifying homologues to human disease genes in organisms suitable for experimentation

S ______ Identifying a large collection of simple tandem repeat or microsatellite sequences to use as DNA markers within one organism

F, C ______ Making gene knockouts and observing the phenotypic changes associated with them

15.14 Genomic analyses of *Mycoplasma genitalium* have not only identified its 517 genes but also allowed identification of the genes that this organism needs for life. The analyses are presented in detail at the website of the Institute for Genomic Research (http://www.tigr.org/minimal/). Explore this website and then address the following questions.

a. How many protein-coding genes have unknown functions?
b. How do we know which genes are required for life?
c. Are any of the protein-coding genes with unknown functions required for life?
d. What functional genomics approaches might be helpful to discern their function?
e. What practical value might there be in identifying the minimal gene set needed for life?
f. Is it possible to synthesize DNA sequences containing all the genes identified as essential for life and assemble these on one chromosome? Would this be enough to have a template for a new life form? If not, what additional sequence information would be needed to generate a new life form?
g. What is the difference between the life form generated in (f) and *Mycoplasma genitalium?*
h. What ethical issues arise from these analyses? Who should decide how to address them?

Answer:

a. There are about 180 genes with no known homologues outside of the mycoplasmas, which makes it difficult to assign their function.
b. Both theoretical and experimental paradigms can be used to provide evidence for which genes are required for life.

   A theoretical approach is to estimate how many genes are conserved across large phylogenetic distances, making the assumption that highly conserved genes provide essential functions. Comparison of the *Haemophilus influenzae* and *M. genitalium* genomes suggest that about 265 genes are needed. Comparison of two *Mycoplasma* species (*M. pneumoniae* and *M. genitalium*) shows that *M. pneumoniae* has all 480 of the *M. genitalium* protein-coding genes, plus an additional 197 genes.

   To test which of these 480 are essential for life, an experimental approach can be used. One paradigm is to ask whether a mutation that "knocks out" a gene causes a visible mutation, affects the organism's ability to grow, or affects its viability (ability to live). Hutchison et al. used transposons (DNA segments capable of moving and inserting themselves in other sites) to estimate the size of the minimal mycoplasma genome. Their work suggested that between 265 and 300 of the 480 protein-coding genes were essential under laboratory conditions for growth. It is unclear whether any of the genes not essential under laboratory conditions might still be essential for growth in a human host. Thus, the approach begs the question—what "quality" of life?—or "life" under what conditions?

c. Yes. There are 180 genes without a functional assignment. While 69 of these could be disrupted by a transposon insertion without affecting the viability of the organism, 111 could not. This suggests that at least some of them are not dispensable, since if they were dispensable, transposon disruptions in them would have been recovered. The inference from this experiment is that at least some of them must encode essential cellular functions.

d. Multiple functional genomic approaches could be helpful to infer their function. Note that while the knock-out approach using transposon disruption suggested that these genes have an essential function, the results do *not* provide clues as to what they are. To understand the functions of individual genes, it might be helpful to understand when they are transcribed and whether they are coordinately transcribed with other genes during the life cycle of the organism. This could provide some information about the cellular pathways utilizing these genes and so indirectly allow inferences as to their function. For this purpose, gene chips could be used to assess their expression profiles during growth or under different experimental conditions. For example, one might glean some information about the function of a gene by learning that its expression was up- or down-regulated during a particular growth phase or in a mutant strain where a known function was disrupted. An alternate approach would be to generate mutations in a single gene (whose function is unknown) that did not disrupt its function entirely (e.g., by using a chemical mutagen) and then assess which genes had altered expression patterns as a result.

e. In principle, this would allow the design of cells that could be used as bioreactors to synthesize specific complex biological products that could be useful in improving health, crops, or some other aspect of human existence. It might also provide a basis for designing cells that could be used in implementing some biological or chemical process (e.g., as agents to mediate gene therapy, as agents that could be designed with features to use in toxic waste cleanup, as agents for germ warfare).

f. It is not likely that the assembly of the sequences, by itself, would be sufficient for generating a template for a new life form. Encoded in even a simple genome would have to be instructions for regulating the expression of its genes. Even in the smallest bacteriophage, there are enough instructions encoded to allow for the ordered synthesis of a set of genes and the propagation of a life cycle. In a "cell" there must be at least these types of instructions, which would occur minimally at the level of transcriptional control.

g. *M. genitalium* is able to respond to its environment and grow as an autonomous cell. It has a set of genes that are "fitted" into regulatory circuits that interact and have the capacity to feedback into each other. The "life form" generated in (f) may make a set of gene products, but this may not "make" a cell. The cell is more than just the sum of the proteins it possesses.

h. There are multiple ethical issues. One can list some of them as a series of unanswered questions.
    1. Is there significant potential for the generation of a biohazard that could adversely affect existing life?
    2. How does the possibility of generating a "living" cell affect our religious and moral beliefs about the origin and sanctity of life?

3. What are the benefits of pursuing this line of inquiry, relative to our goals for improving human health or the environment?
4. Do we know enough about the safety, benefits, and risks to venture into such an undertaking?
5. Since it is difficult to understand what the risks of this venture are, how does one evaluate how to proceed?

It would be important that scientists, ethicists, the public, and representative groups who might be affected by this work be part of a team that discussed how to proceed and how to address ethical and other (legal, economic, biohazard) issues that might arise from these investigations. Even with this varied level of input, it might not be possible to anticipate all of the consequences of these types of investigations. There are both pragmatic as well as more philosophical concerns.

16

REGULATION OF GENE EXPRESSION IN BACTERIA AND BACTERIOPHAGES

CHAPTER OUTLINE

REVIEW OF KEY TERMS, SYMBOLS, AND CONCEPTS

In your own words, write a brief, precise definition of each term in the groups on the following page. Check your definitions using the text. Then develop a concept map using the terms in each list.

THINKING ANALYTICALLY

This chapter presents how the expression of bacterial and bacteriophage genes is regulated. While the regulation of a particular set of genes can be complex, it is invariably based on a general paradigm: The structural organization of a set of genes is related to both their induction by effector molecules and subsequent expression (as diagrammed in Figure 16.1). As you study the regulation of each set of genes, keep this general principle in mind. Reconsider it with each of the operons; analyze how it is used and elaborated upon.

Once you have a solid understanding of the general principle underlying prokaryotic gene expression, the challenge is to understand and retain the complex details of the regulatory circuitry. It will help to first study the text figures and then diagram (eventually, from memory) the structure of each operon and the role of regulatory factors within it. This is especially valuable

1	2	3
regulated genes	*lac, trp* operons	lambda (λ) phage
constitutive genes	*lacA, -Z, -Y, -I* genes	lytic pathway
inducible genes	negative, positive control	lysogenic pathway
inducers and effectors	allolactose	genetic switch
controlling site	partial diploids	lambda repressor
housekeeping genes	cis-dominance	*cro* gene
induction	trans-dominance	*recA* gene
coordinate induction	O^c, I^-, I^s, I^{-d} mutations	antiterminator
operon	superrepressor	P_R, P_{RE}, P_{RM}, P_L, P_I
polycistronic mRNA	allosteric shift	O_R, O_L
polygenic mRNA	catabolite repression	antitermination
effector molecule	gluclose effect	readthrough transcription
operator	CAP-cAMP complex	integrase
repressor molecule	CAP site	*cI, cII, cIII* genes
promoter	effector	Q, Cro proteins
	aporepressor	
	attenuation	
	attenuator site	
	pause signal	
	antitermination	
	feedback inhibition	
	inducible, repressible operon	

when considering the regulation of genes in phage λ. This will take time, as it requires repeated practice and comparative review.

You can strengthen your understanding of the principles underlying prokaryotic gene expression and your detailed knowledge of the regulatory circuitry of operons by examining the consequences of mutations in different operon elements. First, address why mutations in regulatory elements such as the promoter or operator will be cis-dominant. Second, address why some mutations in the genes for diffusible regulatory factors *can* be trans-dominant. Third, consider how the properties of different mutations in a specific operon can provide you with a basis to develop a model of its structure. Finally, relate the regulation of the operon to how it functions within a biochemical or growth pathway.

QUESTIONS FOR PRACTICE

Multiple Choice Questions

1. Genes that respond to the needs of a cell or organism in a controlled manner are known as
   a. inducer genes
   b. effector genes
   c. regulated genes
   d. constitutive genes

2. A mutation that causes a gene always to be expressed, irrespective of what the environmental conditions may be, is known as a(n)
   a. inducer mutation
   b. effector mutation
   c. regulator mutation
   d. constitutive mutation
3. A gene that is stimulated to undergo transcription in response to a particular molecular event that occurs at a controlling site near that gene is said to be
   a. inducible
   b. constitutive
   c. a promoter
   d. an inducer
4. An operon that is inducible may be under
   a. positive control only
   b. negative control only
   c. constitutive control only
   d. both positive and negative control
5. The effector molecule that induces the *lac* protein-coding genes is
   a. lactose
   b. allolactose
   c. glucose
   d. β-galactosidase
6. Before transcription of the *lac* operon can occur, RNA polymerase must bind strongly to the promoter. This happens when
   a. CAP binds to the CAP site in the promoter
   b. a CAP-cAMP complex binds to the CAP site in the promoter
   c. catabolite repression occurs
   d. $lacI^+$ is mutated to $lacI^-$
7. Which of the following is an example of an effector molecule acting via positive control?
   a. lactose inducing the *lac* operon
   b. glucose causing catabolite repression
   c. tryptophan attenuating the *trp* operon
   d. the λ *cI* gene product, at a certain cellular level, favoring the lysogenic pathway
8. Bacterial operons containing protein-coding genes for the synthesis of amino acids, such as the tryptophan operon, are customarily classified as
   a. negatively controlled
   b. positively controlled
   c. repressible operons
   d. inducible operons
9. Ultraviolet light induces integrated phage λ to enter the lytic pathway by
   a. cleaving repressor monomers
   b. initiating transcription of the *cro* gene
   c. converting the RecA protein to a protease
   d. all of the above, directly or indirectly

10. Which of the following is *not* critical to the genetic switch controlling the choice between the lysogenic and lytic pathways in phage λ?
    a. the competition between the products of the *cI* gene and the *cro* gene
    b. transcription of the genes *J, K, L,* and *M*
    c. the concentration of λ repressor protein dimers
    d. whether RNA polymerase binds to P_{RM} or P_R

Answers: 1c, 2d, 3a, 4d, 5b, 6b, 7b, 8c, 9d, 10b

Thought Questions

1. What three proteins are synthesized when lactose is the sole carbon source in *E. coli?* What does each do?
2. What are the roles of *lacA, lacI, lacO, lacY,* and *lacZ* in *E. coli* carbohydrate metabolism? In what order are they arranged in the DNA molecule?
3. Describe the sequence of events that occurs in the *lac* operon when *E. coli* is grown in the presence of both glucose and lactose.
4. Distinguish between cis-dominant mutations and trans-dominant mutations. Why is $lacO^c$ cis-dominant but $lacI^s$ trans-dominant? Would a mutation that led to constitutive expression of the *cI* gene be cis- or trans-dominant?
5. How is attenuation related to the coupling of transcription and translation in prokaryotes?
6. What are the fundamental differences between the *lac* and *trp* operons in *E. coli?*
7. How do structural changes in DNA or RNA conformation play a role in:
    a. catabolite repression by glucose and CAP-cAMP?
    b. antitermination and termination in attenuation of the *trp* operon?
8. What is catabolite repression?
9. In what different or similar ways are allosteric shifts used in the regulation of inducible and repressible operons?
10. Speculate as to what phenotype each of the following λ mutations would have: (a) a mutation in the *cI* gene that resulted in repressor molecules that bound O_L and O_R less well; (b) a mutation in the *cro* gene that resulted in no Cro protein being produced; (c) a mutation in O_R that resulted in decreased affinity for repressor molecules.

Thought Questions for Media

After reviewing the media at the *Essential iGenetics* website, try to answer these questions.

1. In the *lac* operon, is the operator wholly within, wholly outside of, or partially within the promoter?
2. How could you distinguish between promoter and operator mutations?
3. An operon is defined as a cluster of genes along with an adjacent promoter and operator that controls the transcription of those genes.
    a. Must all operons be polycistronic?
    b. Must all of the genes in one operon be transcribed together?
4. How does lactose mediate de-repression of the *lac* operon? Does it act directly?
5. Which type of mutation in the *lac* operon causes the production of β-galactosidase whether or not lactose is present?

6. Why does it make sense for *E. coli* to transcribe the *lac* operon at a very slow rate in the presence of glucose? (Consider the products of β-galactosidase enzymatic activity.)
7. What phenotype would you expect of a mutation in the CAP site in the *lac* promoter?

1. In the iActivity, the strain containing the O^c mutation ($I^+ O^c Z^+$) showed constitutive expression both by itself and as a merozygote of the form $I^+ O^c Z^+/F' I^+ O^+ Z^+$. Would the phenotype be different if the merozygote had a genotype that was either $I^+ O^c Z^+/F' I^+ O^+ Z^-$ or $I^+ O^c Z^+/F' I^- O^+ Z^+$? For each case, why or why not?
2. In the iActivity, the strain containing the I^s mutation ($I^s O^+ Z^+$) showed repressed expression both by itself and as a merozygote of the form $I^s O^+Z^+/F' I^+O^+Z^+$. Would the phenotype be different if the merozygote had a genotype that was either $I^s O^+ Z^+/F' I^+ O^+ Z^-$ or $I^s O^+ Z^+/F' I^- O^+ Z^+$? For each case, why or why not?
3. In the iActivity, you analyzed data showing that the *lac* operon was inducible and under negative control.
   a. In theory, if you can experimentally demonstrate that an operon is inducible, must it also be under negative control?
   b. Can an inducible operon be under both negative and positive control?
   c. Can an inducible operon be under only positive control?

SOLUTIONS TO TEXT PROBLEMS

16.1 How does lactose bring about the induction of synthesis of β-galactosidase, permease, and transacetylase? Why does this event not occur when glucose is also in the medium?

Answer: The addition of lactose to *E. coli* cells brings about a rapid synthesis of these three enzymes by the induction of a single promoter that lies upstream of the genes for these three enzymes. The three genes are part of a *lac* operon that is transcribed as a single unit. When lactose is added, it is metabolized (isomerized) to allolactose, which binds to a repressor protein. Without bound allolactose, the repressor protein blocks transcription from the *lac* promoter. Hence, the *lac* operon is normally under a negative control mechanism. When allolactose is bound, the repressor protein is inactivated and is unable to bind to the operator site to block transcription. As a result, RNA polymerase binds to the promoter and initiates transcription of a single mRNA that encodes all three proteins.

One of the enzymes that is synthesized, β-galactosidase, cleaves lactose to produce glucose and galactose (which is converted to glucose in a subsequent enzymatic step). Consequently, if glucose is present in the medium, it is redundant to induce the *lac* operon. Glucose blocks induction of the *lac* operon by utilizing a positive control mechanism. In this catabolite repression, glucose causes a great reduction in the amount of cAMP in the cell. For normal induction of the *lac* operon, cAMP must complex with a CAP (catabolite gene activator protein) that in turn binds to a CAP site upstream of the *lac* promoter and activates transcription. In the absence of cAMP, the cAMP-CAP complex is absent, and so transcription cannot be activated.

16.2 Operons produce polygenic mRNA when they are active. What is a polygenic mRNA? What advantages, if any, does it confer in terms of the cell's function?

Answer: Polygenic mRNAs contain coding information for more than one protein. These mRNAs are transcribed from operons that contain several genes encoding related functions, such as catalyzing steps of a biosynthetic pathway. One advantage conferred by utilizing such mRNAs is that cells can regulate all of the steps of a pathway coordinately. By using a polygenic mRNA, the synthesis of each of a set of enzymes acting in one pathway can be produced by a single regulatory signal.

16.3 If an *E. coli* mutant strain synthesizes β-galactosidase whether or not the inducer is present, what genetic defect(s) might be responsible for this phenotype?

Answer: One possibility is that the repressor protein bound by the inducer cannot bind to the operator (i.e., the repressor is nonfunctional and the strain is I^-), so that its presence or absence makes no difference. Another possibility is that there are base pair alterations in the operator region that make it unrecognizable by the repressor protein (i.e., the operon is constitutively expressed and the strain is O^c).

16.4 Elucidation of the regulatory mechanisms associated with the enzymes of lactose utilization in *E. coli* was a landmark in our understanding of regulatory processes in microorganisms. In formulating the operon hypothesis as applied to the lactose system, Jacob and Monod found that results from particular partial diploid strains were invaluable. Specifically, in terms of the operon hypothesis, what information did the partial diploids provide that the haploids could not?

Answer: The use of partial diploids allowed observation of the consequences of placing sequences in *trans* and in *cis*. Partial diploids were used to show that some regulatory sequences must lie in *cis* to *lacZ* (the *lacO* region upstream of the *lacZ* gene) to exert a regulatory effect. For example, O^c mutations are *cis*-acting: They cause constitutive activation of the *lac* promoter they lie upstream from, but do not activate any other promoter. Partial diploids also were used to show that the *lacI* gene encoded a *trans*-acting factor (a diffusible product that could bind to the *lacO* region) and that promoter function did not require a diffusible substance. For example, a $lacI^+$ gene on a plasmid could function in *trans* to regulate a *lac* promoter in *cis* to a $lacI^-$ gene.

16.5 For the *E. coli lac* operon, write the partial diploid genotype for a strain that will produce β-galactosidase constitutively and permease by induction.

Answer: To observe such a phenotype, there are three requirements: (1) A functional *lacZ* and a nonfunctional *lacY* gene must both lie downstream of a functional operator and promoter; (2) a nonfunctional *lacZ* and a functional *lacY* gene must lie downstream of an inducible (i.e., O^+) promoter; and (3) a functional repressor gene must be present in the cell (I^+). A genotype that satisfies these requirements is the partial diploid $lacI^+\ lacO^c\ lacP^+\ lacZ^+\ lacY^-/lacI^+\ lacO^+\ lacP^+\ lacZ^-\ lacY^+$. Only one $lacI^+$ gene is required, so one may be $lacI^-$.

16.6 Mutants were instrumental in elaborating the model for regulation of the *lac* operon.

a. Discuss why $lacO^c$ mutants are cis-dominant but not trans-dominant.

b. Explain why *lacI*^{s} mutants are trans-dominant to the wild-type *lacI*$^+$ allele but *lacI*$^-$ mutants are recessive.

c. Discuss the consequences of mutations in the repressor gene promoter as compared with mutations in the structural gene promoter.

Answer:

a. *lacO*^{c} mutants are mutants in the operator region that is normally bound by the repressor protein. *lacO*^{c} mutants result from a DNA alteration that precludes the repressor protein from binding the operator. Since the repressor normally blocks transcription initiation from a downstream promoter, *lacO*^{c} mutants result in constitutive transcription at that promoter. Since the operator acts only on an adjacent, and not any other, promoter, mutations in the operator are only cis-dominant. The *lacO*^{c} has no effect on other lactose operons in the same cell because *lacO*^{c} does not code for a product that could diffuse through the cell and affect other DNA sequences.

b. In wild-type strains, *lacI* encodes a repressor that can block transcription at the *lac* operon by binding to an operator region. By binding the operator, the repressor blocks RNA polymerase from binding to the promoter. This activity of the repressor can be altered if allolactose is present, which binds to the repressor and inhibits it from binding the operator. The superrepressor (*lacI*^{s}) mutation results in a repressor protein that can bind the operator, but cannot bind allolactose. Once it binds to the operator, it cannot leave the operator, and transcription is always blocked. This results in a dominant mutation, since even if normal repressor molecules (made by *lacI*$^+$) are present, the superrepressor molecules do not vacate the operator region, and the operon cannot be induced. On the other hand, *lacI*$^-$ mutants either do not make repressor protein or make repressor that is unable to bind to the operator. In a partial diploid that has both *lacI*$^-$ and *lacI*$^+$ genes, repressor proteins are made (by *lacI*$^+$) and are capable of diffusing to any *lac* operator region to regulate a *lac* operon. Hence, *lacI*$^-$ is recessive to *lacI*$^+$ because the defect caused by the absence of the repressor proteins in *lacI*$^-$ mutants can be overcome by the synthesis of diffusible repressor protein from the *lacI*$^+$ gene.

c. The *lacI* gene is constitutively expressed at a low level. Mutations in the repressor gene promoter result in either an increase or a decrease in the level of expression of the repressor gene. They will not affect the structure of repressor molecule, but will affect its cellular concentration. If the mutation causes a complete loss of expression of the repressor protein, it would have the same phenotype as an I^- mutant, since no functional repressor would be produced. These mutants would result in constitutive expression of *lacZ*, *lacY*, and *lacA* and be recessive to I^+ in partial diploids. If the promoter mutation only partially decreases transcription below normal levels, there will be increased expression of *lacZ*, *lacY*, and *lacA* in the absence of inducer. If the promoter mutation increases transcription above normal levels, a large number of repressor molecules will be produced. Such mutants will reduce the efficiency of induction of the *lac* operon by allolactose, and will be trans-dominant.

16.7 This question involves the *lac* operon of *E. coli*, where I = *lacI* (the repressor gene), $P = P_{lac}$ (the promoter), O = *lacO* (the operator), Z = *lacZ* (the β-galactosidase gene), and Y = *lacY* (the permease gene). Complete Table 16.A (following page), using + to indicate that the enzyme in question will be synthesized and – to indicate that the enzyme will not be synthesized.

Table 16.A

Genotype	Inducer Absent		Inducer Present	
	β-galactosidase	Permease	β-galactosidase	Permease
a. $I^+ P^+ O^+ Z^+ Y^+$				
b. $I^+ P^+ O^+ Z^- Y^+$				
c. $I^+ P^+ O^+ Z^+ Y^-$				
d. $I^- P^+ O^+ Z^+ Y^+$				
e. $I^s P^+ O^+ Z^+ Y^+$				
f. $I^+ P^+ O^c Z^+ Y^+$				
g. $I^s P^+ O^c Z^+ Y^+$				
h. $I^+ P^+ O^c Z^+ Y^-$				
i. $I^- P^+ O^+ Z^+ Y^+$ / $I^+ P^+ O^+ Z^- Y^-$				
j. $I^- P^+ O^+ Z^+ Y^-$ / $I^+ P^+ O^+ Z^- Y^+$				
k. $I^s P^+ O^+ Z^+ Y^-$ / $I^+ P^+ O^+ Z^- Y^+$				
l. $I^+ P^+ O^c Z^- Y^+$ / $I^+ P^+ O^+ Z^+ Y^-$				
m. $I^- P^+ O^c Z^+ Y^-$ / $I^+ P^+ O^+ Z^- Y^+$				
n. $I^s P^+ O^+ Z^+ Y^+$ / $I^+ P^+ O^c Z^+ Y^+$				
o. $I^+ P^- O^c Z^+ Y^-$ / $I^+ P^+ O^+ Z^- Y^+$				
p. $I^+ P^- O^+ Z^- Y^-$ / $I^+ P^+ O^c Z^- Y^+$				
q. $I^- P^- O^+ Z^+ Y^+$ / $I^+ P^+ O^+ Z^- Y^-$				
r. $I^- P^+ O^+ Z^+ Y^-$ / $I^+ P^- O^+ Z^- Y^+$				

Answer to Table 16A:

Genotype	Inducer Absent		Inducer Present	
	β-galactosidase	**Permease**	**β-galactosidase**	**Permease**
a. $I^+ P^+ O^+ Z^+ Y^+$	−	−	+	+
b. $I^+ P^+ O^+ Z^- Y^+$	−	−	−	+
c. $I^+ P^+ O^+ Z^+ Y^-$	−	−	+	−
d. $I^- P^+ O^+ Z^+ Y^+$	+	+	+	+
e. $I^s P^+ O^+ Z^+ Y^+$	−	−	−	−
f. $I^+ P^+ O^c Z^+ Y^+$	+	+	+	+
g. $I^s P^+ O^c Z^+ Y^+$	+	+	+	+
h. $I^+ P^+ O^c Z^+ Y^-$	+	−	+	−
i. $I^- P^+ O^+ Z^+ Y^+$ / $I^+ P^+ O^+ Z^- Y^-$	−	−	+	+
j. $I^- P^+ O^+ Z^+ Y^-$ / $I^+ P^+ O^+ Z^- Y^+$	−	−	+	+
k. $I^s P^+ O^+ Z^+ Y^-$ / $I^+ P^+ O^+ Z^- Y^+$	−	−	−	−
l. $I^+ P^+ O^c Z^- Y^+$ / $I^+ P^+ O^+ Z^+ Y^-$	−	+	+	+
m. $I^- P^+ O^c Z^+ Y^-$ / $I^+ P^+ O^+ Z^- Y^+$	+	−	+	+
n. $I^s P^+ O^+ Z^+ Y^+$ / $I^+ P^+ O^c Z^+ Y^+$	+	+	+	+
o. $I^+ P^- O^c Z^+ Y^-$ / $I^+ P^+ O^+ Z^- Y^+$	−	−	−	+
p. $I^+ P^- O^+ Z^- Y^-$ / $I^+ P^+ O^c Z^- Y^+$	−	+	−	+
q. $I^- P^- O^+ Z^+ Y^+$ / $I^+ P^+ O^+ Z^- Y^-$	−	−	−	−
r. $I^- P^+ O^+ Z^+ Y^-$ / $I^+ P^- O^+ Z^- Y^+$	−	−	+	−

16.8 A new sugar, sugarose, induces synthesis of two enzymes from the *sug* operon of *E. coli*. Some properties of mutations affecting the appearance of these enzymes are as follows (here, + = enzyme induced normally, i.e., synthesized only in the presence of the inducer; C = enzyme synthesized constitutively; 0 = enzyme cannot be detected):

Mutation of	Enzyme 1	Enzyme 2
Gene *A*	+	0
Gene *B*	0	+
Gene C	0	0
Gene *D*	C	C

a. The genes are adjacent in the order *ABCD*. Which gene is most likely to be the structural gene for enzyme 1?

b. Complementation studies using partial diploid (*F′*) strains were made. The extrachromosomal element (*F′*) and chromosome each carried one set of *sug* genes. The results were as follows:

Genotype of F	Chromosome	Enzyme 1	Enzyme 2
$A^+ B^- C^+ D^+$	$A^- B^+ C^+ D^+$	+	+
$A^+ B^- C^- D^+$	$A^- B^+ C^+ D^+$	+	0
$A^- B^+ C^- D^+$	$A^- B^- C^+ D^+$	0	+
$A^- B^+ C^+ D^+$	$A^+ B^- C^+ D^-$	+	+

From all the evidence given, determine whether the following statements are true or false:

1. It is possible that gene *D* is a structural gene for one of the two enzymes.
2. It is possible that gene *D* produces a repressor.
3. It is possible that gene *D* produces a cytoplasmic product required to induce genes *A* and *B*.
4. It is possible that gene *D* is an operator locus for the *sug* operon.
5. The evidence is also consistent with the possibility that gene C could be a gene that produces a cytoplasmic product required to induce genes *A* and *B*.
6. The evidence is also consistent with the possibility that gene C could be the controlling end of the *sug* operon (the end from which mRNA synthesis presumably commences).

Answer: First consider the data in general terms. Mutations in genes *A* or *B* result in a loss of one, but not both, enzyme activities. These are likely to be structural genes for the enzymes (*B* = enzyme 1, *A* = enzyme 2). The mutations in genes C and *D* result in loss of both enzyme activities, suggesting that these genes or regions regulate or affect both genes.

a. Gene *B* is likely to be the gene for enzyme 1, since only a mutation in gene B produced a loss of enzyme 1 activity with no effect on enzyme 2 activity. (By similar logic, gene A codes for enzyme 2.)

b. 1. False. A mutation in D leads to the constitutive synthesis of both enzymes 1 and 2, so D cannot be a structural gene for either enzyme.

2. True. D could encode a repressor. Suppose the repressor acted as the *lac* repressor does in the *lac* operon. If mutations in D inactivated the repressor, then an absence of repressor would lead to constitutive activation of the operon. In this model, D^- mutants would be recessive to D^+ mutants, which is seen in the analysis of partial diploids ($A^-\ B^+\ C^+\ D^+/A^+\ B^-\ C^+\ D^-$ shows inducible activity of both A^+ and B^+).

3. False. If D was needed to induce the *sug* operon, D^- mutants should produce no enzymes. This is not observed.

4. False. One does see that D^- mutants are constitutive, as would be expected if a D^- operator region could not be bound by a repressor to repress transcription. However, not all of the results support this view. If D was an operator, consider what phenotype would be expected in the partial diploid $A^-B^+C^+D^+/A^+\ B^-\ C^+\ D^-$. If D^- was a defective operator, this partial diploid would express the A^+ gene (enzyme 2) constitutively, in a cis-dominant manner, and not in an inducible manner. This is not seen. What is seen is a trans-dominant effect where the A^+ gene is inducible. Thus, D is not an operator.

5. False. Since the products of genes A and B are not inducible in C^- mutants, one might speculate that gene C produced a cytoplasmic product that was required to induce genes A and B. However, consider the partial diploid data. In two of the partial diploids ($A^+\ B^-\ C^-\ D^+/A^-\ B^+\ C^+\ D^+$ and $A^-B^+C^-D^+/A^+\ B^-\ C^+\ D^+$), the wild-type genes on the C^- chromosome are not expressed, even though a C^+ gene is on another chromosome. This indicates that C shows cis-dominance and not trans-dominance. If C^+ encoded a cytoplasmic factor, it would diffuse and be capable of acting in a trans-dominant fashion. Since it does not, C does not encode a cytoplasmic, *trans*-acting factor.

6. True. The cis-dominant effects that C^- mutants show in partial diploids could be explained if the mutations were in the controlling end of the *sug* operon, in a region such as the promoter.

16.9 What consequences would a mutation in the catabolite activator protein (*CAP*) gene of *E. coli* have for the expression of a wild-type *lac* operon?

Answer: The CAP, in a complex with cAMP, is required to facilitate RNA polymerase binding to the *lac* promoter. The RNA polymerase binding occurs only in the absence of glucose, and only if the operator is not occupied by repressor (i.e., lactose is also absent). Suppose the *CAP* gene is mutated so that it could no longer bind to the CAP site. In this case, RNA polymerase would not be able to recognize and bind to the promoter, and the operon would not be expressed. Such a mutation could be distinguished from other mutations that eliminate operon expression (e.g., I^s, P^-) by several means. First, it should be recessive to a normal allele in partial diploids, and not show cis- or trans-dominance. Second, it should map to a location other than the *lac* operon.

16.10 The *lac* operon is an inducible operon, whereas the *trp* operon is a repressible operon. Discuss the differences between these two types of operons.

Answer: Both inducible and repressible operons allow sensitive control of transcription. They differ from each other in the details of how such control is achieved. In the *lac* operon, a repressor protein bound to an operator blocks transcription unless lactose is present. Thus, if lactose is absent, the system is OFF. When lactose is added, it is converted to allolactose and acts as an effector molecule to release the repressor from the operator, so that RNA polymerase can transcribe the operon. In the *trp* operon, the control strategy is the opposite. When tryptophan is abundant in the medium, the operon is turned off, as it is unnecessary to synthesize the enzymes needed to build tryptophan. Tryptophan also acts as an effector molecule. It binds to an aporepressor protein and converts it into an active repressor. This repressor is capable of binding the *trp* operator to reduce transcription of the *trp* operon protein-coding genes by RNA polymerase. Transcription of the *trp* operon is reduced by about 70 percent in the presence of tryptophan, while the aporepressor has no affinity for the operator in the absence of tryptophan.

The *trp* operon also can be regulated by attenuation, a mechanism that controls the ratio of the transcripts that include the five structural genes to those that are terminated before the structural genes. Under conditions where some tryptophan is present in the medium, short 140-bp transcripts are produced, and transcription is attenuated. Under conditions of tryptophan starvation or limitation, full-length transcripts are produced. A model for how attenuation occurs is described in text Figures 16.13 and 16.14.

16.11 In the presence of high intracellular concentrations of tryptophan, only short transcripts of the *trp* operon are synthesized because of attenuation of transcription at a point 5′ to the structural genes. This is mediated by the recognition of two Trp codons in the leader sequence. If these codons were mutated to UAG stop codons, what effect would this have on the regulation of the operon in the presence or absence of tryptophan? Explain.

Answer: For a wild-type *trp* operon, the absence of tryptophan results in antitermination; that is, the structural genes are transcribed and the tryptophan biosynthetic enzymes are made. This occurs because a lack of tryptophan results in the absence of, or at least a very low level of, Trp-tRNA.Trp. In turn, this causes the ribosome translating the leader sequence to stall at the Trp codons (see Figure 16.14a). When the ribosome is stalled at the Trp codons, the RNA being synthesized just ahead of the ribosome by RNA polymerase assumes a particular secondary structure. This favors continued transcription of the structural genes by the polymerase. If the two Trp codons were mutated to stop codons, then the mutant operon would function constitutively in the same way as the wild-type operon in the absence of tryptophan. The ribosome would stall in the same place, and antitermination would result in transcription of the structural genes.

For a wild-type *trp* operon, the presence of tryptophan turns off transcription of the structural genes. This occurs because the presence of tryptophan leads to the accumulation of Trp-tRNA.Trp, which allows the ribosome to read the two Trp codons and stall at the normal stop codon for the leader sequence. When stalled in that position, the antitermination signal cannot form in the RNA being synthesized; instead, a termination signal is formed, resulting in the termination of transcription. In a mutant *trp* operon with two stop codons instead of the Trp codons, the stop codons cause the ribosome to stall, even though tryptophan and Trp-tRNA.Trp are present. This results in an antitermination signal and transcription of the structural genes.

In sum, in both the presence and the absence of tryptophan, the mutant *trp* operon will not show attenuation. The structural genes will be transcribed in both cases, and the tryptophan biosynthetic enzymes will be synthesized. Only if an amber suppressor mutation were present in the cell would attenuation be restored. An amber suppressor mutation results when a tRNA has a mutant anticodon that recognizes the amber (UAG) nonsense codon. In this case, the operon will be attenuated under the control of the amino acid that binds the suppressor tRNA.

16.12 In the bacterium *Salmonella typhimurium,* seven of the genes coding for histidine biosynthetic enzymes are located adjacent to one another in the chromosome. If excess histidine is present in the medium, the synthesis of all seven enzymes is coordinately repressed, whereas in the absence of histidine all seven genes are coordinately expressed. Most mutations in this region of the chromosome result in the loss of activity of only one of the enzymes. However, mutations mapping to one end of the gene cluster result in the loss of all seven enzymes, even though none of the structural genes have been lost. What is the counterpart of these mutations in the *lac* operon system?

Answer: Mutations that cause the loss of activity in just one enzyme are mutations in structural genes, similar to *lacZ* and *lacY* mutations. Mutations that cause loss of all seven enzymes could be mutations in the promoter, similar to P_{lac-} mutations.

16.13 On infecting an *E. coli* cell, bacteriophage λ has a choice between the lytic and lysogenic pathways. Discuss the molecular events that determine which pathway is taken.

Answer: When λ infects a cell and its genome circularizes, phage growth begins when RNA polymerase binds the λ early operon promoters P_L and P_R and transcribes mRNA for the *N* and *cro* genes, respectively. N acts as a transcription antiterminator and extends RNA synthesis to other genes that include the gene *cII.* The cII protein turns on *cI,* the gene for the λ repressor, as well as genes *O* and *P,* whose products are needed for DNA replication, and gene *Q,* whose product is needed for transcription of the late genes used in lysis and to produce phage particle proteins. Once the λ repressor and Cro are produced, there is competition between them to set a genetic switch that determines whether a lysogenic or a lytic pathway will be taken. A lysogenic pathway is taken if the λ repressor dominates, while a lytic pathway is taken if Cro dominates. The events that occur under Cro or repressor domination are diagrammed in text Figure 16.16.

In a cell that will undergo lysogeny, the cII protein (stabilized by the cIII protein) will stimulate transcription of mRNA from P_{RE} and from P_I. This results in synthesis of repressor and of integrase. λ repressor binds to two operators: O_R and O_L. This binding blocks transcription from P_R and P_L and so blocks production of Cro and N. It also stimulates transcription from P_{RM} and the production of additional λ repressor molecules. The stable repression of transcription from P_R together with the integrase-catalyzed integration of the λ chromosome into the *E. coli* host chromosome leads to the establishment of lysogeny. If the λ repressor is not present in high enough concentration or is cleaved and inactivated, the absence of repressor at O_R leads to the binding of RNA polymerase at P_R and transcription of the *cro* gene. When the Cro protein is produced in increasing amounts, it acts to decrease transcription from P_R and P_L, which reduces synthesis of the cII protein and block synthesis of repressor from P_{RM}. By this means, an increase in Cro protein blocks production of the λ repressor. It allows enough transcription from P_R for Q proteins to accumulate and stimulate the late gene transcription that is needed for starting the lytic pathway. In this way, the competition between the Cro and λ repressor proteins for the operator sites determines how the genetic switch is set.

16.14 How do the λ repressor protein and the Cro protein regulate their own synthesis?

Answer: The λ repressor binds to the operator O_R. When the λ repressor is bound, it stimulates synthesis of more repressor mRNA from the promoter P_{RM}. Eventually, however, at very high concentrations of λ repressor, O_R will be bound by λ repressor in a way that blocks further transcription from the promoter P_{RM}. This maintains repressor concentrations in the cell. In the absence of repressor binding to O_R, RNA polymerase transcribes *cro* mRNA from the promoter P_R. As the concentration of Cro protein produced from this mRNA increases, Cro binds to the operator O_R and decreases synthesis from P_R and P_L. Not only does this serve to block synthesis of λ repressor mRNA from P_{RM}, but it will eventually block synthesis of *cro* mRNA as well. Thus, just as the λ repressor regulates the ability of RNA polymerase to transcribe its own *cI* gene from P_{RM}, Cro regulates the ability of RNA polymerase to transcribe *cro* from P_R. See Figure 16.16.

16.15 If a mutation in the phage λ *cI* gene results in a nonfunctional *cI* gene product, what phenotype would you expect the phage to exhibit?

Answer: The *cI* gene product is a repressor protein that acts to keep the lytic functions of the phage repressed when λ is in the lysogenic state. A *cI* mutant strain would lack the repressor and be unable to repress lysis, so that the phage would always follow a lytic pathway.

16.16 Bacteriophage λ can form a stable association with the bacterial chromosome because the virus manufactures a repressor. This repressor prevents the virus from replicating its DNA and making lysozyme and all the other tools used to destroy the bacterium. When you induce the virus with UV light, you destroy the repressor, and the virus goes through its normal lytic cycle. The repressor is the product of a gene called the *cI* gene and is a part of the wild-type viral genome. A bacterium that is lysogenic for λ^+ is full of repressor protein, which confers immunity against any λ virus added to these bacteria. Added viruses can inject their DNA, but the repressor from the resident virus prevents replication, presumably by binding to an operator on the incoming virus. Thus, this system has many elements analogous to the *lac* operon. We could diagram a virus as shown in the following figure. Several mutations of the *cI* gene are known. The c_i mutation results in an inactive repressor.

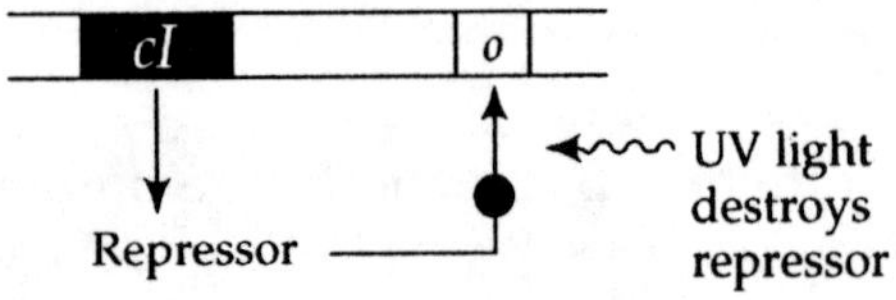

a. If you inject *E. coli* with λ containing a c_i mutation, can it lysogenize (form a stable association with the bacterial chromosome)? Why or why not?

b. If you infect a bacterium simultaneously with a wild-type c^+ and a c_i mutant of λ, can you obtain stable lysogeny? Why or why not?

c. Another class of mutants called c^{IN} makes a repressor that is insensitive to UV destruction. Will you be able to induce a bacterium lysogenic for c^{IN} with UV light? Why or why not?

Answer:

a. No. The repressor is necessary to keep the lytic function of the phage repressed and allow the phage to enter lysogeny. In the presence of a c_i mutation, only the lytic pathway can be taken.

b. Yes. A normal repressor will be made from the wild-type c^+ gene. This repressor is diffusible and will act in a trans-dominant manner to repress lytic growth.

c. No. UV irradiation of lysogenic bacteria destroys repressor function, which in turn leads to induction, including excision of λ and lytic growth. In a c^{IN} mutant, the repressor would not be destroyed following UV irradiation, so that the prophage would not be excised and lysogeny would be retained.

17

REGULATION OF GENE EXPRESSION IN EUKARYOTES

CHAPTER OUTLINE

REVIEW OF KEY TERMS, SYMBOLS, AND CONCEPTS

In your own words, write a brief, precise definition of each term in the groups on the following two pages. Check your definitions using the text. Then develop a concept map using the terms in each list.

THINKING ANALYTICALLY

The complexity of eukaryotic gene regulation has several roots. First, eukaryotic genes and chromosomes are organized differently than their prokaryotic counterparts. Second, the transfer of information from DNA to protein is more complicated in eukaryotes than in prokaryotes, as mRNAs and proteins are processed in additional ways. Third, eukaryotic organisms are capable

1	2	3
transcriptional control	short-term gene regulation	RNA processing control
positive, negative regulatory element/protein	upstream activator sequence	alternative poly(A) sites
promoter, enhancer, silencer	promoter	alternative splice sites
combinatorial gene regulation	positive, regulator, activator	transport control
DNA-binding protein: domain structural motif helix-turn-helix, zinc finger, leucine zipper motif	UAS_G, Gal4, Gal1, Gal10, Gal7	spliceosome retention
transcription initiation	steroid hormone	snRNPs
DNase I–sensitive region	steroid hormone receptor, response element	hnRNA
DNase I–hypersensitive site	chaperone	mRNA translation control
DNase I–hypersensitive region	second messenger	adenylate/uridylate (AU)-rich element (ARE)
histones, nucleosomes	polypeptide hormone	mRNA degradation control
histone acetylation	SHR, HRE, GRE	deadenylation-dependent, deadenylation-independent decay
transcription repression		decapping
DNA methylation		protein degradation control
triplet repeat		proteolysis
FMR-1 gene		ubiquitin
		N-end rule

of substantially more development and differentiation, and the different cell types in a multicellular organism have diverse patterns of gene expression. As you approach this material, keep these roots in mind.

While eukaryotic gene regulation is more complex than in prokaryotes, many of the fundamental principles of gene regulation in bacteria and viruses are retained. For example, the principles of gene regulation used by λ phage to choose between a lytic and lysogenic pathway are similar to many used to control transcription of eukaryotic genes. The genetic switch utilized by λ has parallels in similar genetic switches utilized by eukaryotes to choose between different cell fates and which sets of genes are expressed. It will be helpful to think about how a principle is retained in prokaryotes and eukaryotes while analyzing the differences between them.

The multiple levels in which eukaryotic gene expression is controlled require you to pay very close attention to detail. Consider each level separately, and refer frequently to the text figures, redrawing them to solidify your understanding. In instances where control is at the level of transcription, it will help you to relate the level of control to the organization of the gene or the chromosome. In instances where control is at the level of mRNA processing, translation, and mRNA stability, it will help you to relate the level of control to the stage where information is being transferred from DNA into protein. In instances where control is at the level of posttranslational processing and protein degradation, it will help you to relate the level of control to the fate and function of a protein after it is synthesized.

Once you have understood the levels at which eukaryotic gene expression can be controlled, you will have a foundation to consider how gene expression can be controlled in a temporal and spatial context during development. When genetic pathways and bacterial gene expression was

4	5
development	development
differentiation	*Drosophila*
totipotency	*Caenorhabditis elegans*
regeneration	cell lineage
cloning	*Arabidopsis thaliana*
differential gene activity	polar cytoplasm
α-, β-, γ-, δ-, ϵ- globin genes	syncytium
pseudogene	syncytial, cellular blastoderm
immune system	molecular gradient
DNA rearrangement	parasegment, segment
antigen	maternal effect genes
antibody	segmentation genes
immunoglobin	homeotic genes
clonal selection	gap genes
H, L chains	pair rule genes
C, V, J, D segments	segment polarity genes
somatic recombination	imaginal disks
antibody diversity	serial transplantation
	homeobox
	helix-turn-helix motif
	homeodomain
	colinearity rule

discussed, one considered fairly short timelines in defined situations. In eukaryotic development, timelines are much longer, and the spatial dimension—how do different cell types and tissues differentiate from a common precursor cell—is added. It is essential to place these dimensions in the perspective to the developing organism. Therefore, when you consider the regulation of gene expression during development, be sure you have a thorough understanding of the "biology" of the system that is developing. First understand how the organism (e.g., *Drosophila*) or population of cells (e.g., the immune system) develops, and then consider how genes are regulated in this context. It will be helpful to construct diagrams that illustrate how genes are regulated along a timeline or developmental pathway.

QUESTIONS FOR PRACTICE

Multiple Choice Questions

1. In eukaryotes, cell- and tissue-specific gene expression is achieved via
   a. the use of operons
   b. a cell- and tissue-specific set of activators and repressors
   c. DNase I sensitivity
   d. selective deletion of genes not active in differentiated cells

2. Transcription in eukaryotes is activated if positive regulatory proteins are bound at
   a. the enhancer element
   b. the promoter element
   c. the operator
   d. both a and b
3. Histones act in eukaryotic gene regulation primarily as
   a. enhancers of gene expression
   b. repressors of gene expression
   c. promoters
   d. proteins preventing DNase I digestion
4. A gene lying in chromatin that is actively transcribed by RNA polymerase II is likely to
   a. have DNase I–hypersensitive sites upstream of the protein-coding region
   b. have DNase I–hypersensitive sites within the protein-coding region
   c. be relatively insensitive to DNase I
   d. be heavily methylated
5. The receptors for steroid hormones
   a. lie on the cell surface and act via second messengers
   b. lie on the cell surface and when bound by hormone are transported into the nucleus and bind response elements to activate transcription
   c. lie in the cytoplasm and act via second messengers
   d. lie in the cytoplasm and when bound by hormone are transported into the nucleus and bind response elements to activate transcription
6. Receptors bound by polypeptide hormones can alter gene expression by
   a. acting through a second messenger such as cAMP
   b. directly binding DNA sequences in the 5′ end of genes
   c. directly binding DNA sequences in the 3′ end of genes
   d. affecting a kinase cascade
   e. both a and d
   f. both b and c
7. A wide diversity of immunoglobulin molecules is obtained via
   a. new gene synthesis
   b. combinatorial gene regulation
   c. alternative RNA splicing
   d. somatic recombination of DNA
8. Evidence for the totipotency of nuclei in eukaryotic cells is provided by
   a. the existence of homeotic mutations
   b. nuclear transplantation experiments
   c. transdetermination of imaginal disk after serial transplantation
   d. the transcription of fetal globin genes (γ-globin) only during fetal development and not during adult life
   e. the finding that nuclei in different adult tissues have the same amount of DNA
9. In *Drosophila,* maternal effect genes are genes that
   a. are expressed in the mother during oogenesis and whose products will specify spatial organization in the developing embryo
   b. cause females to lay eggs

c. affect the number of polarity of body segments
d. affect the identity of a segment

Answers: 1b, 2d, 3b, 4a, 5d, 6f, 7d, 8b, 9a

Thought Questions

1. Suppose you have cloned gene *X*. (a) Design an experiment to determine if gene *X is available for transcription* in liver and in brain tissue. (b) Now design an experiment to determine if gene *X is transcribed* in liver and/or in brain tissue. How do these experiments differ?
2. Present evidence that argues for or against the following statement: "Histone and nonhistone proteins function as repressors of gene expression in eukaryotic cells."
3. What is the evidence that methylation plays a role in transcriptional control in some eukaryotes? What evidence is there that methylation does not entirely determine the transcriptional activity of some genes?
4. Consider how hormones act. (a) Since hormones are diffusible, why are certain cells, but not others, targets of a particular hormonal signal? (b) How is the response of a cell to a peptide hormone fundamentally different from the response of a cell to a steroid hormone? (c) Propose a hypothesis to explain why it is that steroid hormone receptors are located inside the target cells, whereas the receptors for peptide hormones are located on the surface of target cells.
5. Suppose Gurdon's experiments on totipotency were repeated using differentiated B lymphocytes as the source of nuclei for transplantation. To what extent would you expect these nuclei to be totipotent? If mature organisms developed from cells with transplanted B cell nuclei, what would be the nature of their immune system?
6. In *Drosophila,* the polar cytoplasm contains factors determining that the nuclei migrating into this region will become germ-line cells. (a) Design an experiment to gather evidence supporting this conclusion. (b) What phenotype would be associated with mutants lacking these factors? (c) Would the mutants from part (b) identify maternal genes?
7. What are homeoboxes, and what is their role in the regulation of development in eukaryotes?
8. In vertebrates, including mammals such as humans, homeotic gene complexes have been identified using homologous probes. Curiously, the gene complexes show similar structural organization, even though vertebrates are not segmented in the same way as invertebrates. What significance might this finding have? How might analyzing mutations in these genes in vertebrates help demonstrate the function of these genes and test this significance?
9. A number of organisms have been studied intensely with the aim of understanding the genetic control of development. Describe the features of *Drosophila, Caenorhabditis,* and *Arabidopsis* that make them useful for such analysis. Why is it important to study more than one model organism?
10. The primary transcripts of many genes undergo tissue-specific alternative mRNA splicing. Indeed, about 15 percent of human disease mutations are associated with mRNA splicing alterations. (a) For genes that are alternatively spliced, do you expect the proteins that are produced to have identical functions? Why or why not? (b) How might spliceosomes process primary mRNAs differently in different tissues to achieve alternative mRNA processing?
11. Dolly is the lamb that was generated by fusing adult mammary epithelial cells with enucleated oocytes. (a) How does this experiment demonstrate totipotency? (b) How could you use

molecular methods to experimentally demonstrate that Dolly had the same genotype as the adult mammary epithelial cells? (c) Suppose you sampled Dolly's DNA at 1 year of age. Do you expect Dolly's simple telomeric repeats to be the same length as other sheep that were 1 year old? (d) What impact might imprinting have on our ability to successfully clone eukaryotic organisms?

Thought Questions for Media

After reviewing the media at the *Essential iGenetics* website, try to answer these questions.

1. In neuronal cells, the calcitonin/CGRP transcript is cleaved, and polyadenylation occurs at the site called pA_2, producing a pre-mRNA with exons 1, 2, 3, 4, and 5. In the thyroid gland, a transcript with exons 1, 2, 3, and 4 is produced and calcitonin, a functional peptide hormone encoded by exon 4, is produced. Why is calcitonin not produced in neurons?

1. In the iActivity, microarrays were used to identify genes that were differentially expressed between differentiating tissues (mesoderm and endoderm) and stem cells. Suppose two genes, *A* and *B*, are upregulated in differentiating mesoderm relative to stem cells. Gene *A* is expressed 13-fold more and gene *B* is expressed 2.5-fold more. Neither gene is upregulated in differentiating endoderm.
   a. From these data alone, can you infer anything about the *relative* importance of these genes' expression for mesodermal differentiation?
   b. Suppose mutants were available that caused the rate of each gene's transcription to be halved. Can you predict which mutation might have a more severe effect on mesodermal differentiation?
   c. Based on the results of the microarray analyses, can you infer which gene produces more mRNA copies (the total number of transcripts per gene) in differentiating mesodermal cells?
2. What is meant by a gene that can *serve as a marker* for endoderm differentiation? What features must such a gene have?
3. A gene that encodes a transcription factor is expressed at high levels in differentiating endoderm and differentiating mesoderm, but not in stem cells. What types of genes might be targets of this transcription factor? Would it invariably act through positive regulation of these targets?
4. A gene that encodes a transcription factor is expressed in stem cells, but not in differentiating cells. What role might this gene play in keeping stem cells from differentiating? Would it invariably act through negative regulation of target genes?

SOLUTIONS TO TEXT PROBLEMS

17.1 Promoters, enhancers, transcription factors, and regulatory proteins that are active for one gene typically share structural similarities with these elements in other genes. Nonetheless, the transcriptional control of a gene can be exquisitely specific: It will be specifically transcribed in some tissues at very defined times. Explore how this specificity arises by addressing the following questions.

a. Distinguish between the functions of promoters and enhancers in transcriptional regulation.

b. What structural features are found in proteins that bind these DNA elements?

c. Can an enhancer bound by a regulatory protein stimulate as well as suppress transcription? If so, how?

d. Given that several different genes may contain the same types of promoter and enhancer elements, and a number of transcription factors contain the same structural features, how is transcriptional specificity generated?

Answer:

a. Promoters and enhancers bind specific regulatory proteins. The proteins that bind to promoters are crucial for determining whether transcription can occur, while those that bind enhancers determine whether transcription will be at a minimal or a maximal level.

b. Structural motifs important for DNA binding include the helix-turn-helix, the zinc finger, the leucine zipper, and the helix-loop-helix.

c. When a protein is bound to an enhancer, its effect on transcription is dependent in part on the regulatory proteins bound at other enhancers and promoters in the region. A negative regulatory protein can bind to an enhancer and silence transcription. However, if a negative regulatory protein is bound to an enhancer, and a positive regulatory protein is bound to a promoter, the net result will depend on their interaction. Either the effect of one protein may modulate the effect of the other, or their interaction may result in an effect that neither alone can cause.

d. Transcriptional specificity is generated through the combinatorial use of a set of transcription factors. Some factors are common to many cells, while others have a more restricted spatial and/or temporal distribution. The transcription of defined arrays of genes in different cells at different times is controlled by the subset of regulatory proteins present in a eukaryotic cell at a particular time.

17.2 A cloned DNA sequence was used to probe a Southern blot. There were two DNA samples on the blot, one from white blood cells and the other from a liver biopsy of the same individual. Both samples had been digested with *Hpa*II. The probe bound to a single 2.2-kb band in the white blood cell DNA but bound to two bands (1.5 and 0.7 kb) in the liver DNA.

a. Is this difference likely to result from a somatic mutation in a *Hpa*II site? Explain.

b. How would it affect your answer if you knew that white blood cell and liver DNA from this individual both showed the two-band pattern when digested with *Msp*I?

Answer:

a. A somatic mutation is rather unlikely. Since humans are diploid, the single 2.2-kb band in the white blood cell DNA indicates that both chromosomes have the same restriction sites. The liver cells have two bands whose sizes add up to the size of the band in white blood cells. Thus, it would appear that the liver has an additional site on *both* chromosomes, not just one chromosome (heterozygous cells would have three bands). For this to occur by a mutational mechanism, two different somatic mutations would have had to occur in the same site, in a cell that gave rise to all cells of the tissue being examined (or in all cells of the tissue), an unlikely event.

b. *Msp*I recognizes the same CCGG sequence as *Hpa*II, but unlike *Hpa*II, it will cleave the methylated sequence $C^{m}CGG$ as well as the unmethylated CCGG. That both liver cells and white blood cells show the same two-band pattern when digested with *Msp*I, but not *Hpa*II, indicates that there is a central methylated site in white blood cells, but not in liver. Increased DNA methylation has, in some instances, been correlated with a decrease in transcriptional activity. However, not all methylated DNA is transcriptionally silent.

17.3 Both fragile X syndrome and Huntington disease are caused by trinucleotide repeat expansion. Individuals with fragile X syndrome have at least 200 CGG repeats at the 5′ end of the *FMR-1* gene. Individuals with Huntington disease have at least 36 CAG repeats within the protein-coding region of the huntingtin gene.

a. How is gene expression affected by these repeat expansions?
b. Based on your answer to (a), why is the fragile X syndrome recessive, whereas Huntington disease is dominant?
c. Why is the number of trinucleotide repeats needed to cause the phenotype different for each disease?

Answer:

a. In fragile X syndrome, the expanded CGG repeat results in hypermethylation and transcriptional silencing. In Huntington disease, the expanded CAG repeat results in the inclusion of a polyglutamine stretch within the huntingtin protein, which causes it to have a novel, abnormal function.
b. A heterozygote with a CGG repeat expansion near one copy of the *FMR-1* gene will still have one normal copy of the *FMR-1* gene. The normal gene can produce a normal product, even if the other is silenced. (The actual situation is made somewhat more complex by the process of X inactivation in females, but in general, one would expect that a mutation that caused transcriptional silencing of one allele would not affect a normal allele on a homologue.) In contrast, a novel, abnormal protein is produced by the CAG expansion in the disease allele in Huntington disease. Since the disease phenotype is due to the presence of the abnormal protein, the disease trait is dominant.
c. Transcriptional silencing may require significant amounts of hypermethylation and so require more CGG repeats for an effect to be seen. In contrast, protein function may be altered by a relatively smaller sized polyglutamine expansion.

17.4 Both peptide and steroid hormones can affect gene regulation of a targeted population of cells.

a. What is a hormone?
b. Distinguish between the mechanisms by which a peptide and a steroid hormone affect gene expression.
c. What role do each of the following have in a physiological response to a peptide or a steroid hormone?
   i. steroid hormone receptor (SHR)
   ii. steroid hormone response element (HRE)
   iii. second messenger
   iv. cAMP and adenylate cyclase

Answer:

a. A hormone is typically a low-molecular-weight chemical messenger synthesized in low concentrations in one tissue or cell and transmitted in body fluids to another part of the organism. Hormones act as effector molecules to produce specific effects after binding to a receptor in target cells that may be remote from the hormone's point of origin. Hormones function to regulate gene activity, physiology, growth, differentiation, or behavior.

b and c. Steroid hormones and peptide hormones employ two fundamentally different mechanisms to regulate gene expression. Steroid hormones traverse the cell membrane and bind to a cytoplasmic steroid hormone receptor (SHR). The complex of the hormone and receptor protein then acts directly to regulate gene expression. The complex binds to specific steroid hormone response elements (HREs) near genes to regulate their transcription. The stability of a gene's mRNA and the processing of precursor mRNAs can be affected as well.

Polypeptide hormones, such as insulin and certain growth factors, bind to receptors that reside on cell surfaces. The activated, bound receptor then transduces a signal via a second-messenger system inside the cell. Some activated receptors increase the activity of the membrane-bound enzyme adenylate cyclase, which results in an increase in cAMP levels. Increased intracellular levels of cAMP send an intracellular signal (a second messenger) to activate other cellular processes that ultimately lead to an alteration in patterns of gene expression.

17.5 The following figure shows the effect of the hormone estrogen on ovalbumin synthesis in the oviduct of 4-day-old chicks. Chicks were given daily injections of estrogen ("Primary stimulation") for 10 days and then the injections were stopped. Two weeks after withdrawal (25 days), the injections were resumed ("Secondary stimulation").

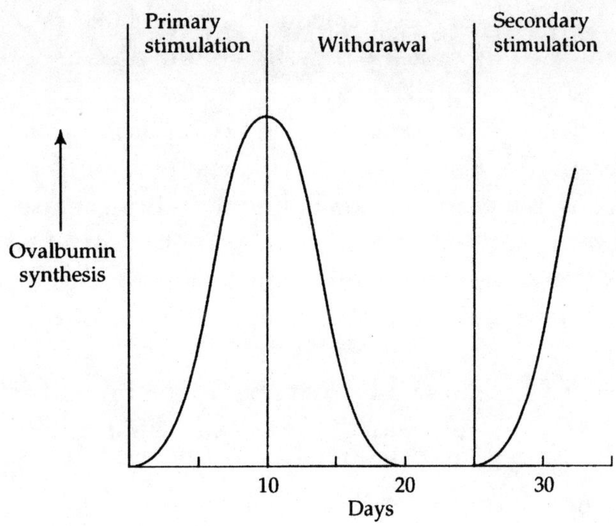

Provide possible explanations of these data.

Answer: The data indicate that the synthesis of ovalbumin is dependent on the presence of the hormone estrogen. These data do not address the mechanism by which estrogen achieves its effects. Theoretically, it could act (1) to increase transcription of the ovalbumin gene by binding to an intracellular receptor that, as an activated complex, stimulates transcription at the ovalbumin gene; (2) to stabilize the ovalbumin precursor mRNA; (3) to increase the processing of the precursor ovalbumin mRNA; (4) to increase the transport of the processed ovalbumin mRNA out of the nucleus; (5) to stabilize the mature ovalbumin mRNA once it has been transported into the cytoplasm; (6) to stimulate translation of the ovalbumin mRNA in the cytoplasm; and/or (7) to stabilize (or process) the newly synthesized ovalbumin protein. Experiments in which the levels of ovalbumin mRNA were measured have shown that the production of ovalbumin mRNA is primarily regulated at the level of transcription.

17.6 Although the primary transcript of a gene may be identical in two different cell types, the translated mRNAs can be quite different. Consequently, in different tissues, distinct protein products can be produced from the same gene. Discuss two different mechanisms by which the production of mature mRNAs can be regulated to this end; give a specific example for each mechanism.

Answer: Two mechanisms are those controlling choice of poly(A) site and choice of splice site. Both mechanisms are illustrated in the generation of tissue-specific products of the human calcitonin gene, *CALC* (Figure 17.7). This gene consists of five exons and four introns. The gene is expressed in certain cells of the thyroid and in certain neurons of the brain. When transcribed in the thyroid cells, polyadenylation occurs at a site next to exon 4, whereas in the neuronal cells, polyadenylation occurs at a site next to exon 5. In thyroid cells, therefore, the fifth exon is absent from the pre-mRNA. Alternative splicing then takes place as the pre-mRNAs are processed in the two tissues. In the thyroid, the pre-mRNA is spliced to remove the introns and bring together exons 1, 2, 3, and 4. Translation of the resulting mRNA produces the precursor polypeptide for the polypeptide hormone calcitonin. Calcitonin's amino acid sequence corresponds to exon 4; the hormone is generated by protease digestion of the precursor polypeptide. In the neuronal cells, the alternative splicing of the pre-mRNA brings together exons 1, 2, 3, and 5; exon 4 is discarded. Translation of the resulting mRNA produces the precursor polypeptide for the polypeptide hormone CGRP (calcitonin gene-related peptide); as with calcitonin, the hormone is generated by protease digestion of the precursor polypeptide.

17.7 Although eukaryotes lack operons like those found in prokaryotes, the exceptional conserved organization of the *ChAT/VAChT* locus is reminiscent of a prokaryotic operon. *ChAT* is the gene for the enzyme choline acetyltransferase, which synthesizes acetylcholine, a neurotransmitter released by one neuron to signal another neuron. *VAChT* is the gene for the vesicular acetylcholine transporter protein, which packages acetylcholine into vesicles before its release from a neuron. Both *ChAT* and *VAChT* are expressed in the same neuron.

Part of the *VAChT* gene is nested within the first intron of the *ChAT* gene, and the two genes share a common regulatory region and a first exon. The structure of a primary mRNA and two processed mRNA transcripts produced by this locus are diagrammed in the following figure. The common regulatory region important for transcription of the locus in neurons is shown in the DNA, black rectangles in RNA represent exons, lines connecting the exons represent spliced intronic regions, and AUG indicates the translation start codons within the *ChAT* and *VAChT* mRNAs. Polyadenylation sites are not shown.

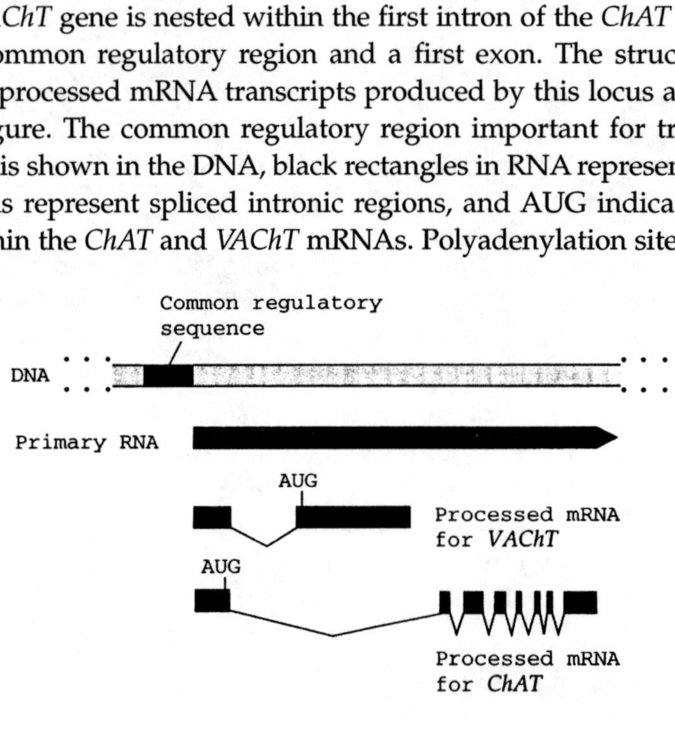

a. In what ways is the organization of the *VAChT/ChAT* locus reminiscent of a bacterial operon?
b. Why is the organization of this locus not structurally equivalent to a bacterial operon?
c. Based on the transcript structures shown, what modes of regulation might be used to obtain two different protein products from the single primary mRNA?

Answer:

a. In bacterial operons, a common regulatory region controls the production of single mRNA from which multiple protein products are translated. These products function in a related biochemical pathway. Here, two proteins that are involved in the synthesis and packaging of acetylcholine are both produced from a common primary mRNA transcript.
b. Unlike the proteins translated from an mRNA synthesized from a bacterial operon, the protein products produced at the *VAChT/ChAT* locus are not translated sequentially from the same mRNA. Here, the primary mRNA appears to be alternatively processed to produce two distinct mature mRNAs. These mRNAs are translated starting at different points, producing different proteins.
c. At least two mechanisms are involved in the production of the different ChAT and VAChT proteins: alternative mRNA processing and alternative translation initiation. After the first exon, an alternative 3′ splice site is used in the two different mRNAs. In addition, different AUG start codons are also employed.

17.8 Distinguish between the terms *development* and *differentiation.*

Answer: Development is a process of regulated growth and cellular change. It results from the interaction of the genome with the cytoplasm and external environment. It involves a programmed sequence of phenotypic changes that are typically irreversible. Differentiation refers to an aspect of cellular change in development and involves the formation of distinct types of cells, tissues, and organs through the regulation of gene expression. Differentiation is thus a part of development and leads to cells that have characteristic structural and functional properties.

17.9 What is totipotency? Give an example of the evidence for the existence of this phenomenon.

Answer: Totipotency refers to the capacity of a nucleus to direct a cell through all of the stages of development. A nucleus taken from the cell of a differentiated tissue is said to be totipotent if, when it is injected into an enucleated egg, it can direct the development of the organism to the adult stage. If the egg is able to develop into an adult, the differentiated nucleus must have retained all of the genetic information needed to direct development again from the start. In 1975, Gurdon demonstrated that the nucleus of a skin cell of an adult frog was at least partly totipotent. When injected into an enucleated egg, such a nucleus could direct development to the tadpole stage. More recently, Wilmut and his colleagues demonstrated that a mammalian nucleus was totipotent. They showed that the nucleus of a mammary epithelium-derived cell could direct the development of a sheep to the adult stage.

17.10 In Woody Allen's 1973 film *Sleeper,* the aging leader of a futuristic totalitarian society has been dismembered in a bomb attack. The government wants to clone the leader from his only remaining intact body part, a nose. The characters Miles and Luna thwart the cloning by abducting the nose and flattening it under a steam roller.

a. In light of the 1996 cloning of the sheep Dolly, how should the cloning have proceeded if Miles and Luna had not intervened?

b. If methods like those used for Dolly had been successful, in what genetic ways would the cloned leader be unlike the original?

c. Suppose that instead of a nose, only mature B cells (B lymphocytes of the immune system) were available. What genetic deficits would you expect in the "new leader"?

d. In the set of experiments used to clone Dolly, six additional live lambs were obtained. Why is the production of Dolly more significant than the production of the other lambs?

e. If the cloning of the leader had succeeded, can you make any prediction about whether the "cloned leader" would be interested in perpetuating the totalitarian state?

Answer:

a. Based on the work of Wilmut and his colleagues, the nose cells would first be dissociated and grown in tissue culture. The cells would be induced into a quiescent state (the G_0 phase of the cell cycle) by reducing the concentration of growth serum in the medium. Then they would be fused with enucleated oocytes from a donor female and allowed to grow and divide by mitosis to produce embryos. The embryos would be implanted into a surrogate female. After the establishment of pregnancy, its progression would need to be maintained.

b. While the nuclear genome would generally be identical to that in the original nose cell, cytoplasmic organelles would presumably derive from those in the enucleated oocyte. Hence, the mitochondrial DNA would not derive from the original leader. In addition, since telomeres in an older individual are "shorter," one might expect the telomeres in the cloned leader to be those of an older individual.

c. In mature B cells, DNA rearrangements at the heavy- and light-chain immunoglobulin genes have occurred. One would expect the cloned leader to be immunocompromised, as he would be unable to make the wide spectrum of antibodies present in a normal individual.

d. The production of Dolly was significant because it demonstrated the apparent complete totipotency of a nucleus from a mature mammary epithelium cell. The other six lambs developed from the transplanted nuclei of embryonic or fetal cells, which one might expect to be less determined and have a higher degree of totipotency due to their younger developmental age.

e. There is no way to predict the psychological profile of the cloned leader based on his genetic identity. Even identical twins, who are genetically more identical than such a clone, do not always share behavioral traits.

17.11 In humans, β-thalassemia is a disease caused by failure to produce sufficient β-globin chains. In many cases, the mutation causing the disease is a deletion of all or part of the β-globin structural gene. Individuals homozygous for certain of the β-thalassemia mutations are able to survive because their bone marrow cells produce γ-globin chains. The γ-globin chains combine with α-globin chains to produce fetal hemoglobin. In these people, fetal hemoglobin is produced by the bone marrow cells throughout life, whereas normally it is produced in the fetal liver. Use your knowledge about gene regulation during development to suggest a mechanism by which this expression of γ-globin might occur in β-thalassemia.

Answer: There are a number of possibilities. One is that the γ-globin genes in bone marrow are under negative regulation by β-globin (or some metabolite of it). When β-globin is not formed, the γ-globin gene is derepressed.

17.12 The following figure shows the percentage of ribosomes found in polysomes in unfertilized sea urchin oocytes (0 hours) and at various times after fertilization.

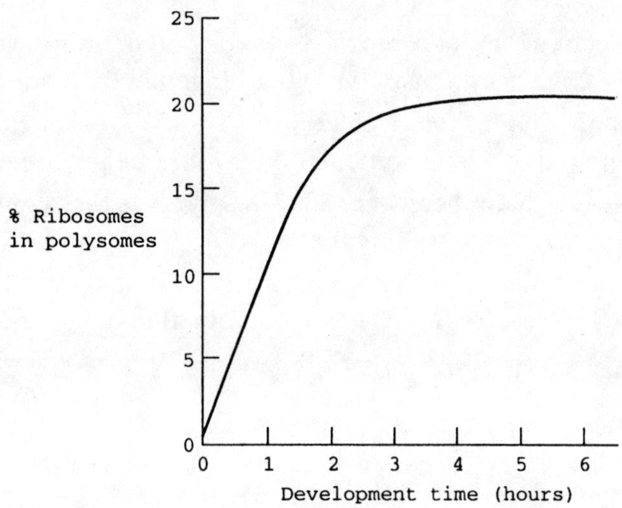

In the unfertilized egg, less than 1 percent of ribosomes are present in polysomes, and at 2 hours postfertilization, about 20 percent of ribosomes are present in polysomes. It is known that no new mRNA is made during the time period shown. How can the data be interpreted?

Answer: Since no new mRNA is synthesized during this time period, preexisting, maternally packaged mRNAs that have been stored in the oocyte must be recruited into polysomes as development begins following fertilization.

17.13 The mammalian genome contains about 10^5 genes. Mammals can produce about 10^6 to 10^8 different antibodies. Explain how it is possible for both of these statements to be true.

Answer: The ability to make 10^6 to 10^8 different antibodies arises from the combinatorial way in which antibody genes are generated in different antibody-producing cells during their development, and not the existence of this many separate antibody genes in each and every mammalian cell. A template that exists in germ-line cells is processed differently during the development of different antibody-producing cells to generate antibody diversity.

Antibody molecules consist of two light (L) chains and two heavy (H) chains. The amino acid sequence of one domain of each type of chain is variable and generates antibody diversity. In the germ-line DNA of mammals, coding regions for these immunoglobulin chains exist in tandem arrays of gene segments. For light chains, there are many variable (V) region gene segments, a few joining (J) segments, and one constant (C) gene segment. Somatic recombination during development results in the production of a recombinant V-J-C DNA molecule which, when transcribed, produces a unique, functional L chain. From a particular gene in one cell, only one L chain is produced. A large number of L chains are obtained by recombining the gene segments in many different ways. Diversity in these L chains results from variability in the sequences of the multiple V segments, variability in the sequences of the four J segments, and variability in the number of nucleotide pairs deleted at the V-J joints. H chains are similar, except that several D (diversity) segments can be used between the V and J segments, increasing the possible diversity of recombinant H chain genes. The type of C gene segment chosen for the constant domains of the H chain determines whether the antibody is IgM, IgD, IgG, IgE, or IgA.

17.14 Antibody molecules (Ig) are composed of four polypeptide chains (two of one light chain type and two of one heavy chain type) held together by disulfide bonds.

a. If for the light chain there were 300 different V_K segments and 4 J_K segments, how many different light chain combinations would be possible?
b. If for the heavy chain there were 200 V_H segments, 12 D segments, and 4 J_H segments, how many heavy chain combinations would be possible?
c. Given the information in parts (a) and (b), what would be the number of possible types of IgG molecules (L + H chain combinations)?

Answer:

a. Not considering the variability in the number of nucleotide pairs deleted at the V_K-J_K joint, there are 300 (V_K segments) $\times$ 4 (J_K segments) = 1,200 different light chain combinations. This number is a lower boundary for an estimate, since significantly more variability can be obtained by using imprecise V-J joining.
b. Not considering the imprecise joining of gene segments that comprise the chain's variable region, there are 200 (V_H segments) $\times$ 12 (D segments) $\times$ 4 (J_H segments) = 9,600 different heavy chain combinations for each of the eight types of Ig (IgM, IgD, IgG_3, IgG_1, IgG_{2b}, IgG_{2a}, IgE, or IgA). If all possible types of Ig are considered, there are 9,600 $\times$ 8 (C_H segments) = 76,800 heavy chain combinations. As in part (a), this is a lower boundary for an estimate of heavy chain variability.
c. IgG has four C segments (γ_3, γ_1, γ_{2b}, γ_{2a}). An estimate of the minimal number of combinations for IgG molecules would be $1200 \times 9{,}600 \times 4 = 4.608 \times 10^7$.

17.15 Define imaginal disk, homeotic mutant, and transdetermination.

Answer: An imaginal disk is a larval structure in *Drosophila* and other insects that undergo a metamorphic transformation from a larva to an adult. During metamorphosis in the pupal period, it will differentiate into an adult structure. It is essentially a sac of cells, a disk, that is set aside early in embryonic development. The disk increases in cell number by mitosis, but remains undifferentiated during larval growth. Although the disk remains undifferentiated up until the time hormonal signals activate the differentiation of adult structures during the pupal period, its fate is *determined* very early in development. Thus, separate imaginal disks are destined to differentiate into different adult structures, such as legs, eyes, antennae, and wings.

Homeotic mutants alter the identity of particular segments, transforming them into copies of other segments. Homeotic mutations thus affect the determination, or fate, of a disk. Mutations in the *Antennapedia* gene complex can, for example, transform the fate of cells of the antenna disk so that a leg is formed where an antenna should be. The general inference from such observations is that homeotic genes normally play a pivotal role in establishing the segmental identity of an undifferentiated cell.

If you research the term independently, you will find that transdetermination is a phenomenon seen in disk transplantation studies. When a disk is transplanted from a larva to an adult abdomen, it can grow and be "incubated" without differentiating. It can be repeatedly transplanted to new abdomens for incubation and remain undifferentiated. If it is later transplanted back to a larva that is allowed to undergo metamorphosis, it will differentiate into an adult structure. In rare instances, the determined fate of the disk is altered so that the disk differentiates to another fate. The disk does not totally dedifferentiate, and its fate is not totally random. Rather, it takes on another determined fate. For example, a leg disk may become an antennal structure or a wing structure, but not a genital structure. This process of selective switching of fate is referred to as transdetermination.

17.16 Imagine that you observed the following mutants (*a–e*) in *Drosophila.* Based on the characteristics given, assign each of the mutants to one of the following categories: maternal effect gene, segmentation gene, or homeotic gene.

a. Mutant *a:* In homozygotes, phenotype is normal, except wings are oriented backward.
b. Mutant *b:* Homozygous females are normal but produce larvae that have a head at each end and no distal ends. Homozygous males produce normal offspring (assuming the mate is not a homozygous female).
c. Mutant *c:* Homozygotes have very short abdomens, which are missing segments A2 through A4.
d. Mutant *d:* Affected flies have wings growing out of their heads in place of eyes.
e. Mutant *e:* Homozygotes have shortened thoracic regions and lack the second and third pair of legs.

Answer:

a. As the *polarity* of an individual, differentiated structure is altered, *a* is likely to be a segmentation gene. In particular, it is likely to be a member of the subclass of segmentation genes known as segment polarity genes.
b. The abnormal appearance of progeny of homozygous mothers, but not homozygous fathers, suggests that homozygous mothers produce abnormal oocytes. This view is bolstered by the phenotype of the abnormal progeny. Progeny with defect(s) at one end could arise if a maternally produced gradient in the oocyte was abnormal. Thus, *b* is likely to be a maternal effect gene.
c. The absence of a set of segments is characteristic of the phenotypes of gap gene mutants, a subclass of segmentation gene mutants.
d. It would appear that the fate of the eye disk has been changed to that of a wing disk. The segmental *identity* of cells is controlled by homeotic genes, and so *d* is likely to be a homeotic mutation.
e. In mutant *e,* a set of segments is missing (T2 and T3). Thus, like *c, e* is likely to be a mutant in a gap gene, a subclass of segmentation gene mutants.

17.17 If actinomycin D, an antibiotic that inhibits RNA synthesis, is added to newly fertilized frog eggs, there is no significant effect on protein synthesis in the eggs. Similar experiments have shown that actinomycin D has little effect on protein synthesis in embryos up until the gastrula stage. After the gastrula stage, however, protein synthesis is significantly inhibited by actinomycin D, and the embryo does not develop any further. Interpret these results.

Answer: Preexisting mRNA that was made by the mother and packaged into the oocyte prior to fertilization is translated up to the gastrula stage. After gastrulation, new mRNA synthesis is necessary for the production of proteins needed for subsequent embryonic development.

17.18 It is possible to excise small pieces of early embryos of the frog, transplant them to older embryos, and follow the course of development of the transplanted material as the older embryo develops. A piece of tissue is excised from a region of the late blastula or early gastrula that would later develop into an eye and is transplanted to three different regions of an older embryo host (see part (a) of Figure 17.A). If the tissue is transplanted to the head region of the host, it will form eye, brain, and other material characteristic of the head region. If the tissue is transplanted to other regions of the host, it will form organs and tissues characteristic of those regions in normal development (e.g., ear, kidney). In contrast, if tissue destined to be an eye is excised from a neurula and transplanted into an older embryo host to exactly the same places as used for the blastula or gastrula transplants, in every case the transplanted tissue differentiates into an eye (see part (b) of Figure 17.A). Explain these results.

Figure 17.A

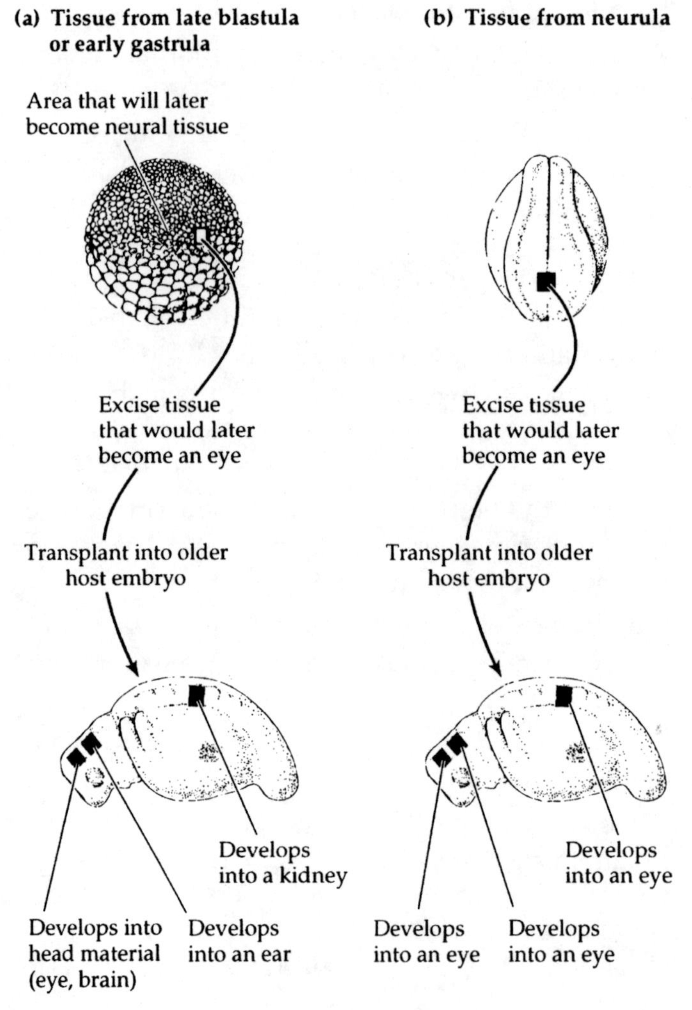

Answer: This experiment demonstrates the phenomenon of *determination* and when it occurs during development. The tissue taken from the blastula/gastrula has not yet been committed to its final differentiated state in terms of its genetic programming; that is, it has not yet been *determined.* Thus, when the tissue is transplanted into the host, it adopts the fate of nearby tissues and differentiates in the same way as they do. Presumably, cues from the tissue surrounding the transplant determine its fate.

In contrast, tissues in the neurula stage are stably determined. By the time the neurula developmental stage has been reached, a developmental program has been set. In other words, the fate of neurula tissue transplants is *determined.* Upon transplantation, they will differentiate according to their own set genetic program. Tissue transplanted from a neurula to an older embryo cannot be influenced by the surrounding tissues. It will develop into the tissue type for which it has been determined, in this case, an eye.

18

GENETICS OF CANCER

CHAPTER OUTLINE

REVIEW OF KEY TERMS, SYMBOLS, AND CONCEPTS

In your own words, write a brief, precise definition of each term in the groups on the following page. Check your definitions using the text. Then develop a concept map using the terms in each list.

THINKING ANALYTICALLY

Although cancer genetics is complex, focusing on two key issues should help put the topics of this chapter in perspective. Since cancer results from unconstrained cellular growth, the first key issue is to understand how each of the processes discussed in this chapter affects cellular growth. It will be helpful to review how cells proceed through a cell cycle. As you read the text, pay particular attention to how progression through the cell cycle is controlled. Notice how mutations in oncogenes and tumor suppressor genes, both of which relate to growth control, can result in cancer through different mechanisms.

A second key issue is that cancer typically occurs in somatic cells, and so it is important to consider the effect of accumulating mutations in somatic cells and not only germ-line cells. To understand the relationship between hereditary and sporadic forms of cancer, it will help to review the relationship between somatic and germ-line cells. To understand how mutations accumulate in cancerous cells, it is necessary to consider the role of DNA repair within the cell cycle, the consequences of errors in DNA repair, and the effects of carcinogens and ionizing radiation.

1	2	3
tumor	tumor virus	multistep nature of cancer
neoplasm	oncogene	cell cycle
transformation	DNA vs. RNA tumor virus	pRB
benign	sarcoma	G_0 phase
malignant	retrovirus	cyclin/Cdk complex
cancer	*gag, pol, env, src* genes	phosophorylation
metastasis	nononcogenic retrovirus	p53
oncogenesis	insertional mutagenesis	p21
terminal differentiation	AIDS	apoptosis
stem cell	helper T cell	cell cycle arrest
germ-line cell	gp120 glycoprotein	mutator gene
somatic cell	transducing retrovirus	tumor suppressor
growth factor	nontransducing retrovirus	oncogene
growth-inhibiting factor	replication-competent	mismatch repair
cell cycle	transformation-competent	HNPCC, herditary FAP
dedifferentiation	viral oncogene, v-*onc*	*hMSH2, hMLH1, hPMS1, hPMS2, mutS, mutL* genes
two-hit model	proto-oncogene	carcinogen
familial cancer	cellular oncogene, c-*onc*	direct-acting carcinogen
sporadic cancer	growth factor	ultimate carcinogen
tumor suppressor gene	protein kinase	procarcinogen
oncogene, proto-oncogene	proviral DNA	UVA
mutator gene	tumor suppressor gene	UVB
multistep nature of cancer		SPF
sporadic vs. hereditary retinoblastoma		ionizing radiation
loss of heterozygosity		
bilateral tumor		

QUESTIONS FOR PRACTICE

Multiple Choice Questions

1. About what percentage of breast cancer is hereditary?
   a. 1%
   b. 5%
   c. 10%
   d. 50%
2. Familial retinoblastoma appears to be inherited as a dominant trait. What factor accounts for this?
   a. A dominant mutation is transmitted through the germ line.
   b. Somatic mutation occurs after a recessive mutation is transmitted through the germ line.
   c. Heterozygous individuals are infected with a retrovirus.
   d. About half of the homozygous, normal individuals in an affected family are infected with a retrovirus at birth.

3. RNA tumor viruses cause cancer
   a. by the same mechanism as DNA tumor viruses
   b. only if their proviral DNA integrates near a proto-oncogene
   c. if they carry an oncogene
   d. if they infect helper T cells
4. Which of the following is *not* a feature of a viral oncogene?
   a. It lacks introns.
   b. Its expression is regulated identically to the corresponding cellular proto-oncogene.
   c. Its structure and location in the viral genome varies.
   d. It is not conserved between viral strains.
5. Which of the following is untrue about the product of the *p53* gene?
   a. It is a transcription factor.
   b. It is a tumor suppressor.
   c. It can induce cell death.
   d. When present, it activates the cell cycle from G_1 to S.
6. What is the cellular function of the unphosphorylated product of the *RB* gene, pRB?
   a. pRB binds to transcription factors E2F and DP1 to maintain cells in G_1 or G_0.
   b. pRB binds to p53 to stimulate apoptosis.
   c. pRB is a growth factor, stimulating cellular proliferation.
   d. pRB is involved in DNA mismatch repair.
7. Which protein phosphorylates the product of the *RB* gene, pRB?
   a. a receptor kinase
   b. phosphorylase
   c. cyclin/cyclin-dependent kinase
   d. p53
8. What is the relationship between a proto-oncogene and its corresponding viral oncogene?
   a. Only the viral oncogene contains introns.
   b. If the viral oncogene encodes a transcription factor, the corresponding proto-oncogene will encode a growth factor.
   c. The proto-oncogene encodes a product essential to the normal development and function of the organism. The viral oncogene encodes an altered product that has aberrant function.
   d. Usually, the proto-oncogene product is produced in greater amounts. The viral oncogene, being under the control of the retroviral promoter, enhancer, and poly(A) signals, produces smaller amounts of a product.
9. What is apoptosis?
   a. the process by which cancerous cells extend cellular processes in an amoeba-like manner
   b. the process of uncontrollable cellular growth caused by overexpression of an oncogene
   c. cell death that results from the multiple mutations that accumulate in a cancerous cell
   d. the process of programmed cell death
10. A genetic predisposition to hereditary nonpolyposis colon cancer (HNPCC) can result from mutations in any of the four genes *hMSH2, hMLH1, hPMS1,* and *hPMS2.* All of these genes are homologous to *E. coli* and yeast genes known to be involved in DNA repair. From this information, one might speculate that HNPCC is caused by a mutation
    a. in a mutator gene
    b. in an oncogene

c. in a proto-oncogene
d. in a tumor suppressor gene

Answers: 1b, 2b, 3c, 4b, 5d, 6a, 7c, 8c, 9d, 10a

Thought Questions

1. What is the relationship of the cell cycle to cancer?
2. What is the two-hit mutation model for cancer? How is it related to the multistep nature of cancer?
3. In all retroviral RNA genomes, specific sequences are found at the left (R and U_5) and right (R and U_3) ends. In proviral DNA, long terminal repeats (LTRs) are found at each end. How are the LTRs produced?
4. What is meant by a nononcogenic retrovirus? Can infection by a nononcogenic retrovirus lead to increased susceptibility to cancer? Give an example.
5. Distinguish between a viral oncogene, a cellular oncogene, and a proto-oncogene. Are each of these always associated with a transducing retrovirus?
6. There are at least six functional classes of oncogene products. What are these functional classes? How can alteration in such diverse functions lead to a similar oncogenic phenotype?
7. What information would you want to know about the UV lights used in tanning salons? Do you think they can be safe?
8. In what ways is programmed cell death a normal process?
9. What is the most lethal common carcinogen you know of?
10. Cancers are staged according to how far they have progressed toward metastasis. This is illustrated in text Figure 18.11 for the development of hereditary adenomatous polyposis, a colorectal cancer. What genetic changes are associated with the staged progression of tumors? Is there a specific order of changes that *always* occurs? Once a tumor has progressed to a certain stage, can it revert to an earlier stage?

Thought Questions for Media

After reviewing the media at the *Essential iGenetics* website, try to answer these questions.

1. How do growth-stimulating factors and growth-inhibiting factors differ, if at all, in terms of being able to
   a. turn on pathways?
   b. bind to specific cell membrane receptors?
   c. transduce signals to the nucleus?
   d. trigger the expression of one or more genes for one or more proteins that stimulate cell division?
2. Many growth-stimulating and growth-inhibiting factors have been identified. Is it the presence or absence of specific factors, the shear number of stimulating or inhibiting factors, or the balance between factors that determines whether a normal, healthy cell in the G_1 phase of the cell cycle is triggered to proceed into the S, G_2, and M phases?
3. After DNA damage occurs, how are the levels of p53 protein altered?
4. What is the immediate consequence of altering the level of p53 protein?
5. Why do normal cells undergo apoptosis after extensive DNA damage? Why don't cells with mutations in each of their *p53* genes undergo apoptosis?

6. How is a cyclin-Cdk complex involved in controlling the ability of a cell to proceed from G_1 into S phase? What is the role of the p21 protein in regulating this process?
7. Why are cells that proceed into S phase with unrepaired DNA damage more likely to become cancerous?
8. The animation illustrated how the p53 protein, following DNA damage, triggers a cellular pathway with inhibitory and stimulatory events.
   a. Which cellular events are inhibited by p53?
   b. Which cellular events are stimulated by p53?
   c. How are these inhibitory and stimulatory events connected?

SOLUTIONS TO TEXT PROBLEMS

18.1 What is the difference between a hereditary cancer and a sporadic cancer?

Answer: Hereditary cancer is associated with the inheritance of a germ-line mutation; sporadic cancer is not. Consequently, hereditary cancer "runs" in families. For some cancers, both hereditary and sporadic forms exist, with the hereditary form being much less frequent. For example, retinoblastoma occurs when both normal alleles of the tumor suppressor gene *RB* are inactivated. In hereditary retinoblastoma, a mutated, inactive allele is transmitted via the germ line. Retinoblastoma occurs in cells of an *RB*/+ heterozygote when an additional somatic mutation occurs. In the sporadic form of the disease, retinoblastoma occurs when both alleles are inactivated somatically.

18.2 Distinguish between a transducing retrovirus and a nontransducing retrovirus.

Answer: Retroviruses are viruses that have two copies of an RNA genome within a protein core surrounded by a membranous envelope. Transducing retroviruses carry an oncogene from a host cell, while nontransducing retroviruses do not. Consequently, transducing retroviruses may be capable of transforming cells to a cancerous state.

18.3 In what ways is the mechanism of cell transformation by transducing retroviruses fundamentally different from transformation by DNA tumor viruses? Even though the mechanisms are different, how are both able to cause neoplastic growth?

Answer: Tumor growth induced by transforming retroviruses results either from the activity of a single viral oncogene or from the activation of a proto-oncogene due to the nearby integration of the proviral DNA. The oncogene can cause abnormal cellular proliferation via the variety of mechanisms discussed in the text. The expression of a proto-oncogene, normally tightly regulated during cell growth and development, can be altered if it comes under the control of the promoter and enhancer sequences in the retroviral LTR.

Although DNA tumor viruses do not carry oncogenes, they are capable of transforming cells through the action of one or more genes within their genomes. For example, in a rare event, the DNA virus can be integrated into the host genome, and the DNA replication of the host cell may be stimulated by a viral protein that activates viral DNA replication. This would cause the cell to move from the G_0 to the S phase of the cell cycle.

For both transducing retroviruses and DNA tumor viruses, abnormally expressed protein leads to the activation of the cell from G_0 to S and abnormal cell growth.

18.4 Although there has been a substantial increase in our understanding of the genetic basis for cancer, the vast majority of cases of many types of cancer are not hereditary.

a. How might studying a hereditary form of a cancer provide insight into a similar, more frequent sporadic form?
b. The incidence of cancer in several members of an extended family might reasonably raise concern as to whether there is a genetic predisposition for cancer in the family. What does the term *genetic predisposition* mean? What might be the basis of a genetic predisposition to a cancer that appears as a dominant trait? What issues must be addressed before concluding that a genetic predisposition for a specific type of cancer exists in a particular family?

Answer:

a. Studies of hereditary forms of cancer have led to insights into the fundamental cellular processes affected by cancer. For example, substantial insights into the important role of DNA repair and the relationship between the control of the cell cycle and DNA repair have come from analyses of the genes responsible for hereditary forms of human colorectal cancer. For breast cancer, studying the normal functions of the *BRCA1* and *BRCA2* genes promises to provide substantial insights into breast and ovarian cancer.
b. Genetic predisposition for cancer refers to the presence of an inherited mutation that, with additional somatic mutations during the individual's life span, can lead to cancer. For diseases such as retinoblastoma, a genetic predisposition has been associated with the inheritance of a recessive allele of the *RB* tumor suppressor gene. Retinoblastoma occurs in *RB*/+ individuals when the normal allele is mutated in somatic cells and the pRB protein no longer functions. Because somatic mutation is likely, the disease appears dominant in pedigrees.

   Though there is a substantial understanding of the genetic basis for cancer and the genetic abnormalities present in somatic cancerous cells, there are also substantial environmental risk factors for specific cancers. Environmental risk factors must be investigated thoroughly when a pedigree is evaluated for a genetic predisposition for cancer.

18.5 Cellular proto-oncogenes and viral oncogenes are related in sequence, but they are not identical. What is the fundamental difference between the two?

Answer: Cellular proto-oncogenes have important roles in regulating normal cellular processes, such as cell division and differentiation. Viral oncogenes are mutated, abnormally expressed forms of proto-oncogenes that cause neoplastic growth. Viral oncogenes lack introns and, as shown in Figure 18.10, are often fused to viral genes.

18.6 Material that has been biopsied from tumors is useful for discerning both the type of tumor and the stage to which a tumor has progressed. It has been known for a long time that biopsied tissues with more differentiated cellular phenotypes are associated with less advanced tumors. Explain this finding in terms of the multistep nature of cancer.

Answer: Tumors result from multiple mutational events that typically involve both the activation of oncogenes and the inactivation of tumor suppressor genes. The analysis of hereditary adenomatous polyposis, an inherited form of colorectal cancer, has shown that the more differentiated cells found in benign, early-stage tumors are associated with fewer mutational events, while the less differentiated cells found in malignant and metastatic tumors are associated with more mutational events. Although the path by which mutations accumulate varies between tumors, additional mutations that activate oncogenes and inactivate tumor suppressor genes generally result in dedifferentiation and increased cellular proliferation.

18.7 An autopsy of a cat that died from feline sarcoma revealed neoplastic cells in the muscle and bone marrow but not in the brain, liver, or kidney. To gather evidence for the hypothesis that the virus FeSV (Figure 18.7c) contributed to the cancer, Southern blot analysis (see Chapter 13, pp. 277–278) was performed on DNA isolated from these tissues and on a cDNA clone of the FeSV viral genome. The DNA was digested with the enzyme *Hin*dIII, separated by size on an agarose gel, and transferred to a membrane. The resulting Southern blot was hybridized with a ^{32}P-labeled probe made from a 1.0-kb *Hin*dIII fragment of the feline *fes* proto-oncogene cDNA. The autoradiogram revealed a 3.4-kb band in each lane, with an additional 1.2-kb band in the lanes with muscle and bone marrow DNA. Only a 1.2-kb band was seen in the lane loaded with *Hin*dIII-cut FeSV cDNA. Explain these results, including the size of the bands seen. Do these results support the hypothesis?

Answer: If FeSV contributed to the feline sarcoma, evidence of the presence of FeSV should be found in the neoplastic tissues (muscle and bone marrow). The Southern blot provides this evidence, as a 1.2-kb DNA fragment hybridizes to the *fes* cDNA probe in the lanes with DNA from muscle and bone marrow as well as in the control lane with FeSV cDNA. The difference in size of the 1.2-kb hybridizing fragment and the size of the 1.0-kb *fes* proto-oncogene *Hin*dIII-cut cDNA probe reflects their different origins. The *fes* proto-oncogene is found normally in a cat, while the FeSV *fes* oncogene is found normally in a cat, while the FeSV *fes* oncogene is found in a retrovirus. The size of the fragment in the retrovirus may reflect a polymorphic *Hin*dIII site and/or a gene rearrangement. Clues to the origin of the 3.4-kb DNA fragment come from observing that it is found in all of the cat tissues and the realization that the *fes* proto-oncogene functions normally in the cat, and so its DNA should also be present in all tissues. Thus, the 3.4-kb DNA fragment is most likely a genomic sequence present in all tissues. Since the *fes* proto-oncogene has a 1.0-kb *Hin*dIII fragment, the mRNA of this gene is very likely spliced to remove an intronic region, 2.4 kb in size.

18.8 The sequences of proto-oncogenes are highly conserved among a large number of animal species. Based on this fact, what hypothesis can you make about the functions of the proto-oncogenes?

Answer: The high degree of conservation of proto-oncogenes suggests that they function in normal, essential, conserved cellular processes. Given the relationship between oncogenes and proto-oncogenes, it also suggests that cancer occurs when these processes are not correctly regulated.

18.9 Explain why HIV-1, the causative agent of AIDS, is considered a nononcogenic retrovirus even though numerous types of cancers are frequently seen in patients with AIDS.

Answer: HIV-1 is a retrovirus that infects and then kills cells of the immune system, particularly helper T cells. This leads to disabling of the immune response and an increase in susceptibility to certain types of cancer. In contrast to the mechanism by which cancer occurs following infection with HIV-1, an oncogenic retrovirus transforms the growth properties of infected cells to neoplastic growth.

18.10 List two ways in which cancer can be induced by a retrovirus.

Answer: One mechanism by which a transducing retrovirus causes cancer is for the viral oncogene it carries to be expressed under the control of retroviral promoters. A second mechanism involves insertional mutagenesis that results when the proviral DNA of a retrovirus integrates near a proto-oncogene. In this event, the expression of the proto-oncogene can be controlled by the promoter and enhancer sequences in the retroviral LTR. As these sequences do not respond to the environmental signals that normally regulate proto-oncogene expression, overexpression of the proto-oncogene is induced.

18.11 Proto-oncogenes produce a diverse set of gene products.

a. What types of gene products are made by proto-oncogenes? Do these gene products share any features?

b. Which of the following mutations might result in an oncogene?
   - i. a deletion of the entire coding region of a proto-oncogene
   - ii. a deletion of a silencer that lies 5′ to the coding region
   - iii. a deletion of an enhancer that lies 3′ to the coding region
   - iv. a deletion of a 3′ splice site acceptor region
   - v. the introduction of a premature stop codon
   - vi. a point mutation (single base pair change in the DNA)
   - vii. a translocation that places the coding region near a constitutively transcribed gene
   - viii. a translocation that places the gene near constitutive heterochromatin

Answer:

a. Proto-oncogenes encode a diverse set of gene products that include growth factors, receptor and nonreceptor protein kinases, receptors lacking protein kinase activity, membrane-associated GTP-binding proteins, cytoplasmic regulators involved in intracellular signaling, and nuclear transcription factors. These gene products all function in intercellular and intracellular circuits that regulate cell division and differentiation.

b. In general, mutations that activate a proto-oncogene convert it into an oncogene. Since (i), (iii), and (viii) cause a decrease in gene expression, they are unlikely to result in an oncogene. Since (ii) and (vii) could activate gene expression, they could result in an oncogene. Mutations (iv), (v), and (vi) cannot be predicted with certainty. The deletion of a 3′ splice site acceptor would alter the mature mRNA and possibly the protein produced and may or may not affect the protein's function and regulation. Similarly, it is difficult to predict the effect of a nonspecific point mutation or a premature stop codon. The text presents examples where these types of mutations have caused the activation of a proto-oncogene and resulted in an oncogene.

18.12 You have a culture of normal cells and a culture of cells dividing uncontrollably (isolated from a tumor). Experimentally, how might you determine whether uncontrolled growth was the result of an oncogene or a mutated pair of tumor suppressor alleles?

Answer: Using the method of Harris, one could experimentally fuse cells from the two cell lines and then test the resultant hybrids for their ability to form tumors. If the uncontrolled growth of the tumor cell line was due to a mutated pair of tumor suppressor alleles, the normal alleles present in the normal cell line would "rescue" the tumor cell-line defect. The hybrid line would grow normally and be unable to form a tumor. If the uncontrolled growth of the tumor cell line was due to an oncogene, the oncogene would also be present in the hybrid cell line. The hybrid line would grow uncontrollably and form a tumor.

18.13 What are the four main ways in which a proto-oncogene can be changed into an oncogene?

Answer: A proto-oncogene can be changed into an oncogene if there is an increase in the activity of its gene product or an increase in gene expression that leads to an increased amount of gene product. This can result from point mutation, deletion, gene amplification, and chromosomal translocation.

18.14 After a retrovirus that does not carry an oncogene infects a particular cell, northern blots indicate that the amount of mRNAs transcribed from a particular proto-oncogene became elevated approximately 13-fold compared with uninfected control cells. Propose a hypothesis to explain this result.

Answer: One hypothesis is that the proviral DNA has integrated near the proto-oncogene, and the expression of the proto-oncogene has come under the control of promoter and enhancer sequences in the retroviral LTR. This could be assessed by performing a whole-genome Southern blot analysis to determine whether the organization of the genomic DNA sequences near the proto-oncogene has been altered.

18.15 Explain how progression through the cell cycle is regulated by the phosphorylation of the retinoblastoma protein pRB. What phenotypes might you expect in cells where

a. pRB was constitutively phosphorylated?
b. pRB was never phosphorylated?
c. a severely truncated pRB protein was produced that could not be phosphorylated?
d. a normal pRB protein was produced at higher than normal levels?
e. a normal pRB protein was produced at lower than normal levels?

Answer: When unphosphorylated, pRB inhibits cell cycle progression by binding to a complex of the transcription factors E2F and DP1. This ensures that the cell remains in G_1 or in the quiescent G_0 state until it is ready to proceed to S. When progression into S phase is signaled, pRB is phosphorylated by a cyclin/Cdk complex. This prevents it from binding to E2F, which in turn allows E2F to activate transcription at genes needed for cell cycle progression. After S phase, pRB is dephosphorylated.

a. If pRB were constitutively phosphorylated, it would never bind to E2F, there would be no inhibition of cell cycle progression at G_1/S, and neoplastic growth would occur.
b. If pRB was never phosphorylated, cells would be stalled in the G_0 phase (and presumably eventually undergo apoptosis).
c. If a truncated protein were produced, it would be unable to bind to E2F and DP1 and (regardless of its phosphorylation state) would be unable to inhibit cell cycle progression. Just as in part (a), this would result in neoplastic growth.

d. Higher levels of normal pRB could result in a longer cell cycle. If more pRB were present in the cell, more unphosphorylated pRB would be available to bind to E2F during G_1. In this event, if an unphosphorylated pRB molecule that was bound to a particular E2F molecule were phosphorylated by a cyclin/Cdk complex and the E2F molecule were released, another unphosphorylated pRB molecule present in the cell could bind to the same E2F molecule. This would prevent this particular E2F molecule from participating in activating the gene expression needed for progression into S phase. Only when enough pRB molecules are phosphorylated will a sufficient number of E2F molecules be released to provide for transcriptional activation. Since the number of pRB molecules is increased relative to the number of cyclin/Cdk complexes, it will take longer for the cyclin/Cdk complexes to phosphorylate enough pRB molecules to effect transcriptional activation. If pRB levels are so high that the cyclin/Cdk complex is unable to phosphorylate enough of the pRB molecules present, cells could be stalled in the G_0 phase, just as in part (b).

e. Lower levels of normal pRB might result in a shorter cell cycle.

18.16 Mutations in the *p53* gene appear to be a major factor in the development of human cancer.

a. Explain what the normal cellular functions of the *p53* gene product are and how alterations in these functions can lead to cancer.

b. Suppose cells in a cancerous growth are shown to have a genetic alteration that results in diminished *p53* gene function. Why can we not immediately conclude that the mutation *caused* the cancer? How would the effect of the mutation be viewed in light of the current, multistep model of cancer?

Answer:

a. The *p53* gene product has multiple functions in tumor suppression. One of its chief roles is to maintain cells in G_0/G_1 arrest until DNA damage has been repaired and, if such damage cannot be repaired, to signal cells to undergo apoptosis. Alterations in *p53* function can lead to cancer in three ways: failure to hold cells in G_0/G_1 arrest, failure to allow DNA repair prior to S (thus allowing mutations to be transmitted to the next generation), or failure to induce apoptosis appropriately.

b. Cancerous cells typically have a number of genetic abnormalities, or lesions. The *p53* mutation may not be the initial genetic abnormality that occurred. Rather, it may have occurred subsequent to a mutation that led to the activation of an oncogene, the introduction of a mutator gene, or the inactivation of alleles at a tumor suppressor gene. Hence, the *p53* mutation may not be the primary genetic lesion. Most cancers develop as a result of the accumulation of mutations in a number of genes, often over extended periods of time (decades). Such mutations accumulate in part due to the lack of successful DNA repair at the G_1/S checkpoint in the cell cycle.

18.17 What is apoptosis? Why is the cell death associated with apoptosis desirable, and how is it regulated?

Answer: Apoptosis is programmed, or suicidal, cell death. Cells targeted for apoptosis are those that have large amounts of DNA damage and so are at a greater risk for neoplastic transformation. (During the development of some tissues in multicellular organisms, cell death via apoptosis is a normal process.) Apoptosis is regulated by p53, among other proteins. In cells with large amounts of DNA damage, p53 is stabilized. This leads to the activation of *WAF1,* whose product, p21, causes cells to arrest in G_1 by binding to cyclin/Cdk complexes and blocking the kinase activity needed for progression from G_1 to S. If the DNA damage cannot be repaired, apoptosis will be induced by p53.

18.18 What mechanisms ensure that cells with heavily damaged DNA are unable to replicate?

Answer: To proceed from G_1 into S, one or more G_1 cyclins bind to the cyclin-dependent kinase CDC28/cdc2 and activate it. The cyclin-dependent kinase then phosphorylates key proteins needed for progression into S. In the presence of heavy DNA damage, p53 is stabilized. The p53 protein acts as a transcription factor to activate *WAF1,* which produces p21. The p21 protein binds to cyclin/Cdk complexes and blocks the kinase activity required to activate the genes needed for the cell to make the transition from G_1 to S. Thus, stabilization of p53 leads to cell cycle arrest at the G_1/S cell cycle checkpoint. This G_1 arrest allows the cell time to induce the necessary repair pathways and repair the DNA or, if damage is too severe, to undergo apoptosis.

18.19 How do radiation and chemical carcinogens induce cancers?

Answer: Radiation and chemical carcinogens increase the amount of DNA damage in cells. If this damage is not repaired or is repaired incorrectly, somatic mutations result. Increasing the amount of DNA damage therefore increases the chance of somatic mutation. Mutations in oncogenes or tumor suppressor genes will contribute to the induction of cancer.

Direct-acting chemical carcinogens bind to DNA and act as mutagens. In contrast, procarcinogens are converted metabolically to active derivatives, so-called ultimate carcinogens, most of which also bind to DNA and act as mutagens. Radiation carcinogens act directly to cause mutation in DNA.

19

DNA MUTATION AND REPAIR

CHAPTER OUTLINE

REVIEW OF KEY TERMS, SYMBOLS, AND CONCEPTS

In your own words, write a brief, precise definition of each term in the groups on the following page. Check your definitions using the text. Then develop a concept map using the terms in each list.

THINKING ANALYTICALLY

Mutation is a fundamental process in genetics and evolution. In this chapter, the conceptual framework used by geneticists to consider the nature of mutation is presented. This is followed by a detailed discussion of the types and causes of mutation, strategies to identify mutants and mutagens, and the biological mechanisms used to repair mutations. Consequently, the material

1	2	3	4
mutation adaptation fluctuation test random mutation base pair mutation, base pair substitution somatic vs. germ-line mutation gene vs. chromosomal mutation point mutation transition mutation transversion mutation missense mutation nonsense mutation neutral mutation silent mutation frameshift mutation forward mutation reverse mutation (reversion) true reversion suppressor mutation, gene second-site mutation intragenic, intergenic suppressor	mutagen carcinogen spontaneous mutation mutation rate vs. mutation frequency tautomer, tautomeric shift depurination, deamination mutational hot spot induced mutation ionizing, nonionizing radiation thymine dimer SOS system base analogue 5BU, AZT, MMS base-modifying agent intercalating agent acridine, proflavin site-specific in vitro mutagenesis Ames test	DNA repair proofreading mutator mutation vs. mutator gene photoreactivation light repair excision (dark) repair alkylating repair O^6-methyl guanine methyltransferase mismatch repair SOS response error-prone repair error-prone bypass synthesis xeroderma pigmentosum ataxia-telangiectasia Fanconi anemia Bloom syndrome Cockayne syndrome hereditary nonpolyposis colon cancer	mutant screen visible mutation replica plating nutritional mutation auxotrophic mutation conditional mutation knockout mutants targeted mutation

presented here is both highly conceptual and quite detailed. Focus first on the concepts, and as in previous chapters, learn to use the terms and definitions to explain the concepts. To follow some of the complicated processes described in this chapter, it is essential to have the precise meaning of each term well in mind.

As you consider the molecular basis of mutation, it will help to review the chemical structure of the bases to envision how each of the mutagens acts. Employ diagrams as you consider the effects of specific mutagens and repair processes on the structure of DNA. It will also be invaluable to use diagrams to follow how a mutant DNA sequence can result in specific alterations in codon usage in an mRNA. It is essential to proceed methodically (and slowly, so as not to overlook a subtle point such as changes in polarity) through any analysis involving base pair changes. Use terms carefully, remembering that mutations affect the bases in DNA and then, indirectly, the bases in transcribed RNA and the amino acids in proteins.

Two topics in this chapter, suppressor mutations and screening procedures used in the isolation of new mutations, warrant particularly careful attention. Proceed by carefully and methodically organizing the information that is presented. Then supplement your reading with diagrams and flowcharts.

QUESTIONS FOR PRACTICE

Multiple Choice Questions

Match the best choice below to each of the descriptions in questions 1–8.

a. missense mutation	e. nonsense mutation
b. transversion mutation	f. transition mutation
c. neutral mutation	g. frameshift mutation
d. forward mutation	h. suppressor mutation

1. A purine-pyrimidine base pair that is mutated to a different purine-pyrimidine base pair
2. A point mutation in the DNA that changes a codon in the mRNA and causes an amino acid substitution that does not alter the function of the translated protein
3. A DNA change in which a GC base pair is replaced by a TA base pair
4. A mutation that results in the addition or deletion of a base pair within the coding region of a gene
5. Any point mutation that is expressed as a change from the wild type to the mutant phenotype
6. A base pair change in the DNA that results in the change of an mRNA codon to either UAG, UAA, or UGA
7. A mutation in the DNA that changes a codon in the mRNA and causes an amino acid substitution that may or may not produce a change in the function of a translated protein
8. Any new mutation that restores some or all of the wild-type phenotype to a previously isolated mutation
9. The Ames test is used to screen for
   a. potential mutagens and carcinogens
   b. the presence of enol forms of thymine and guanine
   c. frameshift mutations
   d. spontaneous mutations
10. In the Ames test, rat liver extracts are used
    a. to chemically alter and detoxify potential mutagens
    b. to chemically alter and toxify potential mutagens
    c. to determine if an environmental chemical that itself is not mutagenic may become mutagenic when processed in the liver
    d. all of the above
11. A population of cells is subjected to UV irradiation and then placed in the dark. There will be significant
    a. production of base tautomers
    b. breakage of phosphodiester bonds
    c. formation of pyrimidine dimers
    d. photolyase activity
12. Single bacterial cells sensitive to the infection of a phage are inoculated into a large number of separate culture dishes and allowed to grow in parallel. After many generations, a sample

is taken from each culture, inoculated with the phage, and separately plated. Which result is expected?

a. All plates will have similar numbers of bacterial colonies.
b. No bacterial colonies will be seen on any of the plates.
c. Only one plate in 10^6 will have bacterial colonies.
d. Some plates will have no bacterial colonies, some will have a few, and some will have many.

Answers: 1f, 2c, 3b, 4g, 5d, 6e, 7a, 8h, 9a, 10d, 11c, 12d

Thought Questions

1. What is the evidence that mutations occur spontaneously and at a low frequency, regardless of selection for or against them?
2. The proofreading ability of DNA polymerase is extremely good, but not quite perfect. What error rates are associated with DNA replication, from whence do they arise, and how might this be important in evolutionary terms?
3. Distinguish between a reversion mutation, an intragenic suppressor mutation, and an intergenic suppressor mutation.
4. Distinguish between a missense and a nonsense mutation and between a missense, a silent, and a neutral mutation.
5. How would you classify the following mutations: (a) a single base change in a promoter region that affects the transcription of a gene; (b) a single base change in a promoter region that has no effect; (c) a RFLP difference between two strains that has no apparent phenotypic effect?
6. What is the basis for the Ames test? What are its specific merits? Are there any kinds of mutagens that it might not detect (e.g., those that specifically cause point mutations, frameshifts, small deletions, or chromosome rearrangements)?
7. In the last several years, significant insight into how DNA is repaired has come from studies in bacteria and yeast, as well as from the molecular cloning of disease genes from humans with defective DNA repair. (a) Speculate as to the phenotype(s) of bacterial and yeast mutations that are defective in DNA repair. (b) How might comparison of these three systems be beneficial?
8. In *Drosophila,* the first mutations isolated (e.g., the first white-eyed mutant) were spontaneous. Currently, most new mutations are isolated by using mutagens. Why are mutagens currently employed, and what benefit do mutagens have for the study of genetics and development? How is site-specific in vitro mutagenesis used to study the function of cloned genes?
9. What spontaneous mutation rates are found in different organisms? Are these rates relatively similar or quite different? Why do you think this is so?
10. Dyes are routinely used to stain nuclei in bacteriological and histological preparations as well as to visualize DNA and RNA molecules separated by size using agarose gel electrophoresis. Some of these dyes bind to the backbone of nucleic acids, while others intercalate between bases. How would you specifically test whether a particular dye is mutagenic? How would you use the results of the test to design general precautions for laboratory workers using these dyes?
11. If a substance is mutagenic, is it necessarily carcinogenic? Why or why not? If a substance is carcinogenic, is it necessarily mutagenic? Why or why not?

Thought Questions for Media

After reviewing the media at *Essential iGenetics* website, try to answer these questions.

1. What are the consequences of a nonsense mutation?
2. What are the consequences of suppressing a nonsense mutation?
3. Suppose the mutation in the animation was an AT-to-GC transition instead of an AT-to-TA transversion.
   a. What would be the result of this mutation on the gene's protein product?
   b. Explain how this mutation could be suppressed using a mechanism similar to that illustrated in the animation.
4. The *Mutagenic Effects of 5BU* animation shows how 5BU can introduce mutations that are either AT-to-GC transition mutations or GC-to-AT transition mutations. For each type of mutation:
   a. How many rounds of replication are involved in the formation of the mutation?
   b. In which round(s) of replication is 5BU required to be in its normal form?
   c. In which round(s) of replication is 5BU required to be in its rare form?
5. If 5BU is incorporated into DNA, but never shifts to its rare form during DNA replication, can it still cause a transition mutation?
6. Based on the information in the animation, is there a way for 5BU to cause a transversion mutation?
7. In principle, a mutagenic compound can cause either forward (new) mutations or reverse mutations.
   a. Does the Ames test assess the frequency of reverse mutations (*his* to *his*$^+$) or the frequency of forward mutations (say, *his*$^+$ to *his*)? How?
   b. Why is it advantageous for the Ames test to assess this type of mutation frequency?
   c. Does this strategy influence which types of mutagenic compounds can be detected using the Ames test?
8. According to the *Ames Test Protocol* animation, why is the S9 liver extract added to the bacteria during the Ames test?

1. In the iActivity, you observed that vinyl chloride causes a missense transversion mutation in the Ames test. Although the allowable EPA concentration of vinyl chloride is 2 μg/L, it is present at 140 μg/L in the groundwater of Russellville.
   a. Can you infer that it is the vinyl chloride in the drinking water that causes missense transversion mutations in the residents of Russellville? Why or why not?
   b. Suppose the concentration of vinyl chloride in a neighboring village is 2 μg/L, the EPA allowable concentration. Is the water safe to drink? What concerns, if any, would you have?
   c. Would your answers to (a) and (b) change if you knew that most of the adult residents of Russellville and the neighboring village smoked three packs of cigarettes a day?
2. In the iActivity, 1,2-dichloroethene was not mutagenic in the Ames test, but was found in the groundwater at about six times the allowable EPA concentration. Since its not mutagenic, why does the EPA bother to regulate how much 1,2-dichloroethene

can be found in groundwater? What other factors might be used to determine the allowable groundwater concentration of a chemical?

3. Suppose that sequence analysis like that performed in the iActivity shows that a chemical causes a silent transition. Since it does not alter the amino acid sequence of a polypeptide, should you still be concerned about the chemical as a potential mutagen?

SOLUTIONS TO TEXT PROBLEMS

19.1 Mutations are (choose the correct answer)

a. caused by genetic recombination
b. heritable changes in genetic information
c. caused by faulty transcription of the genetic code
d. usually but not always beneficial to the development of the individuals in which they occur

Answer: b

19.2 Answer true or false: Mutations occur more frequently if there is a need for them.

Answer: False. Mutations occur spontaneously at a more or less constant frequency, regardless of selective pressure. Once they occur, however, they can be selected for or against, depending on the advantage or disadvantage they confer. It is important not to confuse the frequency of mutations with selection.

19.3 Which of the following is *not* a class of mutation?

a. frameshift
b. missense
c. transition
d. transversion
e. none of the above (meaning all are classes of mutation)

Answer: e. All are classes of mutation.

19.4 Ultraviolet light usually causes mutations by a mechanism involving (choose the correct answer)

a. one-strand breakage in DNA
b. a light-induced change of thymine to alkylated guanine
c. induction of thymine dimers and their persistence or imperfect repair
d. inversion of DNA segments
e. deletion of DNA segments
f. all of the above

Answer: c. The key to this answer is the word "usually." The other choices might apply rarely, but not usually.

19.5 For the middle region of a particular polypeptide chain, the wild-type amino acid sequence and the amino acid sequence of several mutants were determined, as shown in the following table ("..." indicates additional, unspecified amino acids). For each mutant, say what DNA level change has occurred, whether the change is a base pair substitution mutation (transversion or transition, missense or nonsense) or a frameshift mutation, and in which codon the mutation occurred. (Refer to the codon dictionary in Figure 12.7.)

	Codon								
	1	2	3	4	5	6	7	8	9
a. Wild type	... Phe	Leu	Pro	Thr	Val	Thr	Thr	Arg	Trp
b. Mutant 1:	... Phe	Leu	His	His	Gly	Asp	Asp	Thr	Val
c. Mutant 2:	... Phe	Leu	Pro	Thr	Met	Thr	Thr	Arg	Trp
d. Mutant 3:	... Phe	Leu	Pro	Thr	Val	Thr	Thr	Arg	
e. Mutant 4:	... Phe	Pro	Pro	Arg					
f. Mutant 5	... Phe	Leu	Pro	Ser	Val	Thr	Thr	Arg	Trp

Answer: Start methodically and write out the potential codons for the normal protein. Let N represent any nucleotide, R a purine (A or G), and Y a pyrimidine (C or U). Then the codons can be written as:

	1	2	3	4	5	6	7	8	9
Normal:	Phe	Leu	Pro	Thr	Val	Thr	Thr	Arg	Trp
Codon:	UUY	UUR or CUN	CCN	ACN	GUN	ACN	ACN	CGN or AGR	UGG

Mutant 1: The alteration of all of the amino acids after codon 2 suggests that this mutation is a frameshift, either a deletion or an addition of a base pair in the DNA region that codes for the mRNA near codons 2 and 3. His is CAY, which can be generated from CCN by the insertion of a single A between the two Cs. If this is the case, one has:

	1	2	3	4	5	6	7	8	9
Mutant 1:	Phe	Leu	His	His	Gly	Asp	Asp	Thr	Val
Frameshift:	UUY	UUR or CUN	CAC	NAC	NGU	NAC	NAC	NCG or NGR	NUG or RUG

Now identify the unknown bases, given the amino acid sequence:

To code for His, the new fourth codon must be CAC.

To code for Gly, the new fifth codon must be GGU.

To code for Asp, the new sixth codon must be GAC.

To code for Asp, the new seventh codon must be GAC.

To code for Thr, the new eighth codon must be ACG.

To code for Val, the new ninth codon must be GUG.

Thus, one has:

	1	2	3	4	5	6	7	8	9
Mutant 1:	Phe	Leu	His	His	Gly	Asp	Asp	Thr	Val
Codon:	UUY	UUR or CUN	CAC	CAC	GGU	GAC	GAC	ACG	GUG

This analysis indicates that the codons used in the normal protein are:

	1	2	3	4	5	6	7	8	9
Normal:	Phe	Leu	Pro	Thr	Val	Thr	Thr	Arg	Trp
Codon:	UUY	UUR or CUN	CCC	ACG	GUG	ACG	ACA	CGG	UGG

Mutant 2: Codon 5 in mutant 2 encodes Met, instead of Val. This single change could be caused by a point mutation, where the G in codon GUG is changed to an A, leading to an AUG codon. This would occur by a CG-to-TA transition in the DNA.

Mutant 3: Mutant 3 has a normal amino acid sequence, but is prematurely terminated, indicating that a nonsense mutation occurred. The nonsense mutation occurred in the ninth codon. If an A base is substituted (for one of the Gs) or inserted (before or after the first G), one could have either UAG or UGA. Thus, either a frameshift or base substitution (CG-to-TA transition) occurred.

Mutant 4: Mutant 4 shows premature termination at codon 5 (indicating a nonsense codon there) and has missense mutations at codons 2 and 4. Compare the possible sequences of the mutant with the normal sequence to see if one mutational event can account for all of these phenomena.

	1	2	3	4	5	6	7	8	9
Normal:	Phe	Leu	Pro	Thr	Val	Thr	Thr	Arg	Trp
Normal Codon:	UUY	UUR or CUN	CCC	ACG	GUG	ACG	ACA	CGG	UGG
Mutant 4:	Phe	Pro	Pro	Arg	Stop				
Possible Mutant Codons:	UUY	CCN (CCC)	CCN (CCA)	AGR or CGN	UAA UAG UGA				

Consider codon 2 carefully. For Pro to be encoded by CCN in the mutant protein and be obtained by a single change from the second codon, Leu (in the normal protein) must be encoded by CUN. If the U were deleted (by an AT deletion in the DNA) and the N were a C, a CCC Pro codon and a frameshift would result. Codon 3 would become CCA and still code for Pro, codon 4 would become CGG and code for Arg, and codon 5 would become UGA, a nonsense codon.

Mutant 5: Mutant 5 shows an alteration of only the fourth amino acid (Thr to Ser), suggesting a point mutation. If the fourth codon were changed from ACG to UCG (by a TA-to-AT transversion in the DNA), this missense mutation would occur.

19.6 In mutant strain X of *E. coli*, a leucine tRNA that recognizes the codon 5′-CUG-3′ in normal cells has been altered so that it now recognizes the codon 5′-GUG-3′. A missense mutation, which affects amino acid 10 of a particular protein, is suppressed in mutant X cells.

a. What are the anticodons of the two Leu tRNAs, and what mutational event has occurred in mutant X cells?

b. What amino acid would normally be present at position 10 of the protein (without the missense mutation)?

c. What amino acid would be put in at position 10 if the missense mutation is not suppressed (i.e., in normal cells)?

d. What amino acid is inserted at position 10 if the missense mutation is suppressed (i.e., in mutant X cells)?

Answer:

a. If the normal codon is 5′-CUG-3′, the anticodon of the normal tRNA is 5′-CAG-3′. If a mutant tRNA recognizes 5′-GUG-3′, it must have an anticodon that is 5′-CAC-3′. The mutational event was a CG-to-GC transversion.

b. Since a leucine-bearing (mutant) tRNA can suppress the mutation, one presumes that leucine is normally present at position 10.

c. The mutant tRNA recognizes the codon 5′-GUG-3′, which codes for Val. In normal cells, a Val–tRNA. Valine would recognize the codon and insert Valine.

d. Leu

19.7 The mutant *lacZ-1* was induced by treating *E. coli* cells with acridine, whereas *lacZ-2* was induced with 5BU. What kinds of mutants are these likely to be? Explain. How could you confirm your predictions by studying the structure of the β-galactosidase in these cells?

Answer: Acridine is an intercalating agent and so can be expected to induce frameshift mutations. 5BU is incorporated into DNA in place of T. During DNA replication, it is likely to be read as C by DNA polymerase because of a keto-to-enol shift. This results in point mutations, usually TA-to-CG transitions. Considering these expectations, *lacZ-1* would probably result in a completely altered amino acid sequence after some point, although it might be truncated (due to the introduction—out of frame—of a nonsense codon). *lacZ-2* is likely to contain a single amino acid difference due to a missense mutation, although it, too, could contain a nonsense codon.

19.8 a. The sequence of nucleotides in an mRNA is:

5′-AUGACCCAUUGGUCUCGUUAG-3′

Assuming that ribosomes could translate this mRNA, how many amino acids long would you expect the resulting polypeptide chain to be?

b. Hydroxylamine is a mutagen that results in the replacement of an AT base pair for a GC base pair in the DNA; that is, it induces a transition mutation. When applied to the organism that made the mRNA molecule shown in part (a), a strain was isolated in which a mutation occurred at the 11th position of the DNA that coded for the mRNA. How many amino acids long would you expect the polypeptide made by this mutant to be? Why?

Answer:

a. The codons read as

5′-AUG-ACC-CAU-UGG-UCU-CGU-UAG-3′

The last codon is a nonsense (chain termination) codon, while the others are sense codons. The chain would be six amino acids long.

b. The new sequence would be

5′-AUG-ACC-CAU-UAG-. . .

Since UAG is a nonsense (chain termination) codon, the new chain would be only three amino acids long.

19.9 In a series of 94,075 babies born in a particular hospital in Copenhagen, 10 were achondroplastic dwarfs (this is an autosomal dominant condition). Two of these 10 had an achondroplastic parent. The other 8 achondroplastic babies each had two normal parents. What is the apparent mutation rate at the achondroplasia locus?

Answer: There were eight new mutations in 94,073 normal couples. Since the phenotype is dominant, the phenotype is seen when just one of the parental genes is mutated. There were 2 × 94,073 copies of the gene that could have undergone mutation. Therefore, the apparent mutation rate at this locus is

$$8/(2 \times 94{,}073) = 8/188{,}146 = 4 \times 10^{-5} \text{ mutations per locus per generation}$$

19.10 Three of the codons in the genetic code are chain-terminating codons for which no naturally occurring tRNAs exist. Just like any other codons in the DNA, though, these codons can change as a result of base pair changes in the DNA. Confining yourself to single base pair changes at a time, determine which amino acids could be inserted in a polypeptide by mutation of these chain-terminating codons: (a) UAG, (b) UAA, and (c) UGA. (The genetic code is given in Figure 12.7.)

	Codon					
Nucleotide Altered	**UAG**	**Code**	**UAA**	**Code**	**UGA**	**Code**
FIRST	AAG	Lys	AAA	Lys	AGA	Arg
	CAG	Gln	CAA	Gln	CGA	Arg
	GAG	Glu	GAA	Glu	GGA	Gly
SECOND	UUG	Leu	UUA	Leu	UUA	Leu
	UCG	Ser	UCA	Ser	UCA	Ser
	UGG	Trp	UGA	STOP	UAA	STOP
THIRD	UAC	Tyr	UAC	Tyr	UGC	Cys
	UAU	Tyr	UAU	Tyr	UGU	Cys
	UAA	STOP	UAG	STOP	UGG	Trp

19.11 The amino acid substitutions in the following figure occur in the α and β chains of human hemoglobin. Those amino acids connected by lines are related by single nucleotide changes. Propose the most likely codon or codons for each of the numbered amino acids. (Refer to the genetic code in Figure 12.7.)

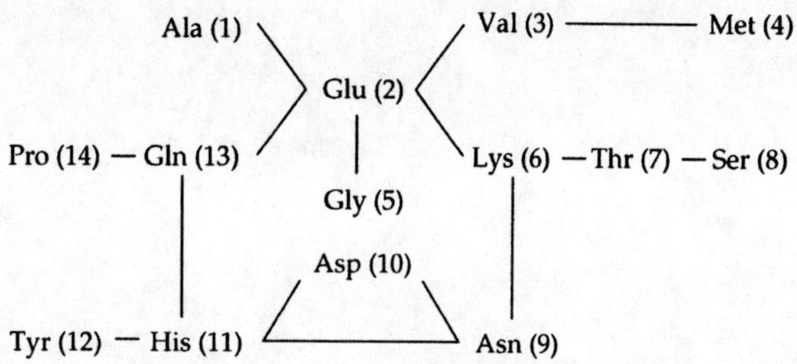

Answer:

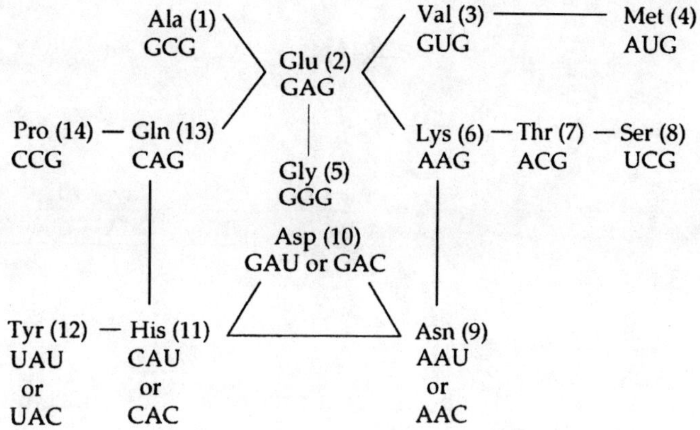

19.12 Yanofsky studied the tryptophan synthetase of *E. coli* in an attempt to identify the base sequence specifying this protein. The wild type gave a protein with a glycine in position 38. Yanofsky isolated two *trp* mutants, *A23* and *A46*. Mutant *A23* had Arg instead of Gly at position 38, and mutant *A46* had Glu at position 38. Mutant *A23* was plated on minimal medium, and four spontaneous revertants to prototrophy were obtained. The tryptophan synthetase from each of the four revertants was isolated, and the amino acids at position 38 were identified. Revertant 1 had Ile, revertant 2 had Thr, revertant 3 had Ser, and revertant 4 had Gly. In a similar fashion, three revertants from *A46* were recovered, and the tryptophan synthetase from each was isolated and studied. At position 38 revertant, 1 had Gly, revertant 2 had Ala, and revertant 3 had Val. A summary of these data is given in the following figure. Using the genetic code in Figure 12.7, deduce the codons for the wild type, for the mutants *A23* and *A46*, and for the revertants, and place each designation in the space provided in the figure on the following page.

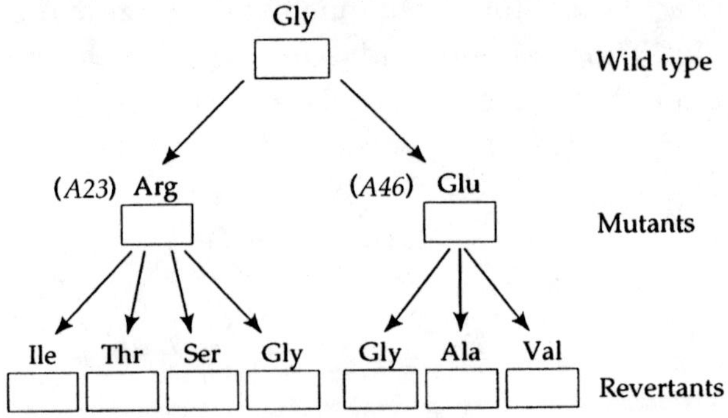

Answer:

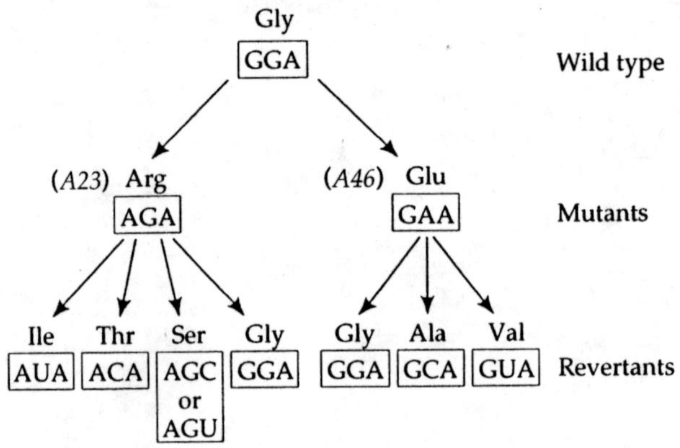

19.13 Consider an enzyme chewase from a theoretical microorganism. In the wild-type cell, the chewase has the following sequence of amino acids at positions 39 to 47 (reading from the amino end) in the polypeptide chain:

```
-Met-Phe-Ala-Asn-His-Lys-Ser-Val-Gly-
 39  40  41  42  43  44  45  46  47
```

A mutant of the organism was obtained; it lacks chewase activity. The mutant was induced by a mutagen known to cause single base pair insertions or deletions. Instead of making the complete chewase chain, the mutant made a short polypeptide chain only 45 amino acids long. The first 38 amino acids were in the same sequence as the first 38 of the normal chewase, but the last seven amino acids were as follows:

```
-Met-Leu-Leu-Thr-Ile-Arg-Val-
 39  40  41  42  43  44  45
```

A partial revertant of the mutant was induced by treating it with the same mutagen. The revertant made a partly active chewase, which differed from the wild-type enzyme only in the following region:

```
-Met-Leu-Leu-Thr-Ile-Arg-Gly-Val-Gly
 39  40  41  42  43  44  45  46  47
```

Using the genetic code given in Figure 12.7, deduce the nucleotide sequences for the mRNA molecules that specify this region of the protein in each of the three strains.

Answer: Using the genetic code, and letting N represent any nucleotide, R a purine, and Y a pyrimidine, the wild-type RNA can be denoted as

```
-Met-Phe-Ala-Asn-His-Lys-Ser-Val-Gly-
 39  40  41  42  43  44  45  46  47
UCN
AUG UUY GCN AAY CAY AAR  or  GUN GGN
AGY
```

Since the mutants were obtained using a mutagen that causes single base pair insertions or deletions, the mutants have frameshift mutations. Considering this fact and comparing the mutant amino acid sequence with that of the wild type, one expects a base pair to be missing or added in codon 40. Try various options to obtain an appropriate amino acid sequence after such a frameshift. If the Y of codon 40 was deleted, an appropriate sequence can be obtained, providing the unknown nucleotides are as specified below:

```
-Met-Leu-Leu-Thr-Ile-Arg-Val-
 39  40  41  42  43  44  45
AUG UUG CNA AYC AYA ARA GYG UNG
N=U Y=C Y=U R=G Y=U N=A
```

This means that the original sequence was

```
-Met-Phe-Ala-Asn-His-Lys-Ser-Val-Gly-
 39  40  41  42  43  44  45  46  47
AUG UUY GCU AAC CAU AAG AGU GUA GGN
```

In the revertant, the reading frame is restored at codon 46, the Val (GUG) at codon 45 is altered to a Gly (GGN), and the chain termination codon at position 46 (UAG) is altered to a Val (GUN). This would have occurred if a G was inserted before or after the first G in codon 45. One would have

```
-Met-Leu-Leu-Thr-Ile-Arg-Gly-Val-Gly-
 39  40  41  42  43  44  45  46  47
AUG UUG CUA ACC AUA AGA GGU GUA GGN
```

19.14 DNA polymerases from different organisms differ in their fidelity of nucleotide insertion; however, even the best DNA polymerases make mistakes. These are usually mismatches. If such mismatches are not corrected, they can become fixed as mutations after the next round of replication.

a. How does DNA polymerase attempt to correct mismatches during DNA replication?
b. What mechanism is used to repair such mismatches if they escape detection by DNA polymerase?

c. How is the mismatched base in the newly synthesized strand distinguished from the correct base in the template strand?

Answer:

a. Many, but not all, DNA polymerases have a proofreading capacity that causes them to stall at mismatches during DNA replication. When present, the 3′→5′ exonuclease activity associated with the polymerase will excise mismatched bases, so that they can be replaced. See text discussion in Chapters 9 and 19 and Tables 9.1.

b. Shortly after DNA replication, mismatches can still be repaired by using a set of enzymes that recognize and excise mismatches in the newly synthesized DNA strand. In *E. coli,* mismatch repair is initiated when single base pair mismatches or small base pair additions or deletions in newly replicated DNA are bound by the MutS protein. The newly replicated strand is distinguished from the parental strand by the presence of an unmethylated A nucleotide in a 5′-GATC-3′ sequence close to the mismatch. The MutL and MutH proteins bring the unmethylated GATC close to the mismatch and form a complex with MutS. MutH nicks the new, unmethylated DNA strand, and a section of this DNA strand, including the mismatch, is excised by an exonuclease. DNA polymerase III and ligase repair the gap, producing the correct base pair. See Figure 19.16.

c. In *E. coli,* the parental strand can be distinguished from the newly replicated strand by methylation of the A in the sequence 5′-GATC-3′. As a newly replicated DNA strand is not methylated until a short time after its synthesis, hemimethylation at this frequently appearing site allows for the identification of which strand is newly synthesized. Since only the strand with the unmethylated A is nicked by the MutH protein and excised by an exonuclease during mismatch repair, only the newly synthesized strand and not the parental strand is resynthesized by DNA polymerase III.

19.15 Two mechanisms in *E. coli* were described for the repair of thymine dimer formation after exposure to ultraviolet light: photoreactivation and excision (dark) repair. Compare these mechanisms, indicating how each achieves repair.

Answer: Photoreactivation requires the enzyme photolyase, which, when activated by a photon of light with a wavelength between 320 and 370 nm, splits the dimers apart. Dark repair does not require light, but requires several different enzymes. First, the UvrABC endonuclease makes two single-stranded nicks, on the 5′ side and the 3′ side of the dimer. Then an exonuclease excises the 12-nucleotide-long segment of one strand between the nicks, including the dimer. Next, DNA polymerase I fills in the single-stranded region in the 5′→3′ direction. Finally, the gap is sealed by DNA ligase.

19.16 DNA damage by mutagens has very serious consequences for DNA replication. Without specific base pairing, the replication enzymes cannot specify a complementary strand, and gaps are left after the passing of a replication fork.

a. What response has *E. coli* developed to large amounts of DNA damage by mutagens? How is this response coordinately controlled?

b. Why is this response itself a mutagenic system?

c. What effects would loss-of-function mutations in *recA* or *lexA* have on this response?

Answer:

a. *E. coli* has developed the SOS response to respond to large amounts of DNA damage. When such damage occurs, the RecA protein becomes activated and stimulates LexA to cleave itself. Since the LexA protein functions as a repressor for about 17 genes, whose products are involved in DNA damage repair (it binds to a 20-nucleotide regulatory sequence called the SOS box), this results in the coordinate transcription of these genes. After the products of these genes "correct" the DNA damage, RecA is inactivated, and newly synthesized LexA represses the expression of the genes.
b. The response is mutagenic because in some instances (e.g., in the repair of C^C dimers), the SOS repair system becomes stalled because of a mismatch. The delay is long enough for C to be deaminated to U. This results in a CG-to-TA transition.
c. If there are loss-of-function mutations in both *recA* and *lexA*, or a loss-of-function mutation only in *lexA*, there would be no functional LexA protein to repress transcription of the 17 genes whose protein products are involved in the SOS response; this would result in constitutive activation of the SOS response. If the loss-of-function mutation is only in *recA*, however, heavy DNA damage would not result in activating the RecA protein, so RecA could not stimulate the LexA protein to cleave itself to induce the SOS response. Instead, the LexA protein would continue to repress the DNA repair genes in the SOS system. *E. coli* with a loss-of-function mutation in *recA* but with a normal *lexA* gene would, therefore, be deficient in SOS repair of heavy DNA damage and consequently would be sensitive to UV irradiation and to X rays. (*E. coli* with a *recA* loss-of-function mutation were observed to have this phenotype by Ann Dee Margulies and Alvin Clark, who discovered the *recA* gene in 1964.)

19.17 A protein contains the amino acid proline (Pro) at one site. Treatment with nitrous acid, which deaminates C to make it U, produces two different mutants. One mutant has a substitution of serine (Ser), and the other has a substitution of leucine (Leu) at the site.

Treatment of the two mutants with nitrous acid produces new mutant strains, each with phenylalanine (Phe) at the site. Treatment of these new Phe-carrying mutants with nitrous acid produces no change. The results are summarized in the following figure:

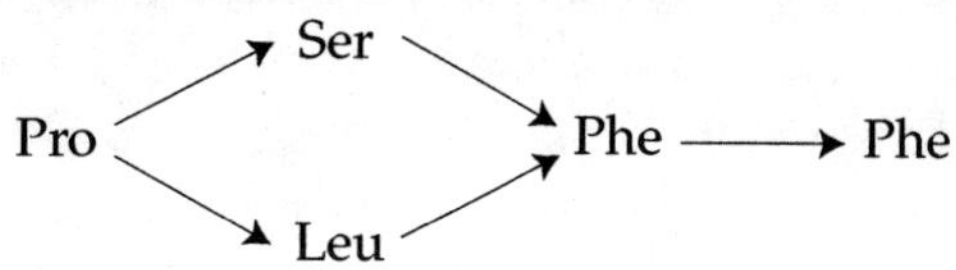

Using the appropriate codons, show how it is possible for nitrous acid to produce these changes and why further treatment has no influence. (Assume that only single nucleotide changes occur at each step.)

Answer: Nitrous acid deaminates C to make it U. U will pair with A, so that treatment with nitrous acid leads to CG-to-TA transitions. Analyze how this treatment would affect the codons of this protein. Use N to represent any nucleotide and Y to represent a pyrimidine (U or C). Then the codons for Pro are CCN, the (relevant) codons for Ser are UCN, the codons for Leu are CUN, and the codons for Phe are UUY. (Nucleotides are written in the 5′→3′ direction unless specifically noted.)

The codon CCN for Pro would be represented by CCN in the nontemplate DNA strand. Deamination of the 5′-C would lead to a nontemplate strand of UCN and a template strand of 3′-AGN-5′. This would produce a UCN codon, encoding Ser. Deamination of the middle C would lead to a nontemplate strand of CUN and a template strand of 3′-GAN-5′. This would produce a CUN codon, encoding Leu.

Further treatment of either mutant would result in deamination of the remaining C and a template strand of 3′-AAN-5′. This would result in a UUN codon. Since we are told that Phe is obtained, N must be C or U, and the template strand must be 3′-AAA-5′ or 3′-AAG-5′.

To explain why further treatment with nitrous acid has no effect, consider that nitrous acid acts via deamination and that T has no amine group. If the template strand was 3′-AAA-5′, the nontemplate strand would have been TTT. Since T cannot be deaminated, nitrous acid will not have any effect on the nontemplate strand.

19.18 In the past few years, many knockout mice that lack specific gene functions have been bred. In addition to the knockout for *p53* described in this chapter, knockouts for many different *Hox* genes (see Chapter 17), lysosomal hexosaminidase (the enzyme missing in Tay-Sachs disease), and many other enzymes have been made.

a. Why is the ability to make knockout mice important for treating and curing human disease?

b. Consider a situation in which a dominant mutation causes a human disease such as Huntington disease. A hypothetical researcher clones the disease gene and discovers that a novel, abnormal protein is responsible for the disease phenotype. Would you expect the phenotype of a knockout mouse to be similar to the disease phenotype? Why or why not?

Answer:

a. The production of knockout mice allows for the development of an animal model system to investigate human disease. It allows for greater investigation into the consequences of not having a particular gene function and also allows for the development of therapeutic approaches in a model system that closely parallels the human disease.

b. Since knockout mice lack a particular gene function, the phenotype of a knockout mouse will not necessarily mimic the phenotype of a dominant mutation that has an additional abnormal function. To design a mouse model in this instance, it is necessary to add a copy of a dominant allele into the normal genetic background or replace a normal gene with the dominant allele. Only in this way would the aberrant protein product of the gene be expressed and the phenotype it causes be observed.

20

TRANSPOSABLE ELEMENTS

CHAPTER OUTLINE

GENERAL FEATURES OF TRANSPOSABLE ELEMENTS

TRANSPOSABLE ELEMENTS IN PROKARYOTES

Insertion Sequences
Transposons

TRANSPOSABLE ELEMENTS IN EUKARYOTES

Transposons in Plants
Ty Elements in Yeast
P Elements in *Drosophila*
Human Retrotransposons

REVIEW OF KEY TERMS, SYMBOLS, AND CONCEPTS

In your own words, write a brief, precise definition of each term in the groups on the following page. Check your definitions using the text. Then develop a concept map using the terms in each list.

THINKING ANALYTICALLY

As the material in this chapter is broadly descriptive, it will be helpful to organize and categorize its information. Distinguish between the various categories of transposable elements by considering transposable elements from four perspectives: (1) structure, (2) mode of transposition, (3) the effect of a transposed element on the structure of a gene or chromosome, and (4) the effect of a transposed element on the function of a gene. As you read the chapter, refer frequently to the text figures to visualize more clearly the different transposon structures and modes of transposition.

QUESTIONS FOR PRACTICE

Multiple Choice Questions

1. How are transposition and recombination fundamentally different?
   a. Recombination requires DNA homology but transposition does not.
   b. Transposition requires DNA homology but recombination does not.
   c. Only during transposition can a piece of DNA be moved from a virus to a cell.
   d. Only during recombination can a deletion occur.

1	2
transposable element	transposon vs. retrotransposon
transposition	replicative vs. conservative transposition
transposition event	homologous recombination
insertion sequence (IS) element	deletion
IS module	inversion
inverted terminal repeat	*Ac*, *Ds* controlling elements
transposase	null mutation
target site	autonomous vs. nonautonomous element
target site duplication	donor site
transposon (Tn)	*Ty* element
illegitimate recombination	long terminal (direct) repeat
composite transposon	delta repeat
noncomposite transposon	target site duplication
Tn*10*, Tn*3*	reverse transcriptase
replicative transposition	hybrid dysgenesis
conservative (nonreplicative) transposition	*P* element
simple insertion	P, M cytotype
cointegrate	LINEs, SINEs
cointegration model	Alu, L1 elements
transposase	
resolvase	
IS element	

2. Which of the following can result from the insertion of a transposon in bacteria?
   a. gene inactivation
   b. an increase or decrease in transcriptional activity of a gene
   c. deletions or insertions
   d. all of the above
3. Which of the following can be consequences of transposition in eukaryotes?
   a. no effect on a nearby gene
   b. production of a mutation by insertional mutagenesis
   c. increased or decreased transcription from a nearby promoter
   d. all of the above
4. In what ways are bacterial IS and Tn elements alike?
   a. Both have inverted repeat sequences at their ends.
   b. Both integrate into target sites and cause a target site duplication.
   c. Both contain a transposase gene.
   d. Both contain antibiotic resistance genes.
   e. Both always use replicative transposition.
   f. a, b, and c
   g. all of the above

5. How might plasmids *X* and *Y* differ? *X* is an *F* factor in an F^+ bacterium sensitive to the antibiotics ampicillin, streptomycin, and tetracycline. *Y* is an *R* plasmid in a bacterium resistant to these antibiotics.
   a. *Y* is likely to be smaller in size.
   b. *X* is likely to lack the *tra* genes.
   c. *Y* is likely to have additional genes between flanking IS modules.
   d. *Y* is likely to lack the *RTF* genes.
6. Reverse transcriptase is used by which two of the following transposable elements?
   a. The *Ty* element in yeast.
   b. The IS element in bacteria.
   c. The TN *10* element in bacteria.
   d. The L1 element in humans.
7. In a bacterial IS element, how does transposase know the boundaries of the DNA to be transposed?
   a. There are special restriction sites at the ends of the IS element.
   b. It recognizes the target site of insertion.
   c. It recognizes the IR sequences.
   d. It recognizes the delta sequences.
8. In corn, the insertion of an autonomous element within a gene results in the production of an allele that mutates at high frequency. Why?
   a. Although an autonomous element cannot excise, it is highly mutable.
   b. An autonomous element has transposase and can excise, thereby causing a mutation.
   c. Autonomous alleles are unstable if nonautonomous elements are also in the genome.
   d. Autonomous alleles generally insert into promoter regions.
9. *Ty* elements in yeast are similar to retroviruses because (choose the correct answer)
   a. both *Ty* elements and retroviruses utilize an RNA transposition intermediate
   b. both *Ty* elements and retroviruses utilize reverse transcriptase
   c. introns are usually found in retroviruses and *Ty* elements
   d. more than one of the preceding
10. Which of the following is true about hybrid dysgenesis in *Drosophila?*
    a. It occurs in the progeny of M females after they are mated to P males.
    b. It occurs in the progeny of P females after they are mated to M males.
    c. It is a result of the mobilization of *P* elements in a P cytotype.
    d. It is a result of the mobilization of *M* elements in a P cytotype.
11. Which of the following statements is *not* true about the L1 family of LINE elements in mammals?
    a. They are retrotransposons.
    b. They comprise about 5 percent of the genome.
    c. All elements in one individual are full-length elements.
    d. They are capable of causing disease by insertional mutagenesis during an individual's lifetime.

Answers: 1a, 2d, 3d, 4f, 5c, 6a,d, 7c, 8b, 9d, 10a, 11c

Thought Questions

1. Contrast conservative and replicative transposition. How can conservative transposition of the *Ac* element result in gene duplication? Can particular IS sequences behave in both ways?
2. Consider how the distinction between direct and inverted repeats is central to the recognition of insertion sequences. The insertion sequence itself contains inverted terminal repeats. Thus, inverted repeats are present as a part of the transposable element both before and after the insertion. On the other hand, the act of insertion results in a staggered cut in the host DNA. When the gaps are filled in after insertion, direct repeats will be generated flanking the inverted terminals of the transposable element. Indeed, insertion sequences have been identified by the presence of a pair of inverted repeats flanked by direct repeats. Using this logic, find the insertion sequence in the following single-stranded length of DNA:

   CGTAGCCATTTGCGATATGCATCCGAATATCGCAAGCCATGCCA

3. In *Drosophila*, hybrid dysgenesis results in the mobilization of transposable elements. This in turn results in progeny flies that have normal somatic cells but are sterile. Why is the somatic tissue unaffected and only the germ line affected? Males from three *Drosophila* strains recently isolated from wild populations in different geographic areas are crossed with females from the same M cytotype laboratory strain. In the first cross, 12 percent of the progeny are sterile. In the second, 64 percent of the progeny are sterile. In the third, 97 percent of the progeny are sterile. Generate a hypothesis to address why different degrees of sterility are seen in these three strains.
4. In what situations can a nonautonomous element behave like an autonomous element? Give specific examples in corn and *Drosophila.*
5. Following the mobilization of *Ty* elements in yeast, two mutants were recovered that affected the activity of enzyme X. One of these mutants lacked enzyme activity and was a null mutation. The other mutant had three times as much enzyme activity. Both mutants had a *Ty* element inserted into the *X* gene. Generate a hypothesis to explain how this is possible.
6. What differences and similarities are there between transposable elements that utilize an RNA intermediate and those that do not?
7. What kinds of repetitive elements are SINEs and LINEs? Support for a hypothesis that a particular repetitive element is a transposon often comes by analyzing its DNA sequence for open reading frames that encode transposase-like products. In humans, Alu sequences do not encode any of the enzymes that are presumably needed for retrotransposition. What aspects of their structure suggest that they might be capable of retrotransposition? What evidence is there that Alu sequences are retrotransposons capable of moving via an RNA intermediate, much like the yeast *Ty* element?
8. How has understanding the basis for hybrid dysgenesis and the transposition of *P* elements allowed for the development of methods for DNA-mediated transformation of *Drosophila?* How are transformants selected for?
9. In humans, insertion of transposable elements has occasionally been associated with disease. An L1 element inserted into the gene for blood-clotting factor VIII resulted in hemophilia in two separate individuals. An Alu element insertion disrupted pre-mRNA processing at the gene for neurofibromatosis. In each case, the insertion was not in either of the affected individual's parents and so appeared to be a result of an insertional event during the affected individual's lifetime. Given that each of these classes of elements represents a substantial fraction of moderately repetitive DNA, how likely is it that most spontaneous mutations in humans are due to insertional mutagenesis by L1 elements? If it is not very likely, why not?

Thought Questions for Media

After reviewing the media at the *Essential iGenetics* website, try to answer these questions.

1. Suppose two different transposable elements are mobilized and insert themselves itself into two different genes. In gene A, transposon IS*1* inserts itself into the translated region. In gene B, transposon IS2 inserts itself at a site 3 base pairs downstream of the TATA box, but before the start of the transcribed region.
   a. What similarities do you expect to find between the sequences of the transposable element inserted in each gene?
   b. In addition to the sequence of the transposable elements, what sequence alterations do you expect to find in the region near the site of insertion?
   c. What features do the sequence alterations in (b) have, and how do they arise?
   d. How might the transposon insertions influence the functions of genes A and B? Are they similar or different? Why?

1. The experiment in the iActivity allowed you to demonstrate that Tn*10* transposition is conservative. Does it allow you to infer whether Tn*10* transposition proceeds through a DNA or RNA intermediate?
2. Suppose you performed an experiment similar to the one described in the iActivity to identify the mode of transposition of an uncharacterized transposon. You observe that none of the colonies exhibit any sectoring and that they are either blue or white. After picking several blue and several white colonies, you use PCR to amplify the region surrounding the *donor* DNA in each colony. You sequence the PCR products and compare the sequences to that of the molecules used to generate the heteroduplexes. What do you find, and how do you explain your findings?

SOLUTIONS TO TEXT PROBLEMS

20.1 Distinguish between prokaryotic insertion elements and transposons. How do composite transposons differ from noncomposite transposons?

Answer: Insertion elements (IS elements) are simpler in structure than transposons and have a transposase gene flanked by perfect or nearly perfect inverted repeat sequences (IR sequences). The more complex transposable elements (Tn elements) exist in two forms, composite transposons and noncomposite transposons. Composite transposons have a central gene-bearing region flanked by IS elements. Transposition of these elements, and the genes they contain (e.g., genes for antibiotic resistance), occurs because of the function of the IS elements. Noncomposite transposons also contain genes, but do not terminate with IS elements. However, they do have inverted terminal repeated sequences that are required for transposition.

Both types of elements can integrate into target sites with which they have no homology, where they cause target site duplications. The transposition of IS elements and composite transposons requires use of the host cell's replicative enzymes and the transposase encoded in the element. Noncomposite transposons can transpose by two different mechanisms. Tn*3*-like noncomposite transposons utilize a replicative transposition mechanism employing a cointegrate—a fusion between a transposable element and the recipient DNA. Genes within Tn3 code for the enzymes transposase and resolvase that are needed for its transposition. Tn*10*-like composite transposons can move by a conservative (non-replicative) transposition.

Upon insertion, all of the elements can cause mutations by a number of different mechanisms, including disruption of a gene's coding sequence or regulatory region.

20.2 What properties do bacterial and eukaryotic transposable elements have in common?

Answer: The structure and, at a general level, the function of eukaryotic and prokaryotic transposable elements are very similar. For example, both Tn and *Ac* elements have genes within them and have inverted repeats at their ends. Both prokaryotic and eukaryotic elements may affect gene function in a variety of different ways, depending on the element involved and how it integrates into or nearby a gene. The integration events of eukaryotic elements, like those of most prokaryotic transposable elements, involve nonhomologous recombination. Some eukaryotic elements, such as *Ty* elements in yeast and retrotransposons, move via an RNA intermediate, unlike the IS and Tn prokaryotic elements.

20.3 An IS element became inserted into the *lacZ* gene of *E. coli.* Later, a small deletion occurred in this gene, which removed 40 base pairs, starting to the left of the IS element. Ten *lacZ* base pairs were removed, including the left copy of the target site, and the 30 leftmost base pairs of the IS element were removed. What will be the consequence of this deletion?

Answer: The left inverted repeat of the IS element has been removed, so that the two ends of this IS element are no longer homologous. The element will not be able to move out of this location and insert into another site.

20.4 An understanding of the molecular structures of *Ac* and *Ds* elements, together with an understanding of the basis for their transposition, has allowed for a clearer, molecular-based interpretation of Barbara McClintock's observations. Ponder the significance of this interpretation on the acceptance of her work, and use your understanding of transposition of *Ac* and *Ds* elements in corn to propose an explanation for the following.

Two different true-breeding strains of corn with colorless kernels A and B are crossed with each of two different true-breeding strains with purple kernels C and D.

The F_1 from each cross is selfed, with the following results:

P:	A × C	A × D
F_1:	all purple	all purple
F_2:	3 purple : 1 colorless	3 purple : 1 colorless
P:	B × C	B × D
F_1:	all purple	all purple
F_2:	3 purple : 1 spotted*	3 purple : 1 colorless

* spotted = kernels with purple spots in a colorless background

Answer:

As is described in Figure 20.6, the spotted-kernel phenotype arises from the transposition of *Ds* elements (incomplete transposons incapable of transposition themselves) out of the *C* gene during kernel development. This means that the mutation in colorless strain B is caused by an insertion of a *Ds* element in the *C* gene. It is able to transpose because the purple strain C has *Ac* elements, full-length transposons, that can activate *Ds* transposition.

Since spotted kernels are not seen in any of the other crosses, transposition of *Ds* elements is not occurring in them. There are two different explanations for this. Since spotted kernels are not seen in the A × C cross, the colorless phenotype in strain A is not caused by the insertion of a *Ds* element. (It could, for example, just be a point mutation in the *C* gene.) Since spotted kernels are not seen in the B × D cross, the D strain lacks *Ac* elements. For these reasons, the crosses A × C, A × D, and B × D all show only expected Mendelian patterns of inheritance.

20.5 Consider two theoretical yeast transposons, A and B. Each contains an intron. Each transposes to a new location in the yeast genome and then is examined for the presence of the intron. In the new locations, you find that A has no intron, but B does. What can you conclude about the mechanisms of transposon movement for A and B from these facts?

Answer: Since introns are spliced out only at the RNA level, a transposition event that results in the loss of an intron (such as that used by *Ty* elements) indicates that the transposition occurred via an RNA intermediate. Thus, A is likely to move via an RNA intermediate. The lack of intron removal during B transposition suggests that it uses a DNA → DNA transposition mechanism (either conservative or replicative, or some other mechanism).

20.6 When certain strains of *Drosophila melanogaster* are crossed, hybrid dysgenesis can result in mutations, chromosomal aberrations, and sterility. A curious feature of hybrid dysgenesis is that it has been seen when females from certain laboratory strains (M cytotype) are mated to males from wild populations (P cytotype), but not when males from the same laboratory strains are mated to females from wild populations.

a. What occurs during the development of the F_1 hybrids to cause hybrid dysgenesis?
b. Why does it occur only in F_1 hybrids from M females × P males and not in F_1 hybrids from P females × M males?
c. Why does hybrid dysgenesis affect only the F_1 germ line and not F_1 somatic cells?
d. Some strains of P cytotype males are more potent at causing hybrid dysgenesis than others. Why might this be the case?
e. How might you test your explanation in (c)? Assume that you have a plasmid vector with an intact *P* element cloned into it.

Answer:

a. The mobilization of *P* elements within the germ line of the F_1 results in a large number of chromosomal breaks and *P* element insertions. These result in mutations (insertions that affect gene function) and chromosomal aberrations in the gametes of the F_1 that affect the phenotypes of the F_2. If the number of chromosome breaks is high enough, there will be sterility in the F_1.
b. In a cross of M females × P males, P males contribute chromosomes bearing *P* elements to an M cytotype. Since the M cytotype lacks a repressor, the *P* elements mobilize, causing hybrid dysgenesis. In the reciprocal cross, *P* elements remain in a P cytotype (contributed by the P female), where repressor exists, and therefore mobilization of *P* elements is repressed.

c. *P* element transposition is restricted to the germ line and does not occur in somatic cells. (This occurs because transposase is produced only in germ-line cells due to the alternative splicing of the transposase pre-mRNA in germ-line cells.)

d. The degree of hybrid dysgenesis that is seen could be due to the number of intact *P* elements in the strain. The mobilization of a greater number of *P* elements would result in more chromosomal damage and cause a greater degree of hybrid dysgenesis.

e. A genomic Southern blot analysis would assess the number of intact *P* elements in a strain. Isolate genomic DNA from a set of strains, cleave it with a restriction enzyme that does not cut within a *P* element, and separate the fragments by size using agarose gel electrophoresis. Blot the size-separated DNA to a membrane and allow the DNA bound to the membrane to hybridize to a labeled probe made from the cloned *P* element DNA. Detect the sites that bind to the probe. Determine the number of different-sized fragments bound by the probe to estimate the number of *P* elements in the strain. Then determine whether there is a correlation between the number of *P* elements in a strain and the degree of hybrid dysgenesis it shows.

20.7 In the experiment described in Figure 20.9, a wild-type ry^+ ($rosy^+$) gene was introduced into a strain homozygous for a mutant *ry* allele by using *P* element–mediated gene transformation.

a. If a transformed male was mated to an M strain female (see problem 20.6), would you expect the *P* element construct used in the experiment to transpose to another site in the genome of the F_1?

b. If a transformed male was mated to a P strain female (see problem 20.6), would you expect the *P* element construct used in the experiment to transpose to another site in the genome of the F_1?

c. Crosses of transformed flies to well-defined laboratory strains are more often useful for experimentation than crosses to animals from wild populations. Based on your answers to (a) and (b), what problems could arise from using transformants obtained by the approach in Figure 20.9?

Answer:

a. Since the transformed male would be transmitting *P* elements into an M cytotype that lacks a repressor, there would be additional transposition events.

b. In a P cytotype, transposition is repressed. No additional transposition events would occur.

c. The answers to (a) and (b) suggest that it may be a problem to maintain stable lines of *P* elements in an M cytotype. This has been circumvented in two ways. First, backcrossing transformed lines repeatedly into strains having a P cytotype will result in animals with sufficient amounts of repressor that a stable line can be maintained. However, crosses must be done with other P cytotype strains or in one direction, P female × M male.

Second, methods have been developed to target DNA for insertion without linking the DNA to the gene for transposase, which is provided separately. In this method, two DNA segments are simultaneously injected into embryos. One DNA segment contains the gene for transposase, but itself is unable to be inserted into the genome (due to a lack of the 31-bp inverted repeats found in *P* elements. The other DNA segment contains the gene targeted for insertion flanked by the 31-bp inverted repeats required for transposition by transposase.

20.8 After the discovery that *P* elements could be used to develop transformation vectors in *Drosophila melanogaster,* attempts were made to use them for the development of germ-line transformation in several different insect species. Savakis and his colleagues have succeeded in using a different transposable element found in *Drosophila,* the *Minos* element, to develop germ-line transformation for both *Drosophila melanogaster* and the medfly, *Ceratitus capitata,* a major agricultural pest present in Mediterranean climates.

a. What is the value of developing a transformation vector for an insect pest?
b. What basic information about the *Minos* element would need to be gathered before it could be used for germ-line transformation?

Answer:

a. The ability to modify a pest species genetically may provide a means to develop better strategies for pest control. For insects harboring parasites that affect human and animal populations (e.g., mosquitoes that are hosts to malaria), the ability to modify the insect host genetically may provide a means to influence the reproduction of a human or animal pathogen.
b. By analogy with the information used to develop *P* element–mediated transformation in *Drosophila,* it is essential to understand the mode of transposition of the element and to have identified the essential features of the element that allow it to transpose (the nature of the transposase and its regulation, the sequences the transposase recognizes to mediate transposition, etc.).

21

CHROMOSOMAL MUTATIONS

CHAPTER OUTLINE

TYPES OF CHROMOSOMAL MUTATIONS

VARIATIONS IN CHROMOSOME STRUCTURE

- Deletion
- Duplication
- Inversion
- Translocation
- Fragile Sites and Fragile X Syndrome

VARIATIONS IN CHROMOSOME NUMBER

- Changes in One or a Few Chromosomes
- Changes in Complete Sets of Chromosomes

REVIEW OF KEY TERMS, SYMBOLS, AND CONCEPTS

In your own words, write a brief, precise definition of each term in the groups on the following page. Check your definitions using the text. Then develop a concept map using the terms in each list.

THINKING ANALYTICALLY

In this chapter, it is crucial to be able to visualize the physical structure of chromosomes in the context of cellular processes such as meiosis and gene expression. Start by developing mental images of chromosomal aberrations. Practice drawing inversions, transpositions, translocations, duplications, and deletions.

To understand fully the impact of chromosomal mutations on gene expression and inheritance of traits, you will need a solid grasp of cellular processes involving chromosome movement. Review how chromosomes move and align during meiosis. Then consider how chromosomal aberrations and mutations affecting chromosome number alter the behavior of

1	2
chromosomal mutation, aberration	euploid
deletion	aneuploid
duplication	polyploid
translocation	diploid
inversion	haploid
position effect	nullisomy
karyotype	monosomy
polytene chromosomes, bands and interbands	trisomy
multigene family	tetrasomy
unequal crossing-over	autopolyploidy
pericentric vs. paracentric inversion	allopolyploidy
dicentric bridge, acentric fragment	allotetraploid
intra-, interchromosomal translocation	odd-number polyploidy
reciprocal vs. nonreciprocal translocation	even-number polyploidy
alternate, adjacent 1, adjacent 2 segregation	chromosome
semisterility	centromere
fragile sites	distal to/from
triplet repeat amplification	promixal to/from
cri-du-chat syndrome	p, q arms
Prader-Willi syndrome	Down syndrome
Drosophila Bar mutation	Patau syndrome
Philadelphia chromosome	Edwards syndrome
Burkitt lymphoma	Turner syndrome
chronic myelogenous leukemia	Klinefelter syndrome
oncogenes, *c-abl, c-myc*	familial Down syndrome
fragile X syndrome	Robertsonian translocation
FMR-1	

chromosomes during meiosis and thereby affect the inheritance of traits. The disruptions in meiosis that result from a particular chromosomal aberration are best understood by drawing out the chromosomal mutation as it proceeds through meiosis. As you do the problems, take the time to draw out the consequences of crossing-over in inversion and translocation heterozygotes. To understand how chromosome aberrations affect gene expression and gene function, consider what happens when the structure of a gene is rearranged and broken into two separate parts by a chromosomal aberration. Ask yourself, What are the possible consequences if a rearrangement breakpoint is in the amino acid–coding region of the gene, if it is in the promoter or other regulatory region, if it rearranges the gene nearby the enhancer of a different gene, or if it fuses two unrelated genes in their intronic regions?

Aneuploidy and polyploidy have direct effects on gene dosage: A normal phenotype requires proper gene dosage. As you examine the consequences of different types of aneuploidy and polyploidy, consider how the different phenotypes that arise result from altering gene dosage.

QUESTIONS FOR PRACTICE

Multiple Choice Questions

1. The number of polytene chromosomes in dipterans, such as *Drosophila*, is characteristically
   a. twice the diploid number
   b. hundreds or more times the diploid number
   c. half the diploid number
   d. the same as the diploid number
2. The chromosomes that become polytene chromosomes are
   a. somatically paired
   b. genetically inert
   c. uniformly pycnotic
   d. acentric
3. Nonhomologous chromosomes that have exchanged segments are the products of a
   a. double deletion
   b. reciprocal translocation
   c. pericentric inversion
   d. paracentric inversion
4. The abbreviated karyotype 2N – 1 describes
   a. nullisomy
   b. monosomy
   c. trisomy
   d. haploidy
5. The abbreviated karyotype 2N + 1 + 1 describes
   a. double trisomy
   b. tetrasomy
   c. double monosomy
   d. none of these
6. All known monosomics in humans have been
   a. lethal
   b. semilethal
   c. treatable
   d. deletion heterozygotes
7. The cultivated bread wheat *Triticum aestivum*, although a polyploid, is fertile because it
   a. has an odd number of chromosome sets
   b. has an even multiple of chromosome sets
   c. is a double hybrid
   d. is propagated by grafting
8. Cultivated bananas and Baldwin apples are
   a. double diploids
   b. tetraploids
   c. double haploids
   d. triploids

9. Down syndrome is associated with
   a. an inversion
   b. a Robertsonian translocation
   c. trisomy-21
   d. both b and c
10. Gene expression can be altered by
    a. an inversion
    b. a duplication
    c. a transposition
    d. any of the above

Answers: 1c, 2a, 3b, 4b, 5a, 6a, 7b, 8d, 9d, 10d

Thought Questions

1. Identify and describe five examples of changes in chromosome structure that alter gene expression.
2. A chromosomal inversion in the heterozygous condition seems to act as a suppressor of crossing-over. Why might this be the case? How might such chromosomes be important for keeping a laboratory stock of three linked recessive mutations?
3. What are two modes by which Down syndrome can be caused? Why might one be more frequent than the other? If a normal female has a sibling with Down syndrome, what information would be critical in the assessment of the chances that she might have a Down syndrome child? How would the assessment be different if the assessed individual were a normal male with a Down syndrome sibling?
4. Consider the data presented in Table 21.2. Given these data, how do you account for the fact that most children with Down syndrome are born to women under 25 years of age? Does this alter the need for caution and prenatal diagnosis in women over 35 years of age?
5. How do you account for the fact that polyploidy is significantly more common in plants than in animals? (Hint: What basic difference between sexual and asexual reproduction might be significant here?)
6. What might be the significance of the following in the evolutionary process of speciation? (1) autopolyploidy, (2) allopolyploidy, (3) inversions, (4) translocations. Provide an explanation that considers how animals might be affected differently than plants.
7. Many small deletions, when heterozygous, display a normal phenotype. Others display a mutant phenotype. For example, a small deletion that removes just the *Notch* gene in *Drosophila,* when heterozygous with a normal chromosome, displays a mutant phenotype of notched (nicked) wings. A nearby deletion that removes just the *white* gene, when heterozygous with a normal chromosome, displays the wild-type phenotype of deep-red eyes. Speculate why some heterozygous deletions might show a phenotype while others do not.
8. Which, if any, of the chromosomal aberrations discussed in this chapter are able to revert to a normal phenotype?

Thought Questions for Media

After reviewing the media at the *Essential iGenetics* website, try to answer these questions.

1. After reviewing the *Crossing-Over in an Inversion Heterozygote* animation, consider the following two situations.
   I. No crossovers occur in the inverted region of a pericentric inversion in an inversion heterozygote.
   II. A single crossover occurs in the inverted region of a pericentric inversion in an inversion heterozygote.

   Are different types of *viable* gametes produced in these two situations?
   a. Are the ratios of different types of *viable* gametes produced in these situations the same or different?
   b. Are *inviable* gametes produced in either of these situations? If so, how are they produced and why are they inviable?
2. In the *Meiosis in a Translocation Heterozygote* animation, the chromosomes in a reciprocal translocation heterozygote were depicted as N_1, N_2, T_1, and T_2. N_1 and N_2 were the normal ordered, nonhomologous chromosomes. T_1 and T_2 were the chromosomes in which a reciprocal translocation occurred. N_1 and T_1 have homologous centromeres, as do N_2 and T_2. In the animation, homologous regions of these chromosomes paired during meiosis I and aligned in a crosslike figure. They were arranged with N_1 in the upper left-hand corner, N_2 in the lower right-hand corner, T_1 in the lower left-hand corner, and T_2 in the upper right-hand corner, as follows:

   $N_1 \rightarrow T_2$

   $T_1 \rightarrow N_2$

   Use this illustration as you answer the following questions.
   a. The animation illustrated three different patterns of segregation of these chromosomes during meiosis I called alternate, adjacent 1 and adjacent 2 segregation. Which chromosomes go to each pole in the different types of segregation?
   b. Which two types of segregation are most frequent? Which one type of segregation is rare?
   c. Which type(s) of segregation pattern produces balanced gametes? Which type(s) produces inviable gametes?
   d. Why are individuals with a reciprocal translocation semisterile?

1. Suppose Sonia and Ramon conceive a fetus that has the T(14;21) translocation. Can you tell, from this information alone, whether the child will have Down syndrome or be healthy? If not, what additional information about the karyotype would you need to know? If the child has Down syndrome, what can you infer about the chromosomes carried by the sperm used in fertilization?
2. Suppose Sonia and Ramon, having had one normal child, are interested in having a second child. They seek genetic counseling for advice. What advice should the counselor give them, and how, if at all, would it differ from any advice they might have received when attempting to have their first child?
3. What is the chance that Sonia and Ramon have a child with the same karyotype as Ramon? With the same karyotype as Sonia?

SOLUTIONS TO TEXT PROBLEMS

21.1 A normal chromosome has the following gene sequence:

A B C D○*E F G H*

Determine the chromosomal mutation illustrated by each of the following chromosomes:

a. *A B C F E* ○ *D G H*

b. *A D* ○ *E F B C G H*

c. *A B C D* ○ *E F E F G H*

d. *A B C D* ○ *E F F E G H*

e. *A B D* ○ *E F G H*

Answer:

a. pericentric inversion (inversion of *D*○*EF*)
b. nonreciprocal translocation (*BC* moved from left to right arm)
c. tandem duplication (*EF* duplicated)
d. reverse tandem duplication (*EF* duplicated)
e. deletion (*C* deleted)

21.2 Distinguish between pericentric and paracentric inversions.

Answer: A pericentric inversion is an inversion that includes the centromere, while a paracentric inversion is an inversion that is wholly within one chromosomal arm. See text Figure 21.6.

21.3 What would be the effect on protein structure if a small inversion were to occur within the amino acid–coding region of a gene?

Answer: In answering this question, it is important to keep in mind that DNA is composed of two strands that have opposite polarity (one is 5′→3′, the other 3′→5′) and that the transcription of DNA into RNA always proceeds so that the RNA transcript is synthesized in the 5′→3′ direction using a 3′→5′ template strand. The sequence of bases in the mRNA transcript are then read as a triplet code and translated into an amino acid sequence.

There are several possible consequences of an inversion in the DNA of a single gene, all of them serious. This is because the polarity of DNA must be maintained to obtain the correct reading frame in the mRNA. Within the region of an inversion, the bases of the nontemplate strand of DNA would be inserted in the template strand (try drawing this out). After the DNA is transcribed, the mRNA transcript will be decoded during translation. The region of the transcript originating from the inverted region will encode an amino acid sequence that is quite different from the normal, uninverted region and may have possible nonsense codons (translation termination signals) within it. Consequently, a nonfunctional or truncated protein may be produced.

21.4 Inversions are known to affect crossing-over. The following homologous with the indicated gene order are given (the filled and open circles are homologous centromeres):

● A B C D E

○ A D C B E

a. Diagram the alignment of these chromosomes during meiosis.

b. Diagram the results of a single crossover between homologous genes *B* and *C* in the inversion.

c. Considering the position of the centromere, what is this sort of inversion called?

Answer:

a.

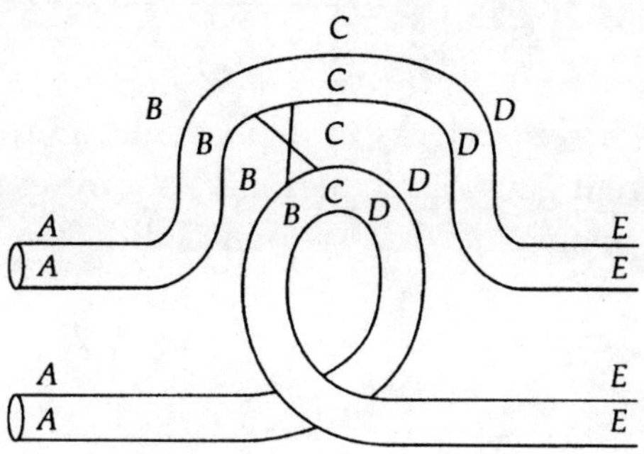

b. A crossover between B and C results in the following chromosomes:

● A B C D E (normal order)

● A B C D A ○ (dicentric, duplication for *A*, deletion for *E*)

E B C D E (acentric, duplication for E, deletion for *A*)

○ A D C B E (inverted order)

c. Paracentric inversion, because the centromere is not included in the inverted DNA segment

21.5 Single crossovers within the inversion loop of inversion heterozygotes give rise to chromatids with duplications and deletions. What happens if, within the inversion loop, there is a two-strand double crossover in such an inversion heterozygote when the centromere is outside the inversion loop?

Answer: If a two-strand double crossover occurs within a paracentric inversion, the four products of meiosis have a complete set of genes, without duplications or deletions. No acentric or dicentric fragments are formed. Consequently, all four meiotic products are viable. This is illustrated in the following example:

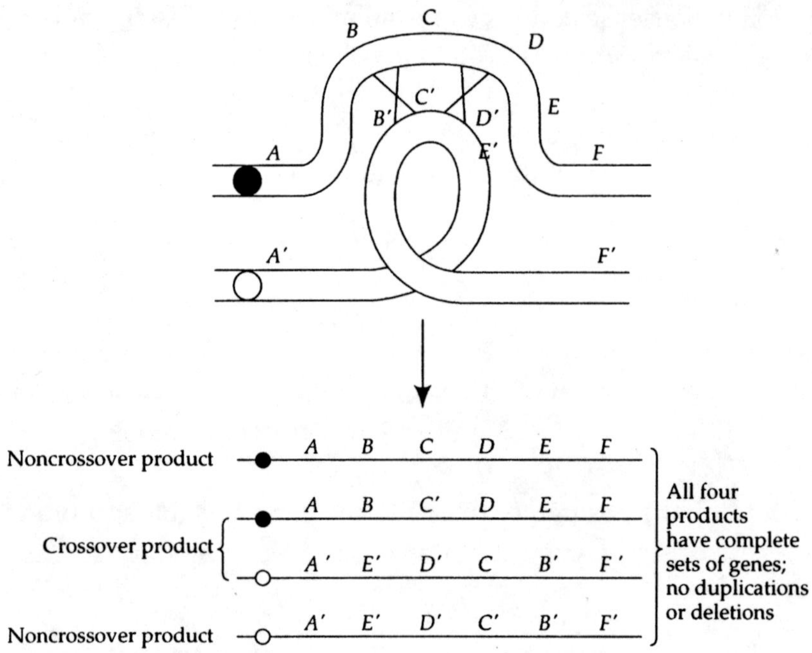

21.6 The following gene arrangements in a particular chromosome are found in *Drosophila* populations in different geographic regions. Assuming that the arrangement in part (a) is the original arrangement, in what sequence did the various inversion types probably arise?

a. *A B C D E F G H I*
b. *H E F B A G C D I*
c. *A B F E D C G H I*
d. *A B F C G H E D I*
e. *A B F E H G C D I*

Answer: One can envision a series of inversions occurring in the following sequence:

a → c → d → e → d
↓
b

The regions inverted in each step are illustrated below.

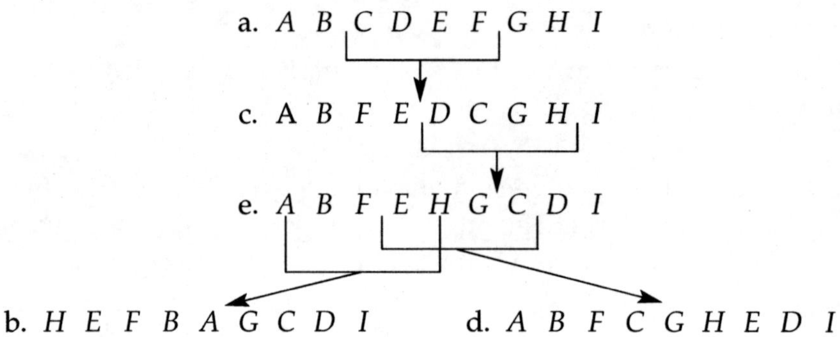

21.7 A particular plant species that had been subjected to radiation for a long time in order to produce chromosome mutations was then inbred for many generations until it was homozygous for all of these mutations. It was then crossed to the original unirradiated plant, and the meiotic process of the F_1 hybrids was examined. It was noticed that the following structures occurred, at low frequency, in anaphase I of the hybrid: a cell with a dicentric chromosome (bridge) and a fragment.

a. What kind of chromosome mutation occurred in the irradiated plant? In your answer, indicate where the centromeres are.

b. Explain in words and with a clear diagram where crossover(s) occurred and how the bridge chromosome of the cell arose.

Answer:

a. The irradiated chromosome has a paracentric inversion, with an inversion within one of its arms. Dicentric chromosomes and fragments arise as the result of single crossovers within paracentric inversions, and dicentric chromosomes with two bridges and two fragments result from a four-strand double crossover within paracentric inversions. An example of such a chromosome is diagrammed below.

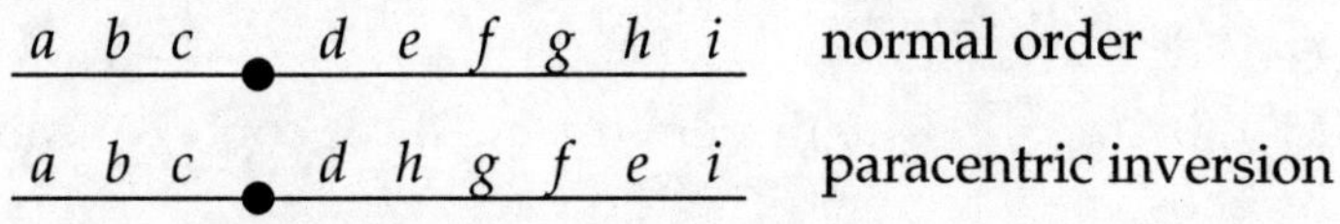

b. The bridge chromosome would arise by a single crossover within an inversion loop during meiosis. In the above example, the crossover would occur between *d* and *i*. See text Figure 21.7b for an illustration of such a crossover.

21.8 Mr. and Mrs. Lambert have not yet been able to produce a viable child. They have had two miscarriages and one severely defective child who died soon after birth. Studies of banded chromosomes of father, mother, and child showed that all chromosomes were normal except for pair number 6. The number 6 chromosomes of mother, father, and child are shown in the following figure:

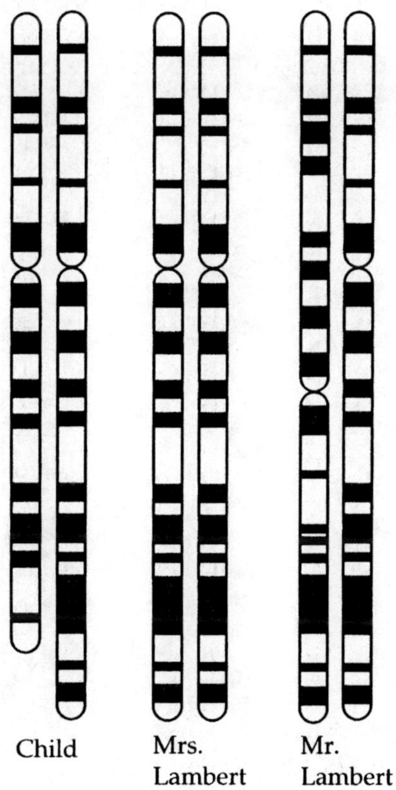

a. Does either parent have an abnormal chromosome? If so, what is the abnormality?
b. How did the chromosomes of the child arise? Be specific as to what events in the parents gave rise to these chromosomes.
c. Why is the child not phenotypically normal?
d. What can be predicted about future conceptions by this couple?

Answer:

a. Mr. Lambert is heterozygous for a pericentric inversion of chromosome 6. Relative to the centromere of the normal chromosome 6, one of the breakpoints is within the fourth light band up from the centromere, while the other is in the sixth dark band below the centromere. Mrs. Lambert's chromosomes are normal.
b. When Mr. Lambert's number 6 chromosomes paired during meiosis, they formed an inversion loop that included the centromere. Crossing-over occurred within the loop and gave rise to the partially duplicated, partially deficient chromosome 6 that the child received.
c. The child's abnormalities stem from having three copies of some, and only one copy of other, chromosome 6 regions. The top part of the short arm is duplicated, and there is a deficiency of the distal part of the long arm in this case.
d. The inversion appears to cover more than half of the length of chromosome 6, so crossing-over will occur in this region in the majority of meioses. In the minority of meioses where there is a two-strand double crossover inside the loop, or where crossing-over occurs outside the loop, and in the cases where a crossover has occurred within the loop but the child receives a noncrossover chromosome, the child can be normal. There is significant risk of abnormality, so fetal chromosomes should be monitored.

21.9 Mr. and Mrs. Simpson have been trying for years to have a child but have been unable to conceive. They consulted a physician, and tests revealed that Mr. Simpson had a markedly low sperm count. His chromosomes were studied, and a testicular biopsy was done. His chromosomes proved to be normal, except for pair 12. The following figure shows Mrs. Simpson's normal pair of number 12 chromosomes and Mr. Simpson's number 12 chromosomes.

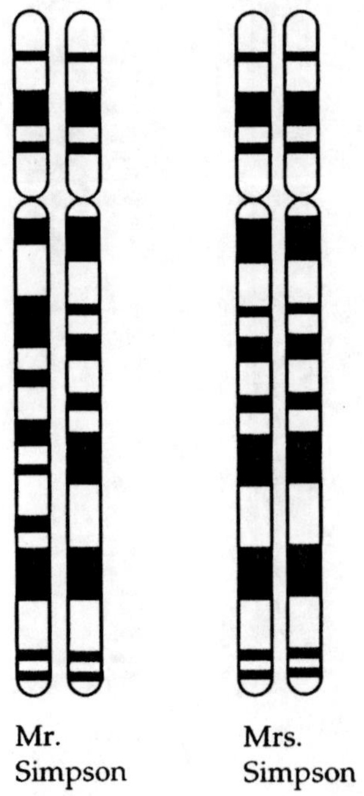

a. What is the nature of the abnormality in Mr. Simpson's chromosomes of pair number 12?

b. What abnormal feature would you expect to see in the testicular biopsy (cells in various stages of meiosis can be seen)?

c. Why is Mr. Simpson's sperm count low?

d. What can be done about Mr. Simpson's low sperm count?

Answer:

a. Mr. Simpson has a paracentric inversion in the long arm of one of his number 12 chromosomes. Moving downward (distally) from the centromere of the normal chromosome, the breakpoints are in the first dark band and in the sixth light band.

b. Crossing-over within the inversion loop will produce dicentric chromatids that will form anaphase bridges. These chromatin bridges joining the two chromatin masses at the beginning of anaphase I will be visible in the testicular biopsy.

c. The inversion is large, so the frequency of crossing-over within it will be significant. Consequently, bridges will be formed in the majority of meioses. Cells in which bridges form do not complete meiosis or form sperm in mammals.

d. Nothing can be done to increase Mr. Simpson's sperm count. This might be an instance to consider in vitro fertilization.

21.10 Chromosome I in maize has the gene sequence *ABCDEF,* whereas chromosome II has the sequence *MNOPQR*. A reciprocal translocation resulted in *ABCPQR* and *MNODEF*. Diagram the expected pachytene (see Chapter 1, p. 20) configuration of the F_1 of a cross of homozygotes of these two arrangements.

Answer:

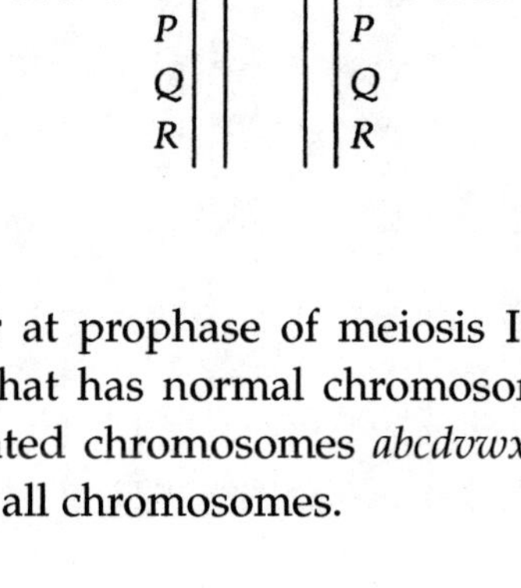

21.11 Diagram the pairing behavior at prophase of meiosis I (see Chapter 1, pp. 18–20) of a translocation heterozygote that has normal chromosomes of gene order *abcdefg* and *tuvwxyz* and has the translocated chromosomes *abcdvwxyz* and *tuefg*. Assume that the centromere is at the left end of all chromosomes.

Answer:

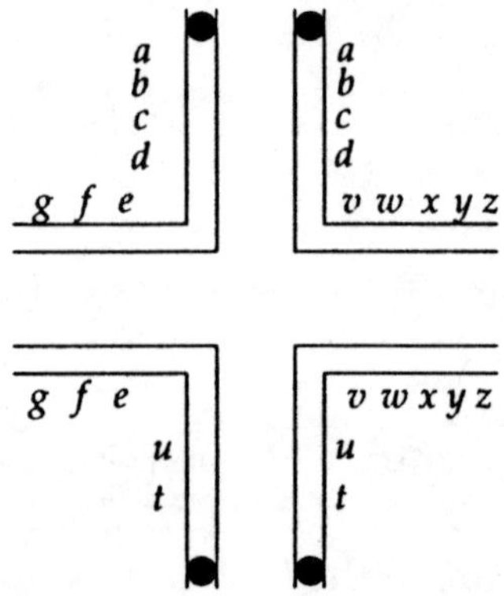

21.12 Mr. and Mrs. Denton have been trying for several years to have a child. They have experienced a series of miscarriages, and last year they had a child with multiple congenital defects. The child died within days of birth. The birth of this child prompted the Dentons' physician to order a chromosome study of parents and child. The results of the study are shown in the following figure. Chromosome banding was done, and all chromosomes were normal in these individuals except some copies of number 6 and number 12. The number 6 and number 12 chromosomes of mother, father, and child are shown in the figure (the number 6 chromosomes are the larger pair).

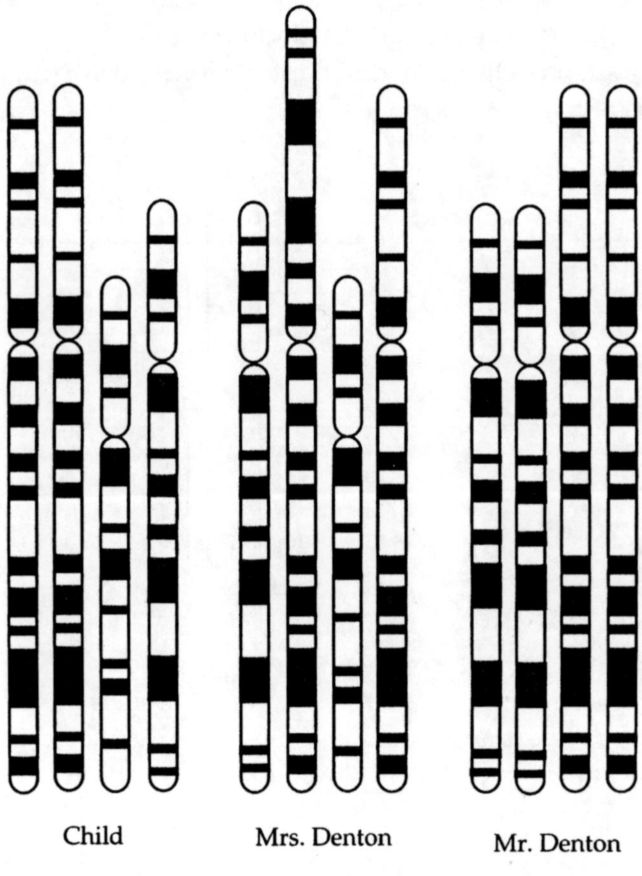

a. Does either parent have an abnormal karyotype? If so, which parent has it, and what is the nature of the abnormality?

b. How did the child's karyotype arise (what pairing and segregation events took place in the parents)?
c. Why is the child phenotypically defective?
d. What can this couple expect to occur in subsequent conceptions?
e. What medical help, if any, can be offered to them?

Answer:

a. Mr. Denton has normal chromosomes. Mrs. Denton is heterozygous for a balanced reciprocal translocation between chromosomes 6 and 12. Most of the short arm of chromosome 6 has been reciprocally translocated onto the long arm of chromosome 12. The breakpoints appear to be in the first thick, dark band just above the centromere of 6 and in the third dark band below the centromere of 12.
b. The child received a normal 6 and a normal 12 from his father. In prophase I of meiosis in Mrs. Denton, chromosomes 6 and 12 and the reciprocally translocated 6 and 12 paired to form a cruciform-like structure. Segregation of adjacent, nonhomologous centromeres to the same pole ensued, so that the child received a gamete containing a normal 6 and one of the translocation chromosomes. See text Figure 21.10 for an illustration of adjacent 1 segregation.
c. The child has a normal 6 and a normal 12 chromosome from Mr. Denton. The child also has a normal 6 chromosome from Mrs. Denton. However, the child also has one of the translocation chromosomes from Mrs. Denton. With this chromosome, the child is partially trisomic as well as partially monosomic. It has three copies of part of the short arm of chromosome 6 and only one copy of most of the long arm of chromosome 12. This abnormality in gene dosage is the cause of its physical abnormality.
d. Segregation of adjacent homologous centromeres to the same pole is relatively rare. The segregation pattern seen in this child (adjacent 1 segregation) and alternate segregation (see text Figure 21.10) are more common. About half of the time, when alternate segregation occurs, the gamete will have a complete haploid set of genes, and the embryo should be normal. However, half of the gametes resulting from alternate segregation will be translocation heterozygotes.
e. Prenatal monitoring of fetal chromosomes could be done, and given the severity of the abnormalities (high probability of miscarriage and multiple congenital abnormalities), therapeutic abortion of chromosomally unbalanced fetuses would be a consideration.

21.13 Irradiation of *Drosophila* sperm produces translocations between the X chromosome and autosomes, between the Y chromosome and autosomes, and between different autosomes. Translocations between the X and Y chromosomes are not produced. Explain the absence of X-Y translocations.

Answer: Mature sperm bear either an X or a Y chromosome, not both. For X-Y translocations to be obtained, both chromosomes must be present in the same sperm cell.

21.14 Define aneuploidy, monoploidy, and polyploidy.

Answer: An aneuploid cell or organism is one in which there is not an exact multiple of a haploid set of chromosomes, or one in which part or parts of chromosomes have been duplicated or deleted. It is one that does not have a *euploid* number of chromosomes. It is a general term used to describe a typically abnormal individual with an "unbalanced" chromosomal set.

A monoploid cell or individual has only one set of chromosomes. In humans, a monoploid cell would have 23 chromosomes instead of the normal diploid number of 46. In this case, a monoploid cell is also a haploid cell. A haploid number of chromosomes is typically the number of chromosomes in a gamete. Thus, in diploid individuals, a haploid gamete is also a monoploid cell. This is not always the case for nondiploid individuals. For example, in a hexaploid plant that has 36 chromosomes, a gamete will have 18 chromosomes (haploid number = 18), but the monoploid number will be 6.

A polyploid cell or individual has multiple sets of chromosomes. It is euploid, having multiple *complete* sets of chromosomes.

21.15 If a normal diploid cell is 2N, what is the chromosome content of the following?

a. a nullisomic
b. a monosomic
c. a double monosomic
d. a tetrasomic
e. a double trisomic
f. a tetraploid
g. a hexaploid

Answer:

a. 2N – 2 (two copies of the same chromosome are missing)
b. 2N – 1 (missing one chromosome)
c. 2N – 1 – 1 (one copy of each of two different chromosomes are missing)
d. 2N + 2 (two extra copies of one chromosome)
e. 2N + 1 + 1 (an extra copy of each of two different chromosomes)
f. 4N
g. 6N

21.16 In humans, how many chromosomes would be typical of nuclei of cells that are

a. monosomic?
b. trisomic?
c. monoploid?
d. triploid?
e. tetrasomic?

Answer:

a. 45
b. 47
c. 23
d. 69
e. 48

21.17 An individual with 47 chromosomes, including an additional chromosome 15, is said to be

a. triplet
b. trisomic
c. triploid
d. tricycle

Answer: b

21.18 A color-blind man marries a homozygous normal woman, and after four joyful years of marriage, they have two children. Unfortunately, both children have Turner syndrome, although one has normal vision and one is color-blind. The type of color blindness involved is a sex-linked recessive trait.

a. For the color-blind child, did nondisjunction occur in the mother or in the father? Explain your answer.
b. For the child with normal vision, in which parent did nondisjunction occur? Explain your answer.

Answer:

a. The cross can be written as $X^+X^+ \times X^cY$. Turner syndrome children are XO. If the child is color-blind, it received its father's X^c via a chromosomally normal X-bearing sperm. Therefore, the egg that was fertilized must have lacked an X, and nondisjunction occurred in the mother.
b. If the child has normal vision, it must have received its mother's normal X^+. To be XO, this must be the only X the embryo received, so that the egg must have been fertilized by a nullo-X, nullo-Y sperm. Nondisjunction occurred in the father.

21.19 Assume that x is a new mutant gene in corn. A female x/x plant is crossed with a triplo-10 individual (trisomic for chromosome 10) carrying only dominant alleles at the x locus. Trisomic progeny are recovered and crossed back to the x/x female plant.

a. What ratio of dominant to recessive phenotypes is expected if the x locus is not on chromosome 10?
b. What ratio of dominant to recessive phenotypes is expected if the x locus is on chromosome 10?

Answer:

a. If the new mutant is not on chromosome 10, the cross can be written as $x/x \times +/+$ (considering only the chromosome carrying the x gene). The progeny will be $x/+$, and the backcross will be $x/+ \times x/x$. A 1:1 ratio of x/x (recessive mutant) to $x/+$ (normal) individuals will be seen.
b. If the new mutant is on chromosome 10, the cross can be written as $x/x \times +/+/+$. Trisomic progeny will be $x/+/+$, and the backcross can be written as $x/+/+ \times x/x$. In the trisomic $x/+/+$ individual, there will be four kinds of gametes produced, depending on how chromosome 10 segregates during meiosis. In $\frac{1}{3}$ of the meioses, x and $+/+$ gametes will be produced, giving $\frac{1}{6}$ x and $\frac{1}{6}$ $+/+$ gametes. In $\frac{2}{3}$ of the meioses, + and $+/x$ gametes will be produced, giving $\frac{1}{3}$ $+/x$ and $\frac{1}{3}$ + gametes. When such gametes

fuse with an x-bearing gamete (from the xx parent) at fertilization, ⅙ of the progeny will have the mutant phenotype (i.e., be x/x) and ⅚ will be normal (i.e., $+/+/x$, $+/x/x$, $+/x$). The phenotypic ratio would be 5 dominant : 1 recessive mutant.

21.20 Why are polyploids with even multiples of the chromosome set generally more fertile than polyploids with odd multiples of the chromosome set?

Answer: Polyploids with even multiples of the chromosome set can better form chromosome pairs in meiosis I than can polyploids with odd multiples. Triploids, for example, will generate an unpaired chromatid pair for each chromosome type in the genome, so that chromosome segregation to the gametes is irregular and the resulting zygotes will not be euploid.

21.21 One plant species (N = 11) and another plant species (N = 19) produced an allotetraploid. Select the correct answer from the key regarding the following statements:

I. The chromosome number of this allotetraploid is 30.

II. The number of linkage groups of this allotetraploid is 30.

Key:

A. Statement I is true, and Statement II is true.

B. Statement I is true, but Statement II is false.

C. Statement I is false, but Statement II is true.

D. Statement I is false, and Statement II is false.

Answer: To form the initial alloploid, gametes from each species fused, so that the initial alloploid had 11 + 19 = 30 chromosomes. A fertile, allotetraploid plant was produced by the doubling of this chromosome set, so that the allotetraploid plant has 60 chromosomes, two sets of 11 and two sets of 19. Thus Statement II is true, but Statement I is not. The allotetraploid has 60 chromosomes and 30 linkage groups (30 pairs of homologues). C is correct.

21.22 According to Mendel's first law, genes A and a segregate from each other and appear in equal numbers among the gametes. But Mendel did not know that his plants were diploid. In fact, because plants are frequently tetraploid, he could have been unlucky enough to have started with peas that were 4N rather than 2N. Let us assume that Mendel's peas were tetraploid, that every gamete contains two alleles, and that the distribution of alleles to the gamete is random. Suppose we have a cross of $AAAA \times aaaa$, where A is dominant, regardless of the number of a alleles present in an individual.

a. What will be the genotype of the F_1?

b. If the F_1 is selfed, what will be the phenotypic ratios in the F_2?

Answer:

a. The F_1 will be $AA\ aa$.

b. Label the four alleles in the F_1 as A^1, A^2, a^1, and a^2. These four gametes can segregate into gametes in six ways: A^1A^2, A^1a^1, A^1a^2, A^2a^1, A^2a^2, a^1a^2. ⅙ of the gametes will be AA, ⅘ of the gametes will be Aa, and ⅙ will be aa. Selfing the F_1 will give the following pairings:

	1/6 *AA*	4/6 *Aa*	1/6 *aa*
1/6 *AA*	1/36 *AAAA*	4/36 *AAAa*	1/36 *AAaa*
4/6 *Aa*	4/36 *AAAa*	16/36 *AAaa*	4/36 *Aaaa*
1/6 *aa*	1/36 *AAaa*	4/26 *Aaaa*	1/36 *aaaa*

This would give a phenotypic ratio of 35 *A–* : 1 *aa*.

21.23 The root tip cells of an autotetraploid plant contain 48 chromosomes. How many chromosomes were contained by the gametes of the diploid from which this plant was derived?

Answer: The somatic cells of an autotetraploid plant have four identical sets of chromosomes. As root tip cells are mitotically dividing somatic cells, they, too, have four identical sets of chromosomes. Therefore, the gametes of the diploid from which this plant was derived had 12 chromosomes.

21.24 How many chromosomes would be found in somatic cells of an allotetraploid derived from two plants, one with N = 7 and the other with N = 10?

Answer: The initial allopolyploid will have 17 chromosomes. After doubling, the somatic cells will have 34 chromosomes.

21.25 Plant species A has a haploid complement of four chromosomes. A related species B has five. In a geographic region where A and B are both present, C plants are found that have some characters of both species and somatic cells with 18 chromosomes. What is the chromosome constitution of the C plants likely to be? With what plants would they have to be crossed to produce fertile seed?

Answer: The C plants are allotetraploids, containing a diploid chromosome set from each of species A and species B. These plants should be fertile, as they will have no abnormal chromosome pairing or unpaired chromosomes. They should be able to produce gametes with 9 chromosomes (the four of species A and the five of species B) and can either be selfed or crossed to other C plants to produce fertile seed.

22

POPULATION GENETICS

CHAPTER OUTLINE

REVIEW OF KEY TERMS, SYMBOLS, AND CONCEPTS

In your own words, write a brief, precise definition of each term in the groups below. Check your definitions using the text. Then develop a concept map using the terms in each list.

1	2	3
transmission genetics	models of genetic variation	Hardy-Weinberg equilibrium
molecular genetics	allele frequency cline	random mating
population genetics	classical model	nonrandom mating
quantitative genetics	neutral mutation model	positive assortative mating
genetic structure	fitness	negative assortative mating
genotypic frequency	polymorphic locus	inbreeding
allelic frequency	proportion of polymorphic loci	mutation
Mendelian population	heterozygosity (*H*)	allelic frequency
gene pool	RFLP	forward, reverse mutation
gene counting	nucleotide substitution	equilibrium frequency
Hardy-Weinberg law	synonymous change	random genetic drift
infinitely large population	nonsynonymous change	effective population size
migration	coding sequence	sampling error
natural selection	noncoding sequence	founder effect
random mating		bottleneck effect
mutation		migration
genetic drift		gene flow
natural selection		population viability analysis
Hardy-Weinberg equilibrium		neutral theory of molecular evolution
		natural selection
		Darwinian fitness
		selection coefficient
		heterosis
		overdominance
		heterozygote superiority

THINKING ANALYTICALLY

Population genetics considers patterns of genetic variation found among individuals within groups and how these patterns vary geographically and evolve over time. As you study population genetics, your perspective will shift away from a direct consideration of the molecular and biochemical mechanisms that underlie the inheritance and expression of traits in a single individual or cell to a statistical evaluation of the *effects* of these processes at the level of the group, population, or species.

At the conceptual core of population genetics lies the Hardy-Weinberg law. Initially, focus your efforts on understanding its assumptions and predictions. Practice the examples given in the text and the text problems to become fluent in analyzing whether a population with two alleles at an autosomal locus is in Hardy-Weinberg equilibrium. Then consider extensions of the law to loci with more than two alleles or loci with sex-linked alleles. Finally, investigate the different ways that the assumptions of the Hardy-Weinberg law may be violated and the consequences of each to the genetic structure of a population.

Be aware that the allelic symbolism used in population genetics is often different from that used in Mendelian or molecular genetics. Consider the following example. In earlier chapters, the genotypes for normal and sickle-cell hemoglobin were written either as $\beta^A\beta^A$, $\beta^A\beta^S$ $\beta^S\beta^S$ or Hb^AHb^A, Hb^AHb^S, Hb^SHb^S. Here they are written as *Hb-A/Hb-A, Hb-A/Hb-S, Hb-S/Hb-S*. Similarly, care should be taken with the letters *p* and *q*, which symbolize the frequency within a population of dominant and recessive members of an allelic pair (but do not symbolize the alleles themselves!).

One of the main concerns of population genetics is how to model changes in allele and genotypic frequencies under a set of specified conditions. The models are often very elegant and employ equations that identify relationships between a set of variables identified in a population. Memorization of the equations used to analyze changes in populations is by itself insufficient to understand the models. It will be difficult, if not impossible, to solve even moderately challenging word problems if you can only plug numbers into equations. While you will find it helpful to recognize by sight several of the equations, you need to do more. Take the time to understand the model that leads to the equation and how the mathematical and statistical analysis in the model is related to a biological question.

A few hints for solving some of the problems are in order. First, you will often find that the frequency of one class of homozygote (usually the recessive homozygotes, given by q^2) is the only piece of hard data available. As you work back from this information, keep track of the assumptions that you are making. Otherwise, you may enter into the realm of circular reasoning. For example, you can legitimately calculate q as the square root of q^2 and, if only two alleles exist, calculate p by equating it with $1 - q$. However, if you then determine that heterozygotes exist at a frequency of $2pq$, you are assuming that random mating is occurring and the conditions of Hardy-Weinberg equilibrium are satisfied. Further work with your calculated values of p^2, $2pq$, and q^2 will continue to reflect that assumption. Often, you will need to look for more information or another approach to prove that the population in question is in Hardy-Weinberg equilibrium. Second, doing the math requires care and, if you are not exceptionally fluent with it, patience. It will sometimes help to factor out common multipliers [e.g., $2pq^2 + q^2 = q^2(2p + 1)$]. At other times, it will be helpful to recognize members of a binomial expansion [e.g., $p^2 + 2pq + q^2 = (p + q)^2$].

After you have mastered the concepts underlying the Hardy-Weinberg law, focus on understanding how genetic variation can be measured. A number of important models have been developed that have been supported by data gathered using a variety of methods, including those of molecular genetics. Models have been proposed to address the substantial amount of genetic variation that exists in a population and the factors that can lead to changes in a population's genetic structure. Some of these employ quantitative analysis and equations. Approach these models just as you did the Hardy-Weinberg law. First become familiar with the conceptual issues of the model, and then relate the variables in an equation to the key factors considered in the model.

QUESTIONS FOR PRACTICE

Multiple Choice Questions

1. In a small population, 30 percent of the individuals have blood type M, 40 percent of the individuals have blood type MN, and 30 percent of the individuals have blood type N. If p equals the frequency of the L^M allele and q equals the frequency of the L^N allele, what are p and q?

a. $p = 0.30, q = 0.30$
b. $p = 0.50, q = 0.50$
c. $p = 0.30, q = 0.70$
d. $p = 0.50, q = 0.30$

2. Is the population described in question 1 above in Hardy-Weinberg equilibrium?
   a. Yes; the calculated genotypic frequencies equal the expected genotypic frequencies.
   b. Yes; at equilibrium you always have equal numbers of recessive and dominant homozygotes.
   c. No; the frequency of heterozygotes is too large and the frequency of homozygotes is too low.
   d. No; the frequency of heterozygotes is too low and the frequency of homozygotes is too large.
3. Which of the following is *not* an assumption about a population in Hardy-Weinberg equilibrium?
   a. The population is isolated.
   b. Random mating occurs in the population.
   c. The population is free from mutation.
   d. The population is free from natural selection.
   e. The population is free from migration.
4. The frequency of one form of X-linked color blindness varies among human ethnic groups. What can be said about whether each ethnic group is in Hardy-Weinberg equilibrium? (Let q = the frequency of the normal allele and p = the frequency of the color-blind allele.)
   a. None of the ethnic populations can be in equilibrium, since all have different values for p and q.
   b. Only the entire human population is in equilibrium.
   c. Some of the ethnic populations may be in equilibrium, provided that the frequency of the trait in males is p and the frequency in females is p^2.
   d. Some of the ethnic populations may be in equilibrium, provided that the frequency of the trait in both sexes is p^2.
   e. All of the ethnic populations will be in equilibrium, since each satisfies the criteria for a population in Hardy-Weinberg equilibrium.
5. In a large, randomly mating population, 80 percent of the individuals have dark hair and 20 percent are blond. Assuming that hair color is controlled by one pair of alleles, is the allele for dark hair *necessarily* dominant to the one for blond hair?
   a. Yes, because otherwise the population would not be dominated by dark-haired individuals.
   b. Yes, because more of something (in this case, hair color) is always dominant to less of that thing.
   c. No, because relative frequencies of alleles in a randomly mating population are unrelated to issues of dominance and recessiveness.
   d. No, because although there is a relationship between dominance and allele frequency, that relationship is not seen in this example.
6. The amounts of genetic variation in a population can be explained in a number of ways. In one model, a large amount of genetic variation is explained by recurrent mutation and random changes in allele frequencies. In this model, natural selection selects against some of the

variation affecting fitness but does not select for or against much of the genetic variation. This model is termed

a. the classical model

b. the neutral mutation model

c. the random mutation model

7. Which of the following populations are in Hardy-Weinberg equilibrium?

Population	Genotypes		
	AA	*Aa*	*aa*
a	0.72	0.20	0.08
b	0.12	0.80	0.08
c	0.08	0.01	0.91
d	0.25	0.50	0.25

8. How is the rate of forward mutation likely to be related to the rate of reverse mutation?
   a. The rate of forward mutation is generally lower because there is mutational pull back to a specific form.
   b. The rate of forward mutation is generally higher because once an allele has changed, it is nearly as likely that a subsequent change will be to yet another new form.
   c. The rate of forward and reverse mutations are codependent and usually equal.
   d. The relative rates of forward and reverse mutation are so highly variable that one cannot formulate an accurate generalization comparing the two.
9. Which of the following can result in genetic drift?
   a. a sampling error
   b. random factors producing unexpected mortality
   c. the establishment of a population by a small number of breeding individuals
   d. a drastic reduction in the size of a population
   e. a change in environmental conditions that affects selection
   f. all of the above
10. What is the effective population size for a population consisting of 10 breeding males and 2 breeding females?
    a. 12
    b. 6
    c. 7
    d. 2
11. Which of the following is *not* true concerning the effect of migration among populations?
    a. Migration tends to increase the effective size of the populations, leading to a reduction in genetic drift.
    b. Migration is associated with gene flow and introduces new alleles to the population.
    c. If the allelic frequencies of migrants and the recipient population differ, migration can lead to the further differentiation of two populations.
    d. Migration and genetic drift have opposite effects on size and variability. Migration effectively increases size and variability, whereas drift acts in opposition.

12. What can one generally state about measuring the fitness associated with a specific genotype?
    a. It is easy to assess and can be based on the number of offspring of an individual.
    b. Since it is based on the reproductive ability of a genotype, it is an absolute term, which requires no assumptions.
    c. It will remain the same from generation to generation.
    d. It is difficult to measure because of pleiotropic effects.
13. Consider a recessive trait that results in a complete lack of reproductive success of homozygotes. How might such a recessive trait be maintained at a high level in a population?
    a. through new mutation
    b. through heterosis
    c. through overdominance
    d. through heterozygote superiority
    e. all of the above
14. For a given eukaryotic gene, which rate of DNA change is expected to be the lowest?
    a. the relative rate of change in nonfunctional pseudogenes
    b. the relative rate of change in introns
    c. the relative rate of change for synonymous substitutions in coding sequences
    d. the relative rate of change for nonsynonymous substitutions in coding sequences
    e. the relative rate of change in leaders and trailers
15. Which of the rates in question 14 is expected to be the highest?

Answers: 1b, 2d, 3a, 4c, 5c, 6b, 7d, 8b, 9f, 10c, 11c, 12d, 13e, 14d, 15a

Thought Questions

1. Consider a population that is not in Hardy-Weinberg equilibrium for a pair of alleles. Show why equilibrium values for genotypic frequencies are reached in one generation after the onset of random mating for autosomal alleles, but more than one generation of random mating is required for sex-linked alleles.
2. Distinguish between heterozygosity and the proportion of polymorphic loci.
3. Mutation pressure is rarely the most important determinant of allelic frequency. What other factors are important?
4. Can a population be in Hardy-Weinberg equilibrium for one, but not another, pair of alleles?
5. What is the most likely explanation if the allelic frequencies agree with those predicted by the Hardy-Weinberg law, but the genotype frequencies do not? (Hint: Consider mating systems and heterosis.)
6. Why do mutation, migration, and drift not necessarily lead to adaptation?
7. What are five different molecular methods that can be used to measure genetic variation? What advantages and disadvantages are there to each?
8. Distinguish between nonrandom mating, positive assortative mating, negative assortative mating, and inbreeding.
9. What are the arguments for and against investing funds and human effort in a population viability analysis for an endangered species?
10. What are the values of preserving and promoting genetic diversity within a species? How does the development of genomics and technologies related to genomics affect our ability to assess and influence genetic diversity, both in our own species and in other species?

Thought Questions for Media

After reviewing the media at the *Essential iGenetics* website, try to answer these questions.

1. What assumptions of the Hardy-Weinberg law were illustrated by the animation?
2. Under these assumptions, what are the predictions of the Hardy-Weinberg law?
3. Which of these assumptions were not valid in the population of Darwin's finches under study by Dr. Grant?
4. What was the purpose of Dr. Grant's study of these finches?
5. The animation indicated that beak shape and size are genetically complex. Some genes will have additive affects, some will be epistatic, and a few will show simple dominance and recessiveness. Why, for the purposes of the study, was it useful to assume the existence of a simple gene that, when homozygous recessive, contributes to a decrease in beak size?
6. Is fitness constant for a given trait such as beak length? If not, according to what conditions does it vary?
7. The animation describes how natural selection for small and large beak size in Darwin's finches is complex. For example, even though birds with larger beaks can crack harder seeds, in some years, there can still be selection for small beak size. Why might this be the case?

SOLUTIONS TO TEXT PROBLEMS

22.1 In the European land snail *Cepaea nemoralis,* multiple alleles at a single locus determine shell color. The allele for brown (C^B) is dominant to the allele for pink (C^P) and to the allele for yellow (C^Y). The dominance hierarchy among these alleles is $C^B > C^P > C^Y$. In one population sample of *Cepaea,* the following color phenotypes were recorded:

Brown	236
Pink	231
Yellow	33
Total	500

Assuming that this population is in Hardy-Weinberg equilibrium (large, randomly mating, and free from evolutionary processes), calculate the frequencies of the C^B, C^Y, and C^P alleles.

Answer: Equate frequency of each color with the frequency expected in Hardy-Weinberg equilibrium, letting $p = f(C^B)$, $q = (f(C^P)$, and $r = f(C^Y)$.

$$\text{Brown: } f(C^BC^B) + f(C^BC^P) + f(C^BC^Y) = p^2 + 2pq + 2pr = 236/500 = 0.472$$
$$\text{Pink: } f(C^PC^P) + f(C^PC^Y) = q^2 + 2qr = 231/500 = 0.462$$
$$\text{Yellow: } f(C^YC^Y) = r^2 = 33/500 = 0.066$$

Now solve for p, q, and r, knowing that $p + q + r = 1$

$$r^2 = 0.066, \text{ so } r = \sqrt{0.066} = 0.26$$

There are two approaches to solve for q. First, since $q^2 + 2qr = 0.462$, one can substitute in $r = 0.26$, giving $q^2 + 2q(0.26) = 0.462$. Recognize this as a quadratic equation and set it equal to 0, and solve for q. That is, solve the equation $q^2 + 0.52q - 0.462 = 0$. Solving the quadratic equation for q, one has:

$$q = \frac{-0.52 \pm \sqrt{(0.52)^2 - 4(1)(-0.462)}}{2\,(1)} = 0.467$$

A second approach to solve for q is to realize that

$$q^2 + 2qr = 0.462 \text{ and } r^2 = 0.066$$

Adding left and right sides of the equations together, one has

$$\begin{aligned} q^2 + 2qr + r^2 &= 0.066 + 0.462 \\ (q + r)^2 &= 0.528 \\ q + r &= 0.726 \\ q &= 0.726 - r = 0.726 - 0.26 = 0.467 \end{aligned}$$

Since $p + q + r = 1$, $p = 1 - (q + r) = 1 - (0.26 + 0.467) = 0.273$.

22.2 Three alleles are found at a locus coding for malate dehydrogenase (MDH) in the spotted chorus frog. Chorus frogs were collected from a breeding pond, and each frog's genotype at the MDH locus was determined with electrophoresis. The following numbers of genotypes were found:

M^1M^1	8
M^1M^2	35
M^2M^2	20
M^1M^3	53
M^2M^3	76
M^3M^3	62
Total	254

a. Calculate the frequencies of the M^1, M^2, and M^3 alleles in this population.

b. Using a chi-square test, determine whether the MDH genotypes in this population are in Hardy-Weinberg proportions.

Answer:

a. The tally for M^1 alleles is as follows:

Genotype	**# Individuals**	**# M¹ Alleles**
M^1M^1	8	16
M^1M^2	35	35
M^1M^3	53	53
Total		104

The total number of individuals is 254; thus, the total number of alleles is 254 × 2, or 508. The frequency of M^1 alleles is 104⁄508 = 0.205. The frequency of the other alleles is obtained similarly. One has $f(M^1) = 0.20 = p$, $f(M^2) = 0.30 = q$, and $f(M^3) = 0.50 = r$.

b. For three alleles with frequencies p, q, and r, a population in Hardy-Weinberg equilibrium will have $p^2 + 2pq + 2pr + q^2 + 2qr + r^2 = 1$. To test the hypothesis that the population is in Hardy-Weinberg equilibrium, calculate the numbers of individuals expected in each class using this relationship and the values for p, q, and r obtained in part (a). Calculate the value of $\chi 2$ as shown in the following table.

Genotype	Observed Value (*o*)	Expected Frequency	Expected Value (*e*)	*d* (*o* – *e*)	d^2/e
M^1M^1	8	$p^2 = 0.04$	10	–2	0.40
M^1M^2	35	$2pq = 0.12$	30	5	0.83
M^2M^2	20	$q^2 = 0.09$	23	–3	0.39
M^1M^3	53	$2pr = 0.20$	51	2	0.08
M^2M^3	76	$2qr = 0.30$	75	1	0.01
M^3M^3	62	$r^2 = 0.25$	64	–2	0.06
					$\chi 2 = 1.77$

Since the six phenotypic classes are completely specified by three allele frequencies, the number of phenotypes (6) minus the number of alleles (3) determines the degree of freedom (6 – 3 = 3). With 3 degrees of freedom, $0.70 < P < 0.50$. The hypothesis is accepted as possible, and it appears that the population is in Hardy-Weinberg equilibrium.

22.3 In a large interbreeding population 81 percent of the individuals are homozygous for a recessive character. In the absence of mutation or selection, what percentage of the next generation would be homozygous recessives? Homozygous dominants? Heterozygotes?

Answer: The conditions of this problem meet the requirements for a population in Hardy-Weinberg equilibrium. In such a population, if p equals the frequency of A, and q equals the frequency of a, one expects $p^2(AA)$, $2pq(Aa)$, and $q^2(aa)$ genotypes after random mating. Here, $q^2 = 0.81$, so $q = 0.9$. Since $p + q = 1$, $p = 1 - 0.9 = 0.1$. In the next generation, one would expect $p^2 = (0.1)^2 = 0.01$ (or 1%) *AA* genotypes, $2pq = 2(0.1)(0.9) = 0.18$ (or 18%) *Aa* genotypes, and $q^2 = (0.9)^2 = 0.81$ (or 81%) *aa* genotypes.

22.4 Let A and a represent dominant and recessive alleles whose respective frequencies are p and q in a given interbreeding population at equilibrium (with $p + q = 1$).

a. If 16 percent of the individuals in the population have recessive phenotypes, what percentage of the total number of recessive genes exist in the heterozygous condition?

b. If 1.0 percent of the individuals are homozygous recessive, what percentage of the recessive genes occur in heterozygotes?

Answer:

a. Since $q^2 = 0.16$, $q = \sqrt{0.16} = 0.40$. Since $p + q = 1$, $p = 0.60$. The frequency of heterozygotes is $2pq = 2(0.40)(0.60) = 0.48$. Each of the 48% of the heterozygotes has one recessive allele, while each of the 16% of the homozygous recessive individuals has two.

Thus, the percentage of the total number of recessive alleles in heterozygotes is (0.48)/[0.48 + 2(0.16)] = 0.48/0.80 = 0.60, or 60%.

b. If $q^2 = 0.01$, then $q = 0.1$, and $p = 0.9$. $2pq = 2(0.1)(0.9) = 0.18$. The percentage of the total number of recessive alleles in heterozygotes is (0.18)/[0.18 + 2(0.01)] = 0.18/0.20 = 0.90, or 90%.

22.5 A population has eight times as many heterozygotes as homozygous recessives. What is the frequency of the recessive gene?

Answer: The frequency of heterozygotes in a population in equilibrium is $2pq$, and the frequency of homozygous recessives is q^2. Here, there are eight times as many heterozygotes as homozygous recessives, so $2pq = 8q^2$. Since $p + q = 1$, $p = 1 - q$, and one can substitute $1 - q$ for p. This gives $2(1 - q)q = 8q^2$. Dividing both sides by q and multiplying through, one has $2 - 2q = 8q$. Thus, $2 = 10q$, and q = the frequency of the recessive allele = 0.20.

22.6 In a large population of range cattle, the following ratios are observed: 49 percent red (*RR*), 42 percent roan (*Rr*), and 9 percent white (*rr*).

a. What percentage of the gametes that give rise to the next generation of cattle in this population will contain allele *R*?

b. In another cattle population, only 1 percent of the animals are white and 99 percent are either red or roan. What is the percentage of *r* alleles in this case?

Answer:

a. In the red, *RR* animals, all of the gametes contain the *R* allele, while in the roan, *Rr* animals, half of the gametes contain the *R* allele. Therefore, [49 + (42/2)] = 70 percent of the gametes will contain the *R* allele. Another way to look at this problem is to realize that the frequency of gametes bearing a certain allele is the same as the frequency of the allele in the population. Let p equal the frequency of *R* in the population. Since there are 49 percent red animals, $p^2 = 0.49$, so $p = 0.70$ (or 70%).

b. If one lets q represent the frequency of *r*, since 1 percent of the animals are white, one has $q2 = 0.01$. Hence, $q = 0.1$.

22.7 In a gene pool, the alleles *A* and *a* have initial frequencies of p and q, respectively. Show that the allelic frequencies and zygotic frequencies do not change from generation to generation as long as there is no selection, mutation, or migration, the population is large, and the individuals mate at random.

Answer: Let the frequency of allele *A* equal p and the frequency of allele *a* equal q, with $p + q = 1$. Then, in the initial generation, the frequency of *AA* genotypes is p^2, the frequency of *aa* genotypes is q^2. The frequency of the remaining genotypes (i.e., *Aa* heterozygotes) must be $1 - (p^2 + q^2)$. Since $p + q = 1$, $(p + q)^2 = 1^2$, and $p^2 + 2pq + q^2 = 1$. Therefore, $1 - (p^2 + q^2) = 2pq$, and the frequency of *Aa* heterozygotes must be $2pq$.

Assume there is no selection, mutation, or migration and that the individuals mate at random. Since there are three different genotypes in the population, nine crosses are possible. The frequency of each type of cross is determined by the frequency of each parental genotype. The types of crosses, the frequency of each cross, the types of progeny, and the frequency of each progeny class are listed in the following table:

Cross	Cross Frequency	Progeny Ratios AA	Aa	aa	Progeny Frequencies AA	Aa	aa
$AA \times AA$	$p^2 \times p^2 =$ p4	all			p^4		
$AA \times Aa$	$p^2 \times 2pq = 2p^3q$	½	½		p^3q	p^3q	
$AA \times aa$	$p^2 \times q^2 = p^2q^2$		all			p^2q^2	
$Aa \times AA$	$2pq \times p^2 = 2p^3q$	½	½		p^3q	p^3q	
$Aa \times Aa$	$2pq \times 2pq = 4p^2q^2$	¼	½	½	p^2q^2	$2p^2q^2$	p^2q^2
$Aa \times aa$	$2pq \times q^2 = 2pq^3$		½	½		pq^3	pq^3
$aa \times AA$	$q^2 \times p^2 = p^2q^2$		all			p^2q^2	
$aa \times Aa$	$q^2 \times 2pq = 2pq^3$		½	½		pq^3	pq^3
$aa \times aa$	$q^2 \times q^2 = q^4$			all			q^4

To determine the frequency of a particular zygotic class, add up the frequency of the progeny in that class. Then factor out a common multiplier, and note that $p + q = 1$ and that $(p + q)^2 = p^2 + 2pq + q^2$. One has:

$$\begin{aligned} \text{frequency }(AA) &= p^4 + 2p^3q + p^2q^2 \\ &= p^2(p^2 + 2pq + q^2) \\ &= p^2(p + q)^2 \\ &= p^2(1)^2 = p^2 \end{aligned}$$

$$\begin{aligned} \text{frequency }(Aa) &= 2p^3q + 4p^2q^2 + 2pq^3 \\ &= 2pq(p^2 + 2pq + q^2) \\ &= 2pq(p + q)^2 \\ &= 2pq(1)^2 = 2pq \end{aligned}$$

$$\begin{aligned} \text{frequency }(aa) &= pq + 2pq^3 + q^4 \\ &= q^2(p^2 + 2pq + q^2) \\ &= q^2(p + q)^2 \\ &= q^2(1)^2 = q^2 \end{aligned}$$

Thus, the zygotic frequencies do not change from one generation to the next.

Since all of the gametes of the *AA* parents and half of the gametes of the *Aa* parents will bear the *A* allele, the frequency of *A* in the gene pool of the next generation is $p^2 + pq = p(p + q) = p$. Since all of the gametes of the *aa* parents and half of the gametes of the *Aa* parents will bear the *a* allele, the frequency of *a* in the gene pool of the next generation is $q^2 + pq = q(q + p) = q$. Thus, the allelic frequencies do not change from one generation to the next.

22.8 The *S*-*s* antigen system in humans is controlled by two codominant alleles, *S* and *s*. In a group of 3,146 individuals, the following genotypic frequencies were found: 188 *SS*, 717 *Ss*, and 2,241 *ss*.

a. Calculate the frequency of the *S* and *s* alleles.

b. Determine whether the genotypic frequencies are in Hardy-Weinberg equilibrium by using the chi-square test.

Answer:

a. Let p equal the frequency of S and q equal the frequency of s. Since homozygotes have two identical alleles and heterozygotes have one recessive and one dominant allele, one has:

$$p = \frac{2(188)[SS] + 717[Ss]}{2(3{,}146)} = \frac{1093}{6292} = 0.1737$$

$$q = \frac{717[Ss] + 2(2{,}241)[ss]}{2(3{,}146)} = \frac{5199}{6292} = 0.8263$$

b. Remember that in a χ^2 test, one uses the actual numbers of progeny observed and expected, and not the frequencies. With a hypothesis that the population is in Hardy-Weinberg equilibrium, one has:

Class	Observed (o)	Frequency	Expected (e)	d ($o - e$)	d^2/e
SS	8	$p^2 = 0.0302$	95	93	91.0
Ss	717	$2pq = 0.287$	903	–186	38.3
ss	2,241	$q^2 = 0.683$	2,148	93	4.0
	3,146	1	3,146	0	133.3

There is only 1 degree of freedom because the three genotypic classes are completely specified by two allele frequencies, namely, p and q. (df. = number of phenotypes – number of alleles = 3 – 2 = 1.) The χ^2 value of 133.3, for 1 degree of freedom, gives $P < 0.0001$. Therefore, the distribution of genotypes differs significantly from that expected if the population were in Hardy-Weinberg equilibrium.

22.9 Refer to problem 22.8. A third allele is sometimes found at the S locus. This allele S^u is recessive to both the S and the s alleles and can be detected only in the homozygous state. If the frequencies of the alleles S, s, and S^u are p, q, and r, respectively, what would be the expected frequencies of the phenotypes $S–$, Ss, $s–$, and S^uS^u?

$$f(S–) = f(SS) + f(SS^u) = p^2 + 2pr$$
$$f(Ss) = 2pq$$
$$f(s–) = f(ss) + f(sS^u) = q^2 + 2qr$$
$$f(S^uS^u) = r^2$$

22.10 In a large interbreeding human population, 60 percent of individuals belong to blood group O (genotype i/i). Assuming negligible mutation and no selective advantage of one blood type over another, what percentage of the grandchildren of the present population will be type O?

Answer: The conditions described in the problem indicate that the population is in Hardy-Weinberg equilibrium. Under equilibrium conditions, neither the allele nor the zygotic frequencies change from one generation to the next. Therefore, in two generations there should still be 60 percent type O individuals.

22.11 A selectively neutral, recessive character appears in 40 percent of the males and in 16 percent of the females in a large, randomly interbreeding population. What is the gene's frequency? What proportion of females are heterozygous for it? What proportion of males are heterozygous for it?

Answer: There are several possible explanations for the difference in the frequency of the trait in males and females. Two are sex linkage and autosomal linkage with sex-influenced expression. Sex linkage can be readily examined. If the population is in Hardy-Weinberg equilibrium (which this one is), the frequency of the recessive allele causing the trait is q, and the gene is X-linked, then the frequency of XY males would be q, while the frequency in XX females would be q^2. Since the frequency in males is 0.4 and the frequency in females is $(0.4)^2 = 0.16$, the data fit a model of sex linkage with $q = 0.4$. The frequency of heterozygous XX (female) individuals is $2pq = 2(0.6)(0.4) = 0.48$. Since the trait appears to be sex-linked, no heterozygous males exist.

22.12 Suppose you found two distinguishable types of individuals in wild populations of some organism in the following frequencies:

	Type 1	Type 2
Females	99%	1%
Males	90%	10%

The difference is known to be inherited. Are these dates compatible with the trait being X-linked?

Answer: As in problem 22.11, there are several explanations for differences in the frequency of a trait between males and females. Test the possibility of sex linkage first. Suppose the trait is sex-linked and recessive and the population is in equilibrium. Let q equal the frequency of the recessive allele, and p equal the frequency of the dominant allele. If the gene is X-linked, one would expect XY males to express the recessive trait at a frequency of q, and XX females to express the recessive trait at a frequency of q^2. By inspecting the data given, one can see that the frequency of type 2 individuals in females is 0.01, which is the square of the frequency of type 2 individuals in males: $(0.10)^2 = 0.01$. Thus, this trait appears to be controlled by a sex-linked pair of alleles occurring with allele frequencies of $q = 0.1$ recessive and $p = 0.9$ dominant.

22.13 Red-green color blindness is caused by a sex-linked recessive gene. About 64 women out of 10,000 are color-blind. What proportion of men would be expected to show the trait if mating is random?

Answer: Let q equal the frequency of the recessive allele, and p equal the frequency of the dominant allele. One expects homozygotes showing the trait to appear at a frequency of q^2 in a population at equilibrium. If $q^2 = 64/10{,}000 = 0.0064$, $q = 0.08$. Thus, one would expect 8 percent of XY, male individuals to show the trait.

22.14 About 8 percent of the men in a population are red-green color-blind (because of a sex-linked recessive gene). Answer the following questions with respect to color blindness, assuming random mating in the population.

a. What percentage of women would be expected to be color-blind?

b. What percentage of women would be expected to be heterozygous?

c. What percentage of men would be expected to have normal vision two generations later?

Answer:

a. Let q equal the frequency of the recessive allele, and p equal the frequency of the dominant allele. The frequency of males with the trait will be q, and the frequency of females will be q^2. Here, $q = 0.08$, so $q^2 = 0.0064$, or 0.64 percent.
b. Since $q = 0.08$, $p = 0.92$. The frequency of heterozygotes is $2pq = 0.1472$, or 14.72 percent. Only women can be heterozygotes, so the frequency of heterozygous women is 14.72 percent.
c. If the population is in Hardy-Weinberg equilibrium, the frequencies of alleles and the frequency of zygotic phenotypes will not change in two generations. Since $p = 0.92$, 92 percent of the (XY) males will have normal vision.

22.15 List some of the basic differences in the classical and neutral mutation models of genetic variation.

Answer: In the classical model, natural populations possess little variation. Within each population, a single allele is strongly favored by natural selection because it "functions the best." This is the wild-type allele, and most individuals are homozygous for this allele. Occasionally, a mutation arises. Usually, this mutation is deleterious, and strong selection occurs against it. In the rare case that a new mutation is advantageous for survival or reproduction, the new allele will increase in frequency and eventually become the new wild type.

In a neutral mutation model, there is significant genetic variation within a population. Many alleles exist in intermediate frequencies, and members of the population are heterozygous at numerous loci. Much of this genetic variation is explained by recurrent mutation and random changes in the frequency of alleles that are physiologically equivalent, and not necessarily by natural selection. Hence, in this model, most variation is neutral with regard to selection.

22.16 Two alleles of a locus, A and a, can be interconverted by mutation:

$$A \underset{v}{\overset{u}{\rightleftarrows}} a$$

where u is a mutation rate of 6.0×10^{-7} and v is a mutation rate of 6.0×10^{-8}. What will be the frequencies of A and a at mutational equilibrium, assuming no selective difference, no migration, and no random fluctuation caused by genetic drift?

Answer: Let q equal the frequency of a, and p equal the frequency of A, with $q + p = 1$. As discussed in the text, when the population is at equilibrium, the frequency of p and q is given by

$$q = \frac{u}{u+v} = \frac{6\times10^{-7}}{(6\times10^{-7})+(6\times10^{-8})} = \frac{6\times10^{-7}}{(6\times10^{-7})+(0.6\times10^{-7})} = \frac{6}{6.6} = 0.91$$

$$p = 1 - q = 1 - 0.91 = 0.09$$

Thus, the frequencies are 0.0081 AA, 0.1638 Aa, and 0.8281 aa.

22.17 The land snail *Cepaea nemoralis* is native to Europe but has been accidentally introduced into North America at several localities. These introductions occurred when a few snails were inadvertently transported on plants, building supplies, soil, or other cargo. The snails subsequently multiplied and established large, viable populations in North America.

Assume that today the average size of *Cepaea* populations found in North America is equal to the average size of *Cepaea* populations in Europe. What predictions can you make about the amounts of genetic variation present in European and North American populations of *Cepaea*? Explain your reasoning.

Answer: Since the gene pool in the present, large North American population is derived from a small number of individuals, a founder effect is likely to have occurred. Thus, genetic drift (random change in allelic frequency due to chance) will influence the North American populations to a greater degree than the European populations. One would expect to see less variation within and greater genetic differentiation among the North American populations. The magnitude of the difference between the native North American population and the introduced population will depend on how many times different European populations were introduced. For example, if relatively large numbers of European snails from multiple localities were introduced many times, the founder effect would not be as marked as if a small number of snails were introduced just once from one location.

22.18 A population of 80 adult squirrels resides on campus, and the frequency of the Est^1 allele among these squirrels is 0.70. Another population of squirrels is found in a nearby woods, and there the frequency of the Est^1 allele is 0.5. During a severe winter, 20 of the squirrels from the woods population migrate to campus in search of food and join the campus population. What will be the allelic frequency of Est^1 in the campus population after migration?

Answer: Let p_{I} equal the frequency of A in population I, and p_{II} equal the frequency of A in population II. If individuals in population I migrate to population II and make up proportion m of population II′, the new frequency of A in population II′ (p'_{II}) is given by

$$p'_{\mathrm{II}} = mp_{\mathrm{I}} + (1 - m)p_{\mathrm{II}}$$

Here, $p'_{\mathrm{II}} = [20/(20 + 80)](0.50) + \{1 - [20/(20 + 80)]\}(0.70) = 0.66$.

22.19 Upon sampling three populations and determining genotypes, you find the following three genotypic distributions. What does each of these distributions imply with regard to selective advantages of population structure?

Population	*AA*	*Aa*	*aa*
1	0.04	0.32	0.64
2	0.12	0.87	0.01
3	0.45	0.10	0.45

Answer: For each population, determine the frequency of the A and a alleles, and compare how the population structure compares with that expected if no selection was occurring.

To do this calculation, suppose there were 100 individuals in each population containing 200 alleles.

In population 1, there would be (0.04)(100)(2) + (0.32)(100)(1) = 40 *A* alleles and (0.64)(100)(2) + (0.32)(100)(1) = 160 *a* alleles. Thus, $f(A) = p = 40/200 = 0.20$, and $f(a) = q = 160/200 = 0.80$. If the population was in Hardy-Weinberg equilibrium, one would expect $p^2 = 0.04$ *AA*, $2pq = 0.32$ *Aa*, and 0.64 *aa* individuals, as is observed. If selection is acting at all, it may be acting to maintain the observed frequencies of the A and a alleles in equilibrium.

In population 2, there would be (0.12)(100)(2) + (0.87)(100)(1) = 111 *A* alleles and (0.01)(100)(2) + (0.87)(100)(1) = 89 *a* alleles. Here, $p = 0.555$ and $q = 0.455$. If the population was in Hardy-Weinberg equilibrium, one would expect $p^2 = 0.308$ *AA*, $2pq = 0.505$ *Aa*, and 0.207 *aa* individuals. The population is not in equilibrium, as there are far more heterozygotes and far fewer homozygotes than would be expected. This suggests that the heterozygote is being selected for in this population.

In population 3, there would be (0.45)(100)(2) + (0.10)(100)(1) = 100 *A* alleles and (0.10)(100)(1) + (0.45)(100)(2) = 100 *a* alleles. Here, $p = q = 0.5$. If the population was in Hardy-Weinberg equilibrium, one would expect $f(AA) = f(aa) = p^2 = q^2 = 0.25$ and $f(Aa) = 2pq = 0.50$. The population is not in equilibrium, as there are far more homozygotes (of either type) and far fewer heterozygotes than expected. This suggests that the heterozygote is being selected against in this population.

22.20 The frequency of two adaptively neutral alleles in a large population is 70 percent *A* and 30 percent *a*. The population is wiped out by an epidemic, leaving only four individuals, who produce many offspring. What is the probability that the population several years later will be 100 percent *AA*? (Assume no mutations.)

Answer: There are a variety of situations in which all individuals would become *AA*. One is that the four "founding" individuals are all *AA*. Since the probability of a single individual being *AA* is $(0.7)^2 = 0.49$, the probability of the four founding individuals being *AA* is $(0.49)^4 = 0.0576$, about 1/17. Even if all four founding individuals are not *AA*, there will be some chance that the subsequent population may become all *AA*. For example, there is a low but distinct probability that only *AA* offspring may be had by *Aa* × *AA* parents. Given that "many" offspring are produced, the likelihood of this will be less than 1/17 [e.g., if ten offspring are produced, the likelihood of all being *AA* in such a cross would be $(1/2)^{10} = 1/1{,}024$]. Since the alleles are adaptively neutral, it would appear that 1/17 is an upper bound for the likelihood of a population being 100 percent *AA*.

22.21 A completely recessive gene, through changed environmental circumstances, becomes lethal in a certain population. It was previously neutral, and its frequency was 0.5.

a. What was the genotype distribution when the recessive genotype was not selected against?
b. What will be the allelic frequency after one generation in the altered environment?
c. What will be the allelic frequency after two generations?

Answer:

a. Let $p = f(A)$, and $q = f(a)$. Initially, $p = q = 0.5$, so $p^2 = f(AA) = q2 = f(aa) = 0.5)^2 = 0.25$. $f(Aa) = 2pq = 0.50$.

b. If *aa* individuals are now lethal, they will not contribute to the gene pool in the next generation. Only *AA* and *Aa* individuals will contribute to the gene pool in the next generation. The allele frequency will be

$$f(A) = \frac{(0.25\times 2)+(0.50\times 1)}{(0.25\times 2)+(0.50\times 2)} = \frac{1.0}{1.5} = 0.67$$

$$f(a) = \frac{(0.50\times 1)}{(0.25\times 2)+(0.50\times 2)} = \frac{0.5}{1.5} = 0.33$$

c. Given the result of (b), the genotype of the progeny are $p^2 = (0.67)^2 = 0.449$ *AA*, $2pq = 2(0.67)(0.33) = 0.442$ *Aa*, and $q^2 = (0.33)2 = 0.109$ *aa*. As in part (b), only *AA* and *Aa* individuals will contribute to the gene pool in the next generation. The allele frequency will be

$$f(A) = \frac{(0.449\times 2)+(0.442\times 1)}{(0.449\times 2)+(0.442\times 2)} = 0.75$$

$$f(a) = \frac{(0.442\times 1)}{(0.449\times 2)+(0.442\times 2)} = 0.25$$

Can you show that the following statement is true under the conditions of this problem? If the last generation in the unaltered environment is considered to be generation number 1, then after n generations, the frequency of A will be $n/(n + 1)$ and the frequency of a will be $1/(n + 1)$.

22.22 Human individuals homozygous for a certain recessive autosomal gene die before reaching reproductive age. Despite this removal of all affected individuals, there is no indication that homozygotes occur less frequently in succeeding generations. To what might you attribute the continued appearance of recessives?

Answer: There are several reasons why recessives may appear in a constant frequency. (1) Perhaps most important is that new mutations of *A* to *a* could occur at a low but constant rate. (2) There could also be a selective advantage to heterozygotes (overdominance). (3) There could be nonrandom mating within the population (e.g., positive assortative mating, inbreeding). (4) There could be a low but steady frequency of migration of heterozygotes into the population. Each of these possibilities violates assumptions made for a population in Hardy-Weinberg equilibrium and allows maintenance of a recessive allele that is deleterious as a homozygote. (Problem 22.21 considers the consequences of such an allele if the assumptions were not violated.) (5) If the homozygous recessive condition is a relatively rare occurrence, the frequency of *a* alleles in the population is very low. As a result, only a very small proportion of the *a* alleles would be found in homozygotes; almost all *a* alleles would be found in heterozygotes (protected polymorphism). The few generations for which data have been recorded may not be enough to see a decline in the number of homozygotes.

22.23 A completely recessive gene (Q^1) has a frequency of 0.7 in a large population, and the Q^1Q^1 homozygote has a relative fitness of 0.6.

a. What will be the frequency of Q^1 after one generation of selection?
b. If there is no dominance at this locus (the fitness of the heterozygote is intermediate to the fitnesses of the homozygotes), what will the allelic frequency be after one generation of selection?
c. If Q^1 is dominant, what will the allelic frequency be after one generation of selection?

Answer: See page 479. For parts (a) and (b), $p = 0.3$, $q = 0.7$, and $s = 1 - W = 0.4$.

a. If there is selection against the recessive homozygotes, after one generation,

$$\Delta q = \frac{-spq^2}{1-sq^2} = \frac{-(0.4)(0.3)(0.7)^2}{1-(0.4)(0.7)^2} = -0.073$$

$$q^1 = 0.7 - 0.073 = 0.627$$

b. If there is selection with no dominance, so that the fitness of the heterozygote is intermediate between the two homozygotes,

$$\Delta q = \frac{-spq/2}{1-sq} = \frac{-(0.4)(0.3)(0.7)/2}{1-(0.4)(0.7)} = -0.0583$$

$$q^1 = 0.7 - 0.0583 = 0.642$$

c. To select against Q^1 as a dominant allele, let p equal the frequency of the dominant allele being selected against: $p = 0.7$, $q = 0.3$, and $s = 0.4$.

$$\Delta p = \frac{-spq^2}{1-s+sq^2} = \frac{-(0.4)(0.7)(0.3)^3}{1-s+sq^2} = \frac{-(0.4)(0.7)(0.3)^2}{1-0.4+(0.4)(0.3)^2} = -0.04$$

$$p^1 = 0.70 - 0.04 = 0.66$$

22.24 As discussed earlier in this chapter, the gene for sickle-cell anemia exhibits heterozygote advantage. An individual who is an *Hb-A*/*Hb-S* heterozygote has increased resistance to malaria and therefore has greater fitness than the *Hb-A*/*Hb-A* homozygote, who is susceptible to malaria, and the *Hb-S*/*Hb-S* homozygote, who has sickle-cell anemia. Suppose that the fitness values of the genotypes in Africa are as presented here:

Hb-A/Hb-A = 0.88
Hb-A/Hb-S = 1.00
Hb-S/Hb-S = 0.14

Give the expected equilibrium frequencies of the sickle-cell gene (*Hb-S*).

Answer: From the text discussion, at equilibrium, $p = f(Hb\text{-}A) = t/(t+s)$, and $q = f(Hb\text{-}S) = s/(s+t)$, where t equals the selection coefficient of *Hb-S*/*Hb-S*, and s equals the selection coefficient of *Hb-A*/*Hb-A*. Since *fitness* = 1 – *selection coefficient*, $f(Hb\text{-}S) = 0.12/(0.12 + 0.86) = 0.122$.

22.25 Achondroplasia, a type of dwarfism in humans, is caused by an autosomal dominant gene. The mutation rate for achondroplasia is about 5×10^{-5}, and the fitness of achondroplastic dwarfs has been estimated to be about 0.2, compared with unaffected individuals. What is the equilibrium frequency of the achondroplasia gene based on this mutation rate and fitness value?

Answer: For a dominant allele, the frequency at equilibrium is $\hat{p} = u/s$, where u equals the mutation rate, and s is the selection coefficient (= 1 – fitness). Here, $\hat{p} = 5 \times 10^{-5}/0.8 = 6.25 \times 10^{-5}$.

22.26 The frequencies of the L^M and L^N blood group alleles are the same in each of the populations I, II, and III, but the genotypes' frequencies are not the same, as shown in the following table. Which of the populations is most likely to show each of the following characteristics: random mating, inbreeding, genetic drift? Explain your answers.

	L^ML^M	L^ML^N	L^NL^N
I	0.50	0.40	0.10
II	0.49	0.42	0.09
III	0.45	0.50	0.05

Answer: First calculate the frequencies of each allele. In population I, one has [(0.5 × 2) + (0.4 × 1)/2] = 0.7 L^M, and 0.3 L^N. Similarly, L^M = 0.7 and L^N = 0.3 in populations II and III. A population in Hardy-Weinberg equilibrium would have $p^2 = (0.7)^2 = 0.49$ L^ML^M, $2pq = 2(0.7)(0.3) = 0.42$ L^ML^N, and $q^2 = (0.3)^2 = 0.09$ L^NL^N. As population II shows these features, it must be in Hardy-Weinberg equilibrium and would be expected to exhibit random mating. Inbreeding results in an increase in the frequency of homozygotes and would be associated with population I (but not III). Genetic drift is random change in allelic frequency due to chance and can be explained by random effects, a small effective population size, founder or bottleneck effects, or sampling error. Genetic drift could be associated with either population I or population III. Because the frequencies in populations I and III are fairly close to those of an equilibrium population, sampling error in a population in Hardy-Weinberg equilibrium could also explain the results seen.

22.27 DNA was collected from 100 people randomly sampled from a given human population and was digested with the restriction enzyme *Bam*HI; the fragments were separated by electrophoresis and then transferred to a membrane filter using the Southern blot technique. The blots were probed with a particular cloned sequence. Three different patterns of hybridization were seen on the blots. Some DNA samples (56 of them) showed a single band of 6.3 kb, others (6) showed a single band at 4.1 kb, and yet others (38) showed both the 6.3- and the 4.1-kb bands.

a. Interpret these results in terms of *Bam*HI sites.

b. What are the frequencies of the restriction site alleles?

c. Does this population appear to be in Hardy-Weinberg equilibrium for the relevant restriction sites?

Answer:

a. One explanation is that a restriction fragment length polymorphism exists in the population. On some chromosomes, an RFLP having a size of 4.1 kb is found, while on others, the 6.3-kb fragment is found. Individuals having just one size band are homozygotes, while individuals with two different-sized bands are heterozygous. The difference in sizes of the fragments could result from a missing site in the 6.3-kb individuals or the insertion of a 2.2-kb piece of DNA between two sites that are normally 4.1 kb apart.

b. Homozygotes have two identical alleles, while heterozygotes have two different alleles. There are $(56 \times 2) + 38 = 150$ 6.3-kb alleles and $(6 \times 2) + 38 = 50$ 4.1-kb alleles. Let q equal the frequency of 6.3-kb alleles, and p equal the frequency of 4.1-kb alleles. Then $q = 0.75$ and $p = 0.25$.

c. For a population in Hardy-Weinberg equilibrium, one would expect $q^2 = 0.5625$ 6.3-kb homozygotes (compared to the 0.56 seen), $2pq = 0.375$ heterozygotes (compared to the 0.38 seen), and $p^2 = 0.0625$ 4.1-kb homozygotes (compared to the 0.06 seen). The population appears to be in Hardy-Weinberg equilibrium.

22.28 Fifty tiger salamanders from one pond in west Texas were examined for genetic variation by using the technique of protein electrophoresis. The genotype of each salamander was determined for five loci (AmPep, ADH, PGM, MDH, and LDH-1). No variation was found at AmPep, ADH, and LDH-1; in other words, all individuals were homozygous for the same allele at these loci. The following numbers of genotypes were observed at the MDH and PGM loci.

MDH Genotypes	Number of Individuals	*PGM* Genotypes	Number of Individuals
AA	11	*DD*	35
AB	35	*DE*	10
BB	4	*EE*	5

Calculate the proportion of polymorphic loci and the heterozygosity for this population.

Answer: Since five loci were examined, and only two have more than one allele, (⅖)(100%) = 40% of the loci are polymorphic. Heterozygosity is calculated by averaging the frequency of heterozygotes for each locus. The frequency of heterozygotes for the AmPep, ADH, and LDH-1 loci is zero. At MDH, 35 out of 50 individuals were heterozygous (0.70). At PGM, 10 out of 50 individuals were heterozygous (0.20). Thus, the average heterozygosity is

$$\frac{0+0+0+0.7+2}{5} = 0.18$$

22.29 What factors cause genetic drift?

Answer: Genetic drift arises from random change in allelic frequency due to chance. Random factors producing mortality in natural populations and sampling error can lead to genetic drift. Its causes include a small effective population size over many generations, a small number of founders (founder effect), and a reduction in population size (bottleneck effect).

22.30 What are the primary effects of the following evolutionary processes on the allelic and genotypic frequencies of a population?

a. mutation
b. migration
c. genetic drift
d. inbreeding

Answer:

a. Mutation will lead to change in allelic frequencies within a population if no other forces are acting. It will introduce genetic variation. If the effective population size is small, mutation may lead to genetic differentiation among populations.
b. Migration will increase the population size, and has the potential to disrupt a Hardy-Weinberg equilibrium. It can increase genetic variation and may influence the evolution of allelic frequencies within populations. Over many generations, migration will reduce divergence among populations and equalize allelic frequencies among populations.
c. Genetic drift produces changes in allelic frequencies within a population. It can reduce genetic variation and increase the homozygosity within a population. Over time, it leads to genetic change. When several populations are compared, genetic drift can lead to increased genetic differences among populations.
d. Inbreeding will increase the homozygosity within a population and decrease its genetic variation.

22.31 Explain how overdominance leads to an increased frequency of sickle-cell anemia in areas where malaria is widespread.

Answer: Overdominance results when a heterozygote genotype has higher fitness than either of the homozygotes. The two alleles of the heterozygote are maintained in a population because both are favored in the heterozygote genotype. In the case of the sickle-cell allele, heterozygotes for *Hb-A/Hb-S* are at a selective advantage because the hemoglobin mixture in these individuals provides an unfavorable environment for the growth of malarial parasites. Thus, heterozygotes have higher fitness than *Hb-A/Hb-A* homozygotes who are susceptible to malaria. Heterozygotes also have higher fitness than *Hb-S/Hb-S* homozygotes who suffer from sickle-cell anemia. The favoring of the sickle-cell allele *Hb-S* in the heterozygote results in its relatively high frequency in areas with malaria.

23

QUANTITATIVE GENETICS

CHAPTER OUTLINE

THE NATURE OF CONTINUOUS TRAITS

Why Some Traits Have Continuous Phenotypes

Questions Studied in Quantitative Genetics

STATISTICAL TOOLS

Samples and Populations

Distributions

The Mean

The Variance and the Standard Deviation

Correlation

POLYGENIC INHERITANCE

Inheritance of Ear Length in Corn

Polygene Hypothesis for Quantitative Inheritance

Quantitative Trait Loci

HERITABILITY

Components of the Phenotypic Variance

Broad-Sense and Narrow-Sense Heritability

Understanding Heritability

RESPONSE TO SELECTION

Estimating the Response to Selection

Genetic Correlations

REVIEW OF KEY TERMS, SYMBOLS, AND CONCEPTS

In your own words, write a brief, precise definition of each term in the groups on the following page. Check your definitions using the text. Then develop a concept map using the terms in each list.

THINKING ANALYTICALLY

Quantitative genetics deals with important but complex issues. Some of the concepts used in the analysis of quantitative traits are subtle. Throughout the chapter, clear and precise thinking is

1	2	3
quantitative genetics	quantitative genetics	heritability
discontinuous trait	population	broad-sense heritability
continuous trait	sample	narrow-sense heritability
polygenic trait	frequency distribution	phenotypic variance (V_P)
norm of reaction	normal distribution	genetic variance (V_G)
multifactorial trait	binomial distribution	environmental variance (V_E)
familial trait	binomial expansion	genetic-environmental interaction ($V_{G \times E}$)
penetrance	mean	genetic-environmental covariance ($COV_{G \times E}$)
expressivity	variance	additive genetic variance (V_A)
pleiotropy	standard deviation	dominance variance (V_D)
epistasis	covariance	interaction variance (V_I)
polygene	correlation vs. cause and effect	general environmental effects (V_{Eg})
contributing alleles	correlation vs. identity	special environmental effects (V_{Es})
noncontributing alleles	phenotypic correlation	maternal effects (V_{Em})
polygene or multiple gene hypothesis for quantitative inheritance	genetic correlation	evolution
	pleitropy	artificial selection
	positive vs. negative correlation	selection response
		selection differential
		narrow-sense heritability

required. To sort out the basis for a quantitative trait, it is important to define the contribution of several different factors. These factors can include the environment as well as multiple genes that themselves may or may not interact. Central to the analysis of a quantitative trait is identifying the degree to which variation in a phenotype can be associated with one or more of these factors. Consequently, understanding how variation is measured and used is of the utmost importance in understanding quantitative genetics.

First, gain a solid understanding of the terms used and the concepts they convey. Geneticists have developed a substantial conceptual framework in which to consider quantitative traits. Some concepts are easy to misunderstand, and to understand what something *is*, it is often important to define what it *is not*. For example, the idea of heritability has important qualifications and limitations. Heritability is a measure of the *proportion of phenotypic variance* that results from genetic differences. It is *not* a measure of the extent to which a trait is genetic, or what proportion of an individual's phenotype is genetic. It is also not fixed for a particular trait or a measurement of genetic differences between populations. The concepts underlying quantitative genetics have significant utility, but only if they are correctly and clearly applied.

Second, concentrate on understanding the statistical methods that are used to analyze data on quantitative traits. Your understanding should consist of more than just knowledge of how to crunch numbers by plugging them into an equation. You need to get a feel for what a statistic tells you about a data set.

Third, relate how measurements are made and analyzed to the concepts that have been developed to explain the inheritance of quantitative traits. For example, relate a measurement of heritability to a selection response. Then explain how you would achieve a particular breeding objective through selection. By doing this, you will obtain a better understanding of the power and utility of quantitative genetics.

QUESTIONS FOR PRACTICE

Multiple Choice Questions

1. Which of the following could be used to describe coat colors in mice?
   a. continuous trait
   b. discontinuous trait
   c. polygenic trait
   d. quantitative trait
   e. both b and c
2. What can generally be said about quantitative traits such as crop yield, rate of weight gain, human birth weight, blood pressure, and the number of eggs laid by *Drosophila?*
   a. They are intractable to molecular genetics.
   b. They are multifactorial.
   c. They are discontinuous.
   d. They are not heritable because they are polygenic.
3. Which statement below best describes the aims of a quantitative genetic analysis of a trait?
   a. to determine whether genes or the environment control a trait
   b. to determine the best breeding strategy to retain a trait
   c. to determine how much of the phenotypic variation associated with a trait in a population is due to genetic variation and how much is due to environmental variation
   d. to determine whether a trait is controlled by nature versus nurture
4. The description of a population in terms of the number of individuals that display varying degrees of expression of a character or range of phenotypes is a
   a. polygraph
   b. polynomial
   c. normal distribution
   d. frequency distribution
5. The general expression for the binomial expansion is
   a. $(p + q)^n$
   b. $(a^2 + b^2)$
   c. $(a + b)^2$
   d. $a^2 + 2ab + b^2$
6. Which *two* of the following accurately describe the term *variance?*
   a. It is a reflection of the accuracy of an estimated measurement.
   b. It is a measure of how much a set of individual measurements is spread out around the mean.
   c. It is the average value of a set of measurements.
   d. It is equal to $(\Sigma x_i/n)$.
   e. It is equal to $\dfrac{\Sigma(x_i - \bar{x})^2}{n-1}$

7. What is heritability?
   a. the proportion of a population's phenotype that is attributable to genetic factors
   b. the proportion of a population's phenotypic variation that is attributable to genetic factors
   c. the degree to which family members resemble one another
   d. the degree to which a continuous trait is controlled by genetic factors
8. The proportion of the phenotypic variance that consists of genetic variance, additive or otherwise, is called
   a. heritability
   b. broad-sense heritability
   c. narrow-sense heritability
   d. phenotypic variance derived from genetic-environmental interactions
9. The narrow-sense heritability of a trait is determined to be very close to 1.0. Which of the following inferences can be made?
   a. The trait will be difficult to select for.
   b. The trait can be readily selected for.
   c. Most of the phenotypic variance results from additive genetic variance.
   d. Most of the phenotypic variance results from environmental variance.
   e. both a and c
   f. both b and c
10. In a population of chickens raised under controlled conditions, body weight is negatively correlated with egg production but positively correlated with egg weight. Because of market conditions, a farmer is interested in producing lots of small eggs. What selection strategy might be beneficial?
    a. Select for smaller hens.
    b. Select for larger hens.
11. Which one of the following statements is true?
    a. Traits shared by family members show high heritability.
    b. Artificial and natural selection can occur in a genetically uniform population.
    c. If heritability is high in each of two populations and the populations differ markedly in a trait, the populations are genetically different.
    d. The selection response depends on (1) the proportion of genetic variance that results from additive genetic variance and (2) the difference between the mean phenotypes of the selected parents and the mean phenotype of the unselected population.

Answers: 1e, 2b, 3c, 4d, 5a, 6b, e, 7b, 8b, 9f, 10a, 11d

Thought Questions

1. Assume that mature fruit weight in pumpkins is a quantitative trait. In the following experiment, environmental factors (weather, soil, etc.) are uniform. Two pumpkin varieties, both of which produce fruit with a mean weight of 20 lb, are crossed. The F_1 produces 20-lb pumpkins. The F_2 plants, however, give the following results:

Mean fruit wt. (lb)	5	12.5	20	27.5	35
Number of plants	19	82	119	79	21

   Explain these results: Postulate how many genes are involved and how much each contributes to fruit weight.

2. Distinguish between heritability, broad-sense heritability, and narrow-sense heritability. In what ways can the latter two quantities be measured, and how can they be put to use by plant and animal breeders?
3. What statistics from a regression analysis are used to estimate narrow-sense heritability?
4. What sources contribute to the phenotypic variance associated with a quantitative trait?
5. Distinguish between a genetic correlation and a phenotypic correlation. In particular, address how a phenotypic correlation might exist when the trait is, or is not, influenced by a common set of genes.
6. Even though a positive correlation exists between alcohol consumption and the number of Baptist ministers, there is thought to be no causal relationship between these two phenomena. Can you think of other correlations that are likely to be true, but which clearly demonstrate that correlations (whether positive or negative) do not imply a cause and effect, and that a correlation is not the same thing as an identity?
7. Defend each of the following statements:
   a. Broad-sense heritability does not indicate the extent to which a trait is genetic.
   b. Heritability does not indicate what proportion of an individual's phenotype is genetic.
   c. Heritability is not fixed for a trait.
   d. If heritability is high in two populations, and the populations differ markedly in a particular trait, one cannot assume that the populations are genetically different.
   e. Familial traits do not necessarily have high heritability.
8. If a population is phenotypically homogeneous for a particular trait, can you predict whether it has a high narrow-sense heritability? If not, what additional information would you need?
9. Why is calculating narrow-sense heritability values for some quantitative human traits especially difficult?
10. Distinguish between a selection differential and a selection response.
11. The mean height in two separate populations A and B is identical and is 5 feet 8 inches. The variance in population A is 14 inches, while the variance in population B is 3 inches. In which population would a 6-foot-tall person be more uncommon?
12. In a single population, is it likely that artificial selection will be equally effective at producing changes in three different traits? Why or why not?

Thought Questions for Media

After reviewing the media at the *Essential iGenetics* website, try to answer these questions.

1. In the *Depicting Quantitative Inheritance* animation, a cross of wheat with red kernels to wheat with white kernels gave wheat with pink kernels. Why does this suggest that this trait shows incomplete dominance?
2. What data given in the animation allow one to rule out incomplete dominance as an explanation for the results illustrated in the animation?
3. An alternate explanation for the results illustrated in the animation was that two independently assorting genes controlled wheat kernel color. Why was this explanation also ruled out?
4. In polygenic inheritance, what is a contributing allele? What is a noncontributing allele? How many contributing alleles can be present at one gene? How many contributing alleles can be present in one organism?

5. How does the explanation of two independently assorting genes with contributing and noncontributing alleles differ from the simpler explanation of two independently assorting genes?
6. Explain why, even though a contributing allele contributes a defined amount of pigment to the red color of the wheat kernel, wheat kernel color still appears to show continuous variation.

SOLUTIONS TO TEXT PROBLEMS

23.1 The following measurements of head width and wing length were made on a series of steamer ducks:

Specimen	Head Width (cm)	Wing Length (cm)
1	2.75	30.3
2	3.20	36.2
3	2.86	31.4
4	3.24	35.7
5	3.16	33.4
6	3.32	34.8
7	2.52	27.2
8	4.16	52.7

a. Calculate the mean and the standard deviation of head width and of wing length for these eight birds.
b. Calculate the correlation coefficient for the relationship between head width and wing length in this series of ducks.
c. What conclusions can you make about the association between head width and wing length in steamer ducks?

Answer:

a. The mean is obtained by summing the individual values and dividing by the total number of values. The mean head width is 25.21/8 = 3.15 cm, and the mean wing length is 281.7/8 = 35.21 cm.

The standard deviation equals the square root of the variance (the square root of s^2). The variance is computed by squaring the sum of the difference between each measurement and the mean value, and dividing this sum by the number of measurements minus 1. One has

$$s_{\text{head width}} = \sqrt{s^2} = \sqrt{\frac{\Sigma(x_i - \bar{x})^2}{n-1}} = \sqrt{\frac{1.70}{7}} = \sqrt{0.24} = 0.49 \text{ cm}$$

$$s_{\text{wing length}} = \sqrt{s^2} = \sqrt{\frac{\Sigma(x_i - \bar{x})^2}{n-1}} = \sqrt{\frac{413.35}{7}} = \sqrt{59.05} = 7.68 \text{ cm}$$

b. The correlation coefficient, r, is calculated from the covariance, cov, of two quantities. Let head width be represented by x and wing length be represented by y. r is defined as

$$r = \frac{\text{cov}_{xy}}{s_x s_y} = \frac{\dfrac{\Sigma x_i y_i - \dfrac{1}{n}(\Sigma x_i \Sigma y_i)}{n-1}}{s_x s_y}$$

The first factor $(\Sigma x_i y_i)$ is obtained by taking the sum of the products of the individual measurements of head width and the corresponding measurements for wing length. The next factor $(\Sigma x_i \Sigma y_i)/n$ is the product of the sums of the two sets of measurements divided by the number of pairs of measurements. The difference between these values is then divided by $(n - 1)$ and then by the products of the standard deviations of each measurement. One has

$$r = \frac{\dfrac{913 - (1/8)(25.21 \times 281.7)}{7}}{0.49 \times 7.68} = \frac{3.61}{3.76} = 0.96$$

c. Head width and wing length show a strong positive correlation, nearly 1.0. This means that ducks with larger heads will almost always have longer wings, and ducks with smaller heads will almost always have smaller wings.

23.2 Answer the following questions.

a. In a family of six children, what is the probability that three will be girls and three will be boys?

b. In a family of five children, what is the probability that one will be a boy and four will be girls?

c. What is the probability that in a family of six children, all will be boys?

Answer:

a. Three boys and three girls can be obtained in a variety of different orders, and each of these must be considered to determine the probability of obtaining this result. One could determine the number of different orders that are possible and then use the sum rule and add together the probability of obtaining three boys and three girls in each possible order. It is easier to use the binomial expansion than to work out all of the ways in which these offspring can be obtained. The binomial expansion is given by

$$P(s,t) = \frac{n!}{s!t!} a^s b^t$$

Here, s equals the number of the first class (here, say the number of girls, so $s = 3$), t equals the number of the other class (here, the number of boys, so $t = 3$), n equals the total number in both classes (here $n = 6$), a is the probability of the first event ($P(girl) = ½$), and b is the probability of the second event ($P(boy) = ½$). Substituting these values into the equation, one has

$$P(\text{3 girls, 3 boys}) = [6!/(3!3!)] \times (½)^3 \times (½)^3 = 20/64 = 0.3125$$

b. Again, use the binomial expansion, with $n = 5$, $s = 4$, and $t = 1$. One has $P(4$ girls, 1 boy$) = [5!/(4!1!)] \times (½)^1 \times (½)^4 = 5/32 = 0.15625$

c. One could again use the binomial expansion ($n = 6$, $s = 0$, $t = 6$, $a = b = ½$; note that $0! = 1$), but since there is only one way of obtaining six boys, it is straightforward to do this calculation without the binomial expansion. The chance of obtaining one boy is ½. The chance of obtaining six boys (use the product rule) is $(½)^6 = 1/64 = 0.015625$.

23.3 In flipping a coin, there is a 50 percent chance of obtaining heads and a 50 percent chance of obtaining tails on each flip. If you flip a coin ten times, what is the probability of obtaining exactly five heads and five tails?

Answer: As in part (a) of problem 23.2, there are a variety of different ways to obtain this result. Use the binomial expansion, with $n = 10$, $a = b = ½$, $s = t = 5$.

$$P = \frac{10!}{5!5!} \times \left(\frac{1}{2}\right)^5 \times \left(\frac{1}{2}\right)^5 = \frac{252}{1024} = 0.246$$

23.4 The F_1 generation from a cross of two pure-breeding parents that differ in a size character usually is no more variable than the parents. Explain.

Answer: The degree of phenotypic variability is related to the degree of genetic variability. Since each pure-breeding parent is homozygous for the genes (however many there are) controlling the size character, each parent is homogeneous in type. A cross of two pure-breeding strains will generate an F_1 heterozygous for those loci controlling the size trait. Since the F_1 consists of all heterozygotes, it is genetically as homogeneous as each of the parents. Therefore, it shows no greater variability than the parents.

23.5 If two pure-breeding strains, differing in a size trait, are crossed, is it possible for F_2 individuals to have phenotypes that are more extreme than either grandparent (i.e., be larger than the largest or smaller than the smallest in the parental generation)? Explain.

Answer: If the two grandparents of the F_2 individuals (i.e., the two parents in the parental generation) represent the two extremes for the size trait that is found in the population, the answer is no. However, if neither of the grandparental types is at an extreme end of the size range, and if each strain differs in the appropriate genes, then it is possible to obtain an F_2 that is more extreme than their grandparents. As an example, consider a quantitative trait that is specified by three allelic pairs: A/a, B/b, and C/c. Suppose each capital allele contributes a certain amount to the quantitative character. Then individuals in the population with six capital alleles (*AA BB CC*) show one extreme, while individuals with six lowercase alleles (*aa bb cc*) show the other extreme. If the parental cross were *AA BB cc* × *aa bb CC*, the F_1 would be *Aa Bb Cc*, and some of the F_2 would be *AA BB CC* and *aa bb cc*. In this scenario, some of the F_2 would have more capital alleles than either grandparent and so exhibit a more extreme phenotype. Similarly, some of the F_2 will have fewer capital alleles than either grandparent and so exhibit a more extreme phenotype at the opposite end of the size range.

23.6 Two pairs of genes with two alleles each, A/a and B/b, determine plant height additively in a population. The homozygote *AA BB* is 50 cm tall, and the homozygote *aa/bb* is 30 cm tall.

a. What is the F_1 height in a cross between the two homozygous stocks?
b. What genotypes in the F_2 will show a height of 40 cm after an $F_1 \times F_1$ cross?
c. What will be the F_2 frequency of the 40-cm plants?

Answer:

a. Since the cross is *AA BB* × *aa bb*, the F_1 genotype will be *Aa Bb*. Since capital alleles determine height additively, and individuals with four capital alleles have a height of 50 cm, while individuals with no capital alleles have a height of 30 cm, each capital allele appears to confer (50 – 30)/4 = 5 cm of height over the 30-cm base. *Aa Bb* individuals with two capital alleles should have an intermediate height of 40 cm.
b. Any individuals with two capital alleles will show a height of 40 cm. Thus, *Aa Bb*, *AA bb*, and *aa BB* individuals will be 40 cm high.
c. In the F_2, 1/16 of the progeny are *AA bb*, 4/16 are *Aa Bb*, and 1/16 are *aa BB*. Thus, 6/16 = 3/8 of the progeny will be 40 cm high.

23.7 Three independently segregating genes (*A*, *B*, *C*), each with two alleles, determine height in a plant. Each capital-letter allele adds 2 cm to a base height of 2 cm.
a. What are the heights expected in the F_1 progeny of a cross between homozygous strains *AA BB CC* (14 cm) and *aa bb cc* (2 cm)?
b. What is the distribution of heights (frequency and phenotype) expected in an $F_1 \times F_1$ cross?
c. What proportion of F_2 plants will have heights equal to the heights of the original two parental strains?
d. What proportion of the F_2 will breed true for the height shown by the F_1?

Answer:

a. The cross *AA BB CC* × *aa bb cc* will produce all *Aa Bb Cc* progeny. The three capital-letter alleles will add 3 × 2 = 6 cm to the 2-cm base height, giving a total of 8 cm.
b. There are a variety of ways to approach this problem. One way is to use a modified Punnett square and another is to use the binomial theorem [use the coefficients of $(a + b)^6$]. Shown here is still another approach, a branch diagram:

Aa* × *Aa	***Bb* × *Bb***	***Cc* × *Cc***	**Proportion**	**# Capital Alleles**	**Ht. (cm)**
		1/4 *CC* →	1/64	6	14 cm
	1/4 *BB*	1/2 *Cc* →	2/64	5	12 cm
		1/4 *cc* →	1/64	4	10 cm
		1/4 *CC* →	2/64	5	12 cm
1/4 *AA*	1/2 *Bb*	1/2 *Cc* →	4/64	4	10 cm
		1/4 *cc* →	2/64	3	8 cm
		1/4 *CC* →	1/64	4	10 cm
	1/4 *bb*	1/2 *Cc* →	2/64	3	8 cm
		1/4 *cc* →	1/64	2	6 cm

(continued)

$Aa \times Aa$	$Bb \times Bb$	$Cc \times Cc$	Proportion	# Capital Alleles	Ht. (cm)
		1/4 *CC* →	2/64	5	12 cm
	1/4 *BB*	1/2 *Cc* →	4/64	4	10 cm
		1/4 *cc* →	2/64	3	8 cm
		1/4 *CC* →	4/64	4	10 cm
1/2 *Aa*	1/2 *Bb*	1/2 *Cc* →	8/64	3	8 cm
		1/4 *cc* →	4/64	2	6 cm
		1/4 *CC* →	2/64	3	8 cm
	1/4 *bb*	1/2 *Cc* →	4/64	2	6 cm
		1/4 *cc* →	2/64	1	4 cm
		1/4 *CC* →	1/64	4	10 cm
	1/4 *BB*	1/2 *Cc* →	2/64	3	8 cm
		1/4 *cc* →	1/64	2	6 cm
		1/4 *CC* →	2/64	3	8 cm
1/4 *aa*	1/2 *Bb*	1/2 *Cc* →	4/64	2	6 cm
		1/4 *cc* →	2/64	1	4 cm
		1/4 *CC* →	1/64	2	6 cm
	1/4 *bb*	1/2 *Cc* →	2/64	1	4 cm
		1/4 *cc* →	1/64	0	2 cm

Add up the proportion of individuals with like heights to obtain:

$1/64 \rightarrow 14$ cm
$6/64 \rightarrow 12$ cm
$15/64 \rightarrow 10$ cm
$20/64 \rightarrow 8$ cm
$15/64 \rightarrow 6$ cm
$6/64 \rightarrow 4$ cm
$1/64 \rightarrow 2$ cm.

c. $\frac{1}{64}$ are 2 cm and $\frac{1}{64}$ are 14 cm, so the answer is $\frac{2}{64}$.

d. None of the F_2 with the height of 8 cm can breed true, because this height depends on a genotype having three capital-letter alleles, which in turn requires heterozygosity for at least one gene locus.

23.8 Repeat problem 23.7, but assume that each capital-letter allele doubles the existing height; for example, *Aa bb cc* = 4 cm, *AA bb cc* = 8 cm, *AA Bb cc* = 16 cm, and so on.

Answer:

a. Because the base height is 2 cm, the plant height is 2^{n+1}, where n = the number of capital-letter alleles. The *Aa Bb Cc* F_1 height is $2^4 = 16$ cm.

b. 1/64 → 128 cm
6/64 → 64 cm
15/64 → 32 cm
20/64 → 16 cm
15/64 → 8 cm
6/64 → 4 cm
1/64 → 2 cm.

c. 2/64

d. The 16-cm F_1 height is determined by the presence of three dominant alleles. This can only occur if one allelic pair is heterozygous. Consequently, the 16-cm strain could not breed true.

23.9 Assume that three equally and additively contributing pairs of alleles control flower length in nasturtiums. A completely homozygous plant with 10-mm flowers is crossed with a completely homozygous plant with 30-mm flowers. F_1 plants all have flowers about 20 mm long. F_2 plants show a range of lengths from 10 to 30 mm, with about 1/64 of the F_2 having 10-mm flowers and 1/64 having 30-mm flowers. What distribution of flower length would you expect to see in the offspring of a cross between an F_1 plant and the 30-mm parent?

Answer: Use the clue that 1/64 of the F_2 progeny show either extreme trait. Note that (as in problems 23.7 and 23.8) in an F_2 resulting from a trihybrid cross, the proportion of one kind of homozygote is 1/64. This suggests that three allelic pairs control this quantitative trait. The completely homozygous 10-mm plant would have six lowercase alleles at three loci, *aa bb cc*, while the completely homozygous 30-mm plant would have six capital alleles at three loci, *AA BB CC*. If each capital allele contributes 1/6 of the 20-mm difference between the two extremes, or 20/6 = 3.33 mm, the F_1 trihybrid individual would be 20 mm (= 10 mm base + 3 × (3.33 mm/capital allele) = 20 mm). Thus, the trait fits a model in which three capital alleles contribute additively to this quantitative trait.

If an F_1 plant is crossed to the 30-mm parent, one would have *Aa Bb Cc* × *AA BB CC*. This cross would produce 1/8 each of *Aa Bb Cc*, *Aa Bb CC*, *AA Bb Cc*, *Aa BB Cc*, *AA BB Cc*, *AA Bb CC*, *Aa BB CC*, and *AA BB CC*. Thus, there will be 1/8 that are 30 mm long, 1/8 that are 20 mm long, 3/8 that are 23.33 mm long, and 3/8 that are 26.67 mm long.

23.10 In a particular experiment, the mean internode length in spikes (the floral structures) of the barley variety *asplund* was found to be 2.12 mm. In the variety *abed binder*, the mean internode length was found to be 3.17 mm. The mean of the F_1 of a cross between the two varieties was approximately 2.7 mm. The F_2 gave a continuous range of variation from one parental extreme to the other. Analysis of the F_3 generation showed that in the F_2, 8 out of the total 125 individuals were of the *asplund* type, giving a mean of 2.19 mm. Eight other individuals were similar to the parent *abed binder*, giving a mean internode length of 3.24 mm. Is the internode length in spikes of barley a discontinuous or a quantitative trait? Why?

Answer: Internode length shows the characteristics of a quantitative trait. These characteristics include F_1 progeny that show a phenotype intermediate between the two parental phenotypes, and an F_2 showing a range of phenotypes with extremes in the range of the two parents.

23.11 Assume that the difference between a type of oat yielding about 4 g per plant and a type yielding 10 g is the result of three equal and cumulative multiple-gene pairs, *AA BB CC*. If you cross the type yielding 4 g with the type yielding 10 g, what will be the phenotypes of the F_1 and the F_2? What will be their distribution?

Answer: If six equal and cumulatively acting capital alleles result in the addition of 6 g to the base weight of 4 g, each allele contributes 1 g. The F_1 will have the genotype *Aa Bb Cc* and so have a yield of 4 + 3 = 7 g. The F_2 will appear in a distribution with relative proportions according to the seven coefficients of the binomial expansion $(a + b)^6$ (see page 499). One expects $\frac{1}{64}$ = 4 g, $\frac{6}{64}$ = 5 g, $\frac{15}{64}$ = 6 g, $\frac{20}{64}$ = 7 g, $\frac{15}{64}$ = 8 g, $\frac{6}{64}$ = 9 g, and $\frac{1}{64}$ = 10 g.

23.12 Assume that in squashes the difference in fruit weight between a 3-lb type and a 6-lb type results from three allelic pairs, *A/a, B/b,* and *C/c*. Each capital-letter allele contributes a half pound to the weight of the squash. From a cross of a 3-lb plant (*aa bb cc*) with a 6-lb plant (*AA BB CC*), what will be the phenotypes (weights) of the F_1 and the F_2? What will be their distribution?

Answer: The F_1 will have the genotype *Aa Bb Cc* and so have a weight of 4.5 lb. The F_2 is expected to show a distribution predicted by the seven coefficients of the binomial expansion $(a + b)^6$ (see page 499). One should see $\frac{1}{64}$ = 3 lb, $\frac{6}{64}$ = 3.5 lb, $\frac{15}{64}$ = 4 lb, $\frac{20}{64}$ = 4.5 lb, $\frac{15}{64}$ = 5 lb, $\frac{6}{64}$ = 5.5 lb, and $\frac{1}{64}$ = 6 lb.

23.13 Refer to the assumptions stated in problem 23.12 to determine the range in fruit weight of the offspring in the following squash crosses.

a. *AA Bb CC* × *aa Bb Cc*

b. *AA bb Cc* × *Aa BB cc*

c. *aa BB cc* × *AA BB cc*.

Answer: In each case, one needs to consider the maximum and minimum number of capital-letter alleles that can be contributed to the progeny, noting that each contributes 0.5 lb above the 3-lb base weight.

a. *AA Bb CC* × *aa Bb Cc*. At least two and at most five capital alleles can be contributed to the progeny. The progeny weights will range from 4.0 to 5.5 lb.

b. *AA bb Cc* × *Aa BB cc*. At least two and at most four capital alleles can be contributed to the progeny. The progeny weights will range from 4.0 to 5.0 lb.

c. *aa BB cc* × *AA BB cc*. Exactly three capital alleles will be contributed to the progeny. The progeny weight will be 4.5 lb.

23.14 Assume that the difference between a corn plant 10 dm (decimeters) high and one 26 dm high results from four pairs of equal and cumulative multiple alleles, with the 26-dm plants being *AA BB CC DD* and the 10-dm plants being *aa bb cc dd*.

a. What will be the size and genotype of an F_1 from a cross between these two true-breeding types?

b. Determine the limits of height variation in the offspring from the following crosses.

   i. *Aa BB cc dd* × *Aa bb Cc dd*

ii. *aa BB cc dd* × *Aa Bb Cc dd*
iii. *AA BB Cc DD* × *aa BB cc Dd*
iv. *Aa Bb Cc Dd* × *Aa bb Cc Dd*

Answer: Since four pairs of alleles are equally responsible for the 16-dm height increase, each allele contributes 2 dm.

a. A cross of *AA BB CC DD* × *aa bb cc dd* will produce *Aa Bb Cc Dd* progeny. With four capital-letter alleles, the F_1 height will be 18 dm.
b. (i) *Aa BB cc dd* × *Aa bb Cc dd.* The minimum number of capital alleles contributed to the progeny would be one, and the maximum number of capital alleles contributed to the progeny would be four. The height range would be 12 to 18 dm.
   (ii) *aa BB cc dd* × *Aa Bb Cc dd.* At least one and at most four capital alleles would be contributed to the progeny. The height range would be 12 to 18 dm.
   (iii) *AA BB Cc DD* × *aa BB cc Dd.* At least four and at most six capital alleles would be contributed to the progeny. The height range would be 18 to 22 dm.
   (iv) *Aa Bb Cc Dd* × *Aa bb Cc Dd.* A minimum of none and a maximum of seven capital alleles would be contributed to the progeny. The height range would be 10 to 24 dm.

23.15 Refer to the assumptions given in problem 23.14. For this problem, two 14-dm corn plants, when crossed, give nothing but 14-dm offspring (case A). Two other 14-dm plants give one 18-dm, four 16-dm, six 14-dm, four 12-dm, and one 10-dm offspring (case B). Two other 14-dm plants, when crossed, give one 16-dm, two 14-dm, and one 12-dm offspring (case C). What genotypes for each of these 14-dm parents (cases A, B, and C) would explain these results? Would it be possible to get a plant taller than 48 dm by selection in any of these families?

Answer: In each of cases A, B, and C, both 14-dm parents must have two capital-letter alleles. Consideration of the possible arrangement of these alleles leads to the solution of cases A, B, and C.

Case A: For two 14-dm plants only to give rise to more 14-dm plants, each of the plants must be homozygous for one of the capital alleles. Both plants could be homozygous for the same capital allele (e.g., *AA bb cc dd* × *AA bb cc dd*), or each could be homozygous for a different capital allele (e.g., *AA bb cc dd* × *aa bb CC dd*). The progeny plants would have two capital alleles and be 14 dm high.

Case B: The 1 four capital alleles : 4 three capital alleles : 6 two capital alleles : 4 one capital allele : 1 no capital allele proportions reflect the coefficients in the binomial expansion of $(a + b)^4$. This result suggests that the parents were heterozygous at the two (identical or different) loci (e.g., *Aa Bb cc dd* × *Aa Bb cc dd* or *Aa Bb cc dd* × *aa bb Cc Dd*).

Case C: The 1 one capital allele : 2 two capital alleles : 1 three capital alleles proportions reflect the coefficients in the binomial expansion of $(a + b)^2$. One would expect this ratio if one of the parents was homozygous at one locus and the other heterozygous at two loci. For example, the parents could be *aa BB cc dd* × *Aa Bb cc dd* or *aa BB cc dd* × *Aa bb Cc dd.*

Since each capital allele contributes 2 dm, plants would have to have at least five capital alleles to be taller than 18 dm. To be taller than 48 dm is simply not possible (they would need $>(48 - 10)/2 = >17$ capital alleles at only four genes!). At most, the plants can have eight capital alleles and be 10 + 16 = 26-dm tall. In case B, 1/16 of the progeny of the cross

Aa Bb cc dd × *aa bb Cc Dd* would have four capital alleles and be 18-dm tall. If this is selfed, the F_2 would have 56/256 20-dm, 28/256 22-dm, 8/256 24-dm, and 1/256 26-dm tall plants.

23.16 Pigmentation in the imaginary river bottom dweller *Mucus yuccas* is a quantitative character controlled by a set of five independently segregating polygenes with two alleles each: *A/a, B/b, C/c, D/d,* and *E/e.* Pigment is deposited at three different levels, depending on the threshold of gene products produced by the capital-letter alleles. Greyish-brown pigmentation is seen if at least four capital alleles are present; light tan pigmentation is seen if two or three capital alleles are present; and whitish-blue pigmentation is seen if these thresholds are not met. If an *AA BB CC DD EE* animal is crossed with a true-breeding *aa bb cc dd ee* animal and the progeny are selfed, what kinds of phenotypes are expected in the F_1 and F_2?

Answer: The F_1 is *A/a B/b C/c D/d E/e* and so it is greyish-brown. The F_2 phenotypes are determined by the number of capital alleles contributed from each F_1 parent. Thus, determine the chance of obtaining zero, one, two, or three capital alleles from each F_1 parent: Since the parent is heterozygous at all five loci, the chance of obtaining any specified set of five alleles in one gamete is $(\frac{1}{2})^5$. The chance of obtaining a particular number of capital alleles from an F_1 is the number of ways in which that number of alleles can be obtained multiplied by $(\frac{1}{2})^5$. Consideration of all the possibilities reveals that there is one way to obtain zero capital alleles, five ways to obtain one capital allele, ten ways to obtain two capital alleles, and ten ways to obtain three capital alleles. With this information, tabulate the ways in which the progeny can be formed:

F_1 Gamete #1		**F_1 Gamete #2**		**F_2 Progeny**		
Capital Alleles	**Gamete Fraction**	**Capital Alleles**	**Gamete Fraction**	**Capital Alleles**	**F_2 Fraction**	**Phenotype**
0	$(\frac{1}{2})^5$	0	$(\frac{1}{2})^5$	0	$(\frac{1}{2})^{10}$	whitish-blue
0	$(\frac{1}{2})^5$	1	$5(\frac{1}{2})^5$	1	$5(\frac{1}{2})^{10}$	whitish-blue
1	$5(\frac{1}{2})^5$	0	$(\frac{1}{2})^5$	1	$5(\frac{1}{2})^{10}$	whitish-blue
0	$(\frac{1}{2})^5$	2	$10(\frac{1}{2})^5$	2	$10(\frac{1}{2})^{10}$	light tan
1	$5(\frac{1}{2})^5$	1	$5(\frac{1}{2})^5$	2	$25(\frac{1}{2})^{10}$	light tan
2	$10(\frac{1}{2})^5$	0	$(\frac{1}{2})^5$	2	$10(\frac{1}{2})^{10}$	light tan
0	$(\frac{1}{2})^5$	3	$10(\frac{1}{2})^5$	3	$10(\frac{1}{2})^{10}$	light tan
1	$5(\frac{1}{2})^5$	2	$10(\frac{1}{2})^5$	3	$50(\frac{1}{2})^{10}$	light tan
2	$10(\frac{1}{2})^5$	1	$5(\frac{1}{2})^5$	3	$50(\frac{1}{2})^{10}$	light tan
3	$10(\frac{1}{2})^5$	0	$(\frac{1}{2})^5$	3	$10(\frac{1}{2})^{10}$	light tan

In the F_2, $11(\frac{1}{2})^{10}$ = 11/1,204 will be whitish-blue and $165(\frac{1}{2})^{10}$ = 165/1,024 will be light tan. The remaining $[1 - 176(\frac{1}{2})^{10}]$ = 848/1,024 F_2 progeny will be greyish-brown.

A faster method to solve this problem is to use the coefficients of the binomial expansion to determine the proportion of progeny with different numbers of capital and lowercase alleles. Let n = total number of alleles, s = number of capital alleles, t = number of lowercase alleles, a = chance of obtaining a capital allele, b = chance of obtaining a lowercase allele, and $x! = (x)(x-1)(x-2)\ldots(1)$, with $0! = 1$. Then the chance P of obtaining progeny with a specified number of each type of allele is given by

$$P(s,t) = \frac{n!}{s!t!}a^s b^t$$

$$\left.\begin{aligned} P(0,10) &= \frac{10!}{0!10!}\left(\frac{1}{2}\right)^0\left(\frac{1}{2}\right)^{10} = \frac{1}{1{,}024} \\ P(1,9) &= \frac{10!}{1!9!}\left(\frac{1}{2}\right)^1\left(\frac{1}{2}\right)^9 = \frac{10}{1{,}024} \end{aligned}\right\} \frac{11}{1{,}024} \text{ whitish-blue}$$

$$\left.\begin{aligned} P(2,8) &= \frac{10!}{2!8!}\left(\frac{1}{2}\right)^2\left(\frac{1}{2}\right)^8 = \frac{45}{1{,}024} \\ P(3,7) &= \frac{10!}{3!7!}\left(\frac{1}{2}\right)^3\left(\frac{1}{2}\right)^7 = \frac{120}{1{,}024} \end{aligned}\right\} \frac{165}{1{,}024} \text{ light tan}$$

$$1 - \frac{11+165}{1{,}024} = \frac{848}{1{,}024} \text{ greyish-brown}$$

23.17 Since monozygotic twins share all their genetic material and dizygotic twins share, on average, half of their genetic material, twin studies sometimes can be useful for evaluating the genetic contribution to a trait. Consider the following two instances.

An intelligence quotient (IQ) assesses intellectual performance on a standardized test that involves reasoning ability, memory, and knowledge of an individual's language and culture. IQ scores are transformed so that the population mean score is 100 and 95 percent of the individuals have scores in the range between 70 and 130. Observations in the United States and England found that monozygotic twins had an average difference of 6 IQ points, dizygotic twins had an average difference of 11 points, and random pairs of individuals had an average difference of 21 points.

In a large sample of pairs of twins in the United States where one twin was a smoker, 83 percent of monozygotic twins both smoked, while 62 percent of dizygotic twins both smoked.

From these data, can you infer the genetic determination of IQ or smoking?

Answer: In neither of these cases is any information provided about the environments in which the twins were raised. While it would initially appear that there is some genetic component in each case, there is no way to evaluate the role of environment. It could be that the identical twins sampled in each case were raised in an identical environment, and this could account for the greater similarity in smoking behavior and IQ results for monozygotic twins than for dizygotic twins.

23.18 A quantitative geneticist determines the following variance components for leaf width in a population of wildflowers growing along a roadside in Kentucky:

Additive genetic variance (V_A) = 4.2
Dominance genetic variance (V_D) = 1.6
Interaction genetic variance (V_I) = 0.3

Environmental variance (V_E) = 2.7
Genetic-environmental variance ($V_{G \times E}$) = 0.0

a. Calculate the broad-sense heritability and the narrow-sense heritability for leaf width in this population of wildflowers.
b. What do the heritabilities obtained in part (a) indicate about the genetic nature of leaf width variation in this plant?

Answer:

a. The broad-sense heritability of a trait represents the proportion of the phenotypic variance in a particular population that results from genetic differences among individuals, while the narrow-sense heritability measures only the proportion of the phenotypic variance in that population that results from additive genetic variance.

Broad-sense heritability
$= \text{genetic variance}/\text{phenotypic variance} = V_G/V_P$
$= (4.2 + 1.6 + 0.3)/(4.2 + 1.6 + 0.3 + 2.7 + 0.0)$
$= 6.1/8.8 = 0.69$

Narrow-sense heritability
$= \text{additive genetic variance}/\text{phenotypic variance}$
$= V_A/V_P$
$= 4.2/8.8 = 0.48$

b. About 69 percent of the phenotypic variation in leaf width observed in this population is due to genetic differences among individuals. About 48 percent of the phenotypic variation is due to additive genetic variation. The phenotypic variation due to additive genetic variation represents the part of the phenotypic variance that responds to natural selection in a predictable manner.

23.19 Assume that all genetic variance affecting seed weight in beans is genetically determined and is additive. From a population in which the mean seed weight was 0.88 g, a farmer selected two seeds, each weighing 1.02 g. He planted these and crossed the resulting plants with each other, then collected and weighed their seeds. The mean weight of their seeds was 0.96 g. What is the narrow-sense heritability of seed weight?

Answer: In this example, there was 0.96 – 0.88 = 0.08 g of change after one generation of selection. Thus, the selection response was 0.08 g. This selection response is dependent on both the narrow-sense heritability and the selection differential. The selection differential is the difference between the mean phenotype of the selected parents and the mean phenotype of the unselected population. In this case, the selection differential is 1.02 – 0.88 = 0.14 g. The narrow-sense heritability is given by

Selection response = narrow-sense heritability × selection differential
$0.08 = h^2 \times 0.14$
$h^2 = 0.08/0.14 = 0.57$

23.20 Members of the inbred rat strain SHR are salt-sensitive: They respond to a high-salt environment by developing hypertension. Members of a different inbred rat strain, TIS, are not salt-sensitive. Imagine that you placed a population consisting only of SHR rats in an environment that was variable in regard to distribution of salt, so that some rats would be exposed to more salt than others. What would be the heritability of blood pressure in this population?

Answer: SHR rats will continue to respond to salt by developing hypertension. Since the strain is inbred, any variation in blood pressure will result from the amount of exposure to salt, and not from genetic variation. Therefore, heritability for this population will be zero. Similarly, the inbred TIS rats would also have a heritability of zero (and retain a low blood pressure).

23.21 In Kansas, a farmer is growing a variety of wheat called TK138. He calculates the narrow-sense heritability for yield (the amount of wheat produced per acre) and finds that the heritability of yield for TK138 is 0.95. The next year he visits a farm in Poland and observes that a Russian variety of wheat growing there, UG334, has only about 40 percent as much yield as TK138 grown on his farm in Kansas. Since he found the heritability of yield in his wheat to be very high, he concludes that the American variety of wheat (TK138) is genetically superior to the Russian variety (UG334), and he tells the Polish farmers that they can increase their yield by using TK138. What is wrong with his conclusion?

Answer: Heritability is a measurement of the genetic variance of a *particular* population in a *specific* environment. Heritability is not fixed for a specific trait. It depends on genetic makeup *as well as* the specific environment of the population in which it is measured. Consequently, heritability cannot be used to make inferences about the basis for the differences between two distinct populations. As the environmental conditions on the two farms differ, the heritability calculated for a population in Kansas cannot be used to infer the future performance of the same strain in another environment. The yield of TK138 grown in Poland would most likely be different than when grown in Kansas, perhaps even lower than the yield of the UG334 variety.

23.22 A scientist wants to determine the narrow-sense heritability of tail length in mice. He measures tail length among the mice of a population and finds a mean tail length of 9.7 cm. He then selects the ten mice in the population with the longest tails; mean tail length in these selected mice is 14.3 cm. He interbreeds the mice with the long tails and examines tail length in their progeny. The mean tail length in the F_1 progeny of the selected mice is 13 cm. Calculate the selection differential, the response to selection, and the narrow-sense heritability for tail length in these mice.

Answer: The selection differential equals $14.3 - 9.7 = 4.6$ cm. The response to selection equals $13 - 9.7 = 3.3$ cm. The narrow-sense heritability equals $3.3/4.6 = 0.72$.

23.23 Suppose that the narrow-sense heritability of wool length in a breed of sheep is 0.92, and the narrow-sense heritability of body size is 0.87. The genetic correlation between wool length and body size is –0.84. If a breeder selects for sheep with longer wool, what will be the most likely effects on wool length and on body size?

Answer: The strong narrow-sense heritability of both wool length and body size indicates that these traits will respond to selection. The negative correlation coefficient between wool length and body size indicates that if longer wool is selected for, smaller body size will also be obtained.

23.24 The heights of nine college-age men and the heights of their fathers are presented below.

Height of Son (inches)	Height of Father (inches)
70	70
72	76
71	72
64	70
66	70
70	68
74	78
70	74
73	69

a. Calculate the mean and the variance of height for the sons and do the same for the fathers.

b. Calculate the correlation coefficient for the relationship between the height of father and height of son.

c. Determine the narrow-sense heritability of height in this group by regression of the son's height on the height of the father.

Answer:

a. Sons: mean = 70.0, variance = 10.25

Fathers: mean = 71.9, variance = 11.6

b and c. Let y equal the height of the sons and x equal the height of the fathers. One can calculate $\Sigma x_i y_i = 45{,}333$, $\Sigma x_i = 647$, $\Sigma y_i = 630$, and $n = 9$, so that

$$r = \frac{\text{cov}_{xy}}{s_x s_s} = \frac{\dfrac{\Sigma x_1 y_1 - \frac{1}{n}(\Sigma x_i \Sigma y_i)}{n-1}}{\sqrt{(s_x)^2 (s_y)^2}}$$

$$= \frac{\dfrac{45333 - [(630)(647)/9]}{8}}{\sqrt{11.6 \times 10.25}} = \frac{5.375}{10.904} = 0.493$$

$$b = \frac{\text{cov}_{xy}}{(s_x)^2} = \frac{5.375}{11.6} = 0.463$$

When, as in this case, the mean phenotype of the offspring is regressed against the phenotype of only *one* parent, the narrow-sense heritability is $2b = 2(0.46) = 0.92$. Since h^2 is very close to 1, most of the phenotypic variation is determined by genes with additive effects. Nonadditive factors (genes with dominance, genes with epistasis, environmental factors) contribute little to the phenotypic variation.

23.25 The narrow-sense heritability of egg weight in a particular flock of chickens is 0.60. A farmer selects for increased egg weight in this flock. The difference between the mean egg weight of the unselected chickens and the mean egg weight of the selected chickens is 10 g. How much should egg weight increase in the offspring of the selected chickens?

Answer:

Selection response = narrow-sense heritability × selection differential
Selection response = 0.60 × 10 g = 6 g

24

MOLECULAR EVOLUTION

CHAPTER OUTLINE

REVIEW OF KEY TERMS, SYMBOLS, AND CONCEPTS

In your own words, write a brief, precise definition of each term in the groups on the following page. Check your definitions using the text. Then, develop a concept map using the terms in each list.

THINKING ANALYTICALLY

Perhaps the most striking element in the field of molecular evolution is the time scale in which the mechanisms of evolution operate. Almost throughout this text, we have considered genetic mechanisms that act over relatively short time frames. The replication of the genetic material, its transcription, and the production of gene products through translation occur in time scales that are on the order of seconds and minutes. Genetic crosses may take several generations, but even for organisms that have relatively long life spans, the time scales are on the order of years or tens of years. Molecular evolution occurs over many, many generations, and the time scales being considered are on the order of tens of thousands to millions of years. In this space of time, much

1	2	3
molecular evolution	nucleotide substitution	molecular phylogeny
population, species	homologous proteins, genes	morphological phylogeny
gene	sequence alignment	evolutionary relationship
evolutionary time, history	computer algorithm	convergent evolution
population genetics	optimal alignment	phenotype
Hardy-Weinberg equilibrium	indel	phylogenetic tree
mutation vs. substitution	Jukes-Cantor model	branch, node
gene duplication	substitution rate	rooted, unrooted trees
gene conversion	evolutionary rate	taxa
transposition	divergence time	gene vs. species tree
unequal crossing over	transition, transversion	inferred tree
phylogenetic relationship	coding, noncoding sequence	distance matrix
natural selection	leader, trailer region	parsimony
multigene family	pseudogene	unweighted pair group method with arithmetic mean (UPGMA)
globin, myoglobin genes	synonymous site, codon	transformed distance method
ancestral gene	nonsynonymous site, codon	neighbor-joining method
domain shuffling	mutation vs. substitution	clustering
exon shuffling	comparative genome analysis	informative site
internal duplication	genome project	noninformative site
exon, intron	codon usage bias	tree of maximum parsimony
Bacteria, Archaea, Eukarya	stochastic factor	bootstrap test
	natural selection	confidence level
	diversifying selection	tree of life
	McDonald-Kreitman test	"out of Africa" theory
	functional constraint	"mitochondrial Eve"
	major histocompatibility complex (MHC)	"Y chromosome Adam"
	mitochondrial DNA	
	error-prone replication	
	proofreading ability	
	molecular clock	
	molecular clock hypothesis	
	relative rate test	
	outgroup	

can happen. Some say that if you can think of it, nature has had time to try it. With genomes serving as historical records of molecular evolution, it may be possible to find evidence of it as well. Having said this, consider that small alterations in events that occur quickly can, over a long period of time, result in major effects. For example, variation in the bonding energy associated with the instantaneous interaction between codons and the anticodons of isoacceptor tRNAs can effect the efficiency of translation. Over many, many generations, this may lead to the selection of a species' codon usage bias.

It is therefore helpful to keep in mind the time scales associated with molecular evolution. The text presents data documenting that rates of evolutionary change are not constant between regions of a gene, between different positions of codons within the coding region of a gene, between genes, between species, and between mitochondrial and nuclear genomes. Where possible, relate differences in the rate of evolutionary change to the time frames associated with molecular evolution. Focus on why rates are different—what functional differences or constraints exist that preclude uniform rates. As you do this, draw on everything you have learned about the mechanisms of inheritance and gene expression, since these can not only influence molecular evolution, but have also been derived through molecular evolution and therefore represent the tuning of existing biological instruments.

QUESTIONS FOR PRACTICE

Multiple Choice Questions

1. Why does the molecular clock run at different rates in different proteins?
   a. Neutral substitution rates differ among proteins.
   b. Some proteins have resulted from gene duplication followed by sequence divergence.
   c. Some species have shorter generation times.
   d. Codon usage varies between proteins.
2. Two homologous proteins are found in two different species. These proteins
   a. perform identical functions in the different species
   b. share a common ancestor
   c. arose by gene duplication
   d. have only synonymous substitutions in their genes
3. An indel is
   a. a gap in a sequence alignment caused by either by an insertion or a deletion
   b. a gap in a sequence alignment caused by a deletion in just one of the sequences
   c. a gap in a sequence alignment caused by an insertion in just one of the sequences
   d. a deletion that removes an intron
4. What issues does the Jukes-Cantor model address?
   a. the rate of the molecular clock in different proteins
   b. whether nucleotide substitutions are adaptive or random
   c. the relationship between the observed and actual number of substitutions
   d. whether an inferred phylogenetic tree is robust
5. What is generally true about synonymous and nonsynonymous substitutions?
   a. In coding regions, synonymous substitutions are more frequent than nonsynonymous substitutions.
   b. In coding regions, synonymous and nonsynonymous substitutions are equally frequent.
   c. Synonymous substitutions in coding regions and substitutions in intronic regions are equally frequent.
   d. Both a and c are true.
6. Which one of the following sequences of a gene exhibits the highest rate of evolution?
   a. the 3′ untranslated but transcribed region of the gene
   b. introns within the gene

c. the 5′ flanking region of the gene
d. nonfunctional pseudogenes related to the gene

7. Why might codon usage bias have evolved?
   a. The tRNA for one codon is more abundant than an isoacceptor tRNA for another codon.
   b. The bonding energy between pairs of codons and anticodons differ, leading to increased translation efficiency of certain pairs.
   c. The required abundant expression of some genes favors selection of codons that provide efficient translation.
   d. all of the above
8. Which of the following statements is true?
   a. An outgroup is a species that a phylogenetic analysis using maximum parsimony shows is a common ancestor to two existing species.
   b. The molecular clock typically runs at the same rate in different proteins.
   c. The synonymous substitution rate in mammalian mitochondrial genes is about tenfold greater than that in nuclear genes.
   d. The length of a species' generation time is unlikely to affect its evolutionary rate.
9. What is the best description of convergent evolution?
   a. mutations that introduce the same polymorphism into two species after they diverged from a common ancestor
   b. the evolution of similar phenotypes in distantly related species
   c. the finding that there are parallel gene duplications in two unrelated species
   d. the slowing of the molecular clock in an existing species such as humans
10. In phylogenetic analysis, how do rooted and unrooted trees differ?
    a. When a given number of taxa are being analyzed, there are more possible unrooted than rooted trees.
    b. A rooted tree is based on relationships between genes, while an unrooted tree is based on relationships between species.
    c. Outgroups can be used only in unrooted trees.
    d. A rooted tree specifies one internal node as representing a common ancestor to all the other nodes on the tree; an unrooted tree specifies only the relationship between the nodes and does not indicate the evolutionary path that was taken.
11. Which approach to phylogenetic reconstruction first determines the distance between taxa and then repeatedly clusters the two taxa with the smallest distance separating them into a single, composite taxa until all the taxa are grouped together?
    a. the maximum parsimony method
    b. the UPGMA method
    c. the bootstrap method
    d. the neighbor-joining method

Answers: 1a, 2b, 3a, 4c, 5a, 6d, 7d, 8c, 9b, 10d, 11b

Thought Questions

1. Gene duplication provides a mechanism that allows for the evolution of new protein functions. How does this mechanism affect our view of which proteins are homologous? How can one infer which proteins encoded by a multigene family present in each of two organisms are homologous?

2. What are the relationships between the fields of comparative genomics and molecular evolution?
3. How do the results of the *Arabidopsis* genome project add to our understanding of molecular evolution and how new gene functions evolve?
4. Suppose you have identified a sequence that can be translated and appears to encode an open reading frame (ORF). How could you use an analysis of codon usage bias to infer whether it was likely to be translated?
5. What problems in dating the times of existence of common ancestors are posed by variable molecular clocks in different taxonomic groups?
6. Why is the parsimony approach to phylogenetic tree reconstruction able to give some insight into the nature of long-dead organisms? Are other methods, such as distance matrix-based approaches, also able to do this?
7. What role does gene conversion play in the evolution of new gene functions?
8. Does the proposition that domain shuffling provides a mechanism for the generation of new protein function require that introns once existed in primitive life forms? If so, why might they no longer exist in the simpler Bacteria and Archaea?
9. Do phylogenetic reconstructions tell us anything about the mechanisms of speciation? If so, what?
10. What are the general types of functional constraints that restrict proteins from undergoing rapid evolution? Consider in your answer the histones, which evolve so slowly that the human histone H4 protein can functionally replace its homologue in yeast.

Thought Questions for Media

After reviewing the media at the *Essential iGenetics* website, try to answer these questions

1. The iActivity illustrates how the sequence of ancient DNA is obtained by using PCR. Fragments averaging 100 to 200 base pairs in length are cloned, the cloned segments are sequenced and aligned, and a *consensus nucleotide sequence* is determined from the aligned sequences.
   a. What is a *consensus nucleotide sequence,* and why is it necessary to use one in this type of analysis?
   b. In this analysis, sequences from the control region of mitochondrial DNA were amplified. Why is mitochondrial DNA used, and why might the control region be specifically chosen for analysis?
   c. What types of errors might be introduced when amplifying ancient DNA templates? How might you ensure that a sequence accurately reflects the sequence in the ancient DNA and is not an artifact?
2. Suppose a bone sample from a new, non-Neanderthal, extinct hominid species called *ML* (for *missing link*) was made available to you.
   a. Outline how you would test if *ML* was a direct ancestor of modern humans.
   b. Suppose you undertook an analysis similar to that of the iActivity and compared DNA sequences in the control region of mitochondrial DNA. You find 12 differences in mitochondrial DNA between *ML* and modern European sequences, 11 differences between *ML* and modern African sequences, and 7 differences between the modern European and modern African DNA sequences. What would you conclude? Be as specific as possible.

3. In the iActivity, you were able to estimate the time of divergence of modern Europeans and Africans and the time of divergence of Neanderthals and modern humans.
   a. What assumptions did you make in this analysis?
   b. How did you estimate the rate of nucleotide evolution? That is, what was the basis for relating the number of observed mutations to the times of divergence between the European, African, and Neanderthal populations?
   c. How would your estimates of the times of divergence between the European, African, and Neanderthal populations differ if the rate of nucleotide evolution in hominid species is not constant, but tripled, relative to a previously constant rate, in the last 300,000 years?

SOLUTIONS TO TEXT PROBLEMS

24.1 The following sequence is that of the first 45 codons from the human gene for prepro-insulin:

```
ATG GCC CTG TGG ATG CGC CTC CTG CCC CTG CTG GCG CTG CTG GCC CTC
TGG GGA CCT GAC CCA GCC GCA GCC TTT GTG AAC CAA CAC CTG TGC GGC
TCA CAC CTG GTG GAA GCT CTC TAC CTA GTG TGC GGG GAA
```

Using the genetic code (Figure 12.7), determine what fraction of mutations at the first, second, and third positions of these 45 codons will be synonymous. At which position is natural selection likely to have the greatest effect and are nucleotides most likely to be conserved?

Answer: A synonymous mutation or substitution is a change in nucleotide sequence that does not affect the amino acid sequence in a polypeptide chain. Due to the redundancy of the genetic code, more than one codon can code for one amino acid. Inspection of a table of the genetic code shows that at one extreme, some amino acids, such as methionine and tryptophan, are only coded by one codon (AUG and UGG, respectively), while at the other extreme, some amino acids, such as leucine and serine, are coded for by six codons. Other amino acids are coded for by two, three, or four codons. Say the codon in question codes for leucine and is CUG. Then, since the codons for leucine are UAA, UGG, CUU, CUC, CUA, and CUG, any of three coding strand changes in the third nucleotide (G to T, G to A, or G to C) and a change of C to T in the first nucleotide would still result in leucine being inserted into the polypeptide chain. Hence, any of these substitutions would be synonymous.

To get some sense of which substitutions will be synonymous, inspect the genetic code and look for some general principles. For amino acids coded for by four codons, variation in the third position results in a synonymous substitution; for amino acids coded for by two codons, variation in the third position by one base only (a purine or a pyrimidine) results in a synonymous mutation. For amino acids coded for by either two or four codons, variation in the first or second positions results in a nonsynonymous substitution. Similar but not identical patterns are seen for amino acids coded for by either three or six codons. The sense that synonymous changes are mostly associated with substitutions in the third position is confirmed by a more systematic analysis. The following table lists the number of possible nonsynonymous (NS) and synonymous (S) mutations at each position of each codon for the amino acids in question.

Coding Strand	mRNA Codon	Amino Acid	Position 1		Position 2		Position 3	
			NS	S	NS	S	NS	S
ATG	AUG	methionine	3	0	3	0	3	0
GCG	GCC	alanine	3	0	3	0	0	3
CTG	CUG	leucine	2	1	3	0	0	3
TGG	UGG	tryptophan	3	0	3	0	3	0
ATG	AUG	methionine	3	0	3	0	3	0
CGC	CGC	arginine	3	0	3	0	0	3
CTC	CUC	leucine	3	0	3	0	0	3
CTG	CUG	leucine	2	1	3	0	0	3
CCC	CCC	proline	3	0	3	0	0	3
CTG	CUG	leucine	2	1	3	0	0	3
CTG	CUG	leucine	2	1	3	0	0	3
GCG	GCG	alanine	3	0	3	0	0	3
CTG	CUG	leucine	2	1	3	0	0	3
CTG	CUG	leucine	2	1	3	0	0	3
GCC	GCC	alanine	3	0	3	0	0	3
CTC	CUC	leucine	3	0	3	0	0	3
TGG	UGG	trytophan	3	0	3	0	3	0
GGA	GGA	glycine	3	0	3	0	0	3
CCT	CCU	proline	3	0	3	0	0	3
GAC	GAC	aspartate	3	0	3	0	2	1
CCA	CCA	proline	3	0	3	0	0	3
GCC	GCC	alanine	3	0	3	0	0	3
GCA	GCA	alanine	3	0	3	0	0	3
GCC	GCC	alanine	3	0	3	0	0	3
TTT	UUU	phenylalanine	3	0	3	0	2	1
GTG	GUG	valine	3	0	3	0	0	3
AAC	AAC	asparagine	3	0	3	0	2	1
CAC	CAC	histidine	3	0	3	0	2	1
CTG	CUG	leucine	2	1	3	0	0	3
TGC	UGC	cysteine	3	0	3	0	2	1
GGC	GGC	glycine	3	0	3	0	0	3
TCA	UCA	serine	3	0	3	0	0	3
CAC	CAC	histidine	3	0	3	0	2	1
CTG	CUG	leucine	2	1	3	0	0	3
GTG	GUG	valine	3	0	3	0	0	3
GAA	GAA	glutamate	3	0	3	0	2	1
GCT	GCU	alanine	3	0	3	0	0	3
CTC	CUC	leucine	3	0	3	0	0	3

(continued)

Coding Strand	mRNA Codon	Amino Acid	Position 1		Position 2		Position 3	
			NS	S	NS	S	NS	S
TCA	UCA	serine	3	0	3	0	0	3
CAC	CAC	histidine	3	0	3	0	2	1
CTG	CUG	leucine	2	1	3	0	0	3
GTG	GUG	valine	3	0	3	0	0	3
GAA	GAA	glutamate	3	0	3	0	2	1
GCT	GCU	alanine	3	0	3	0	0	3
CTC	CUC	leucine	3	0	3	0	0	3
TAC	UAC	tyrosine	3	0	3	0	2	1
CTA	CUA	leucine	2	1	3	0	0	3
GTG	GUG	valine	3	0	3	0	0	3
TGC	UGC	cysteine	3	0	3	0	2	1
GGG	GGG	glycine	3	0	3	0	0	3
GAA	GAA	glutamate	3	0	3	0	2	1
		TOTAL	124	11	135	0	34	101
		Percent	91.8	8.2	100	0	25.2	74.8

From this analysis, 8.2 percent of mutations in the first position, none of the mutations in the second position, and 74.8 percent of mutations in the third position will be synonymous. Since none of the mutations in the second position will be synonymous, all mutations in this position will affect the amino acid inserted into a polypeptide chain. Natural selection is therefore likely to have the greatest effect on mutations in the second position, and nucleotides in this position are the most likely to be conserved.

24.2 The following sequences represent an optimum alignment of the first 50 nucleotides from the human and sheep preproinsulin genes:

```
Human: ATGGCCCTGT GGATGCGCCT CCTGCCCCTG CTGGCGCTGC TGGCCCTCTG
Sheep: ATGGCCCTGT GGACACGCCT GGTGCCCCTG CTGGCCCTGC TGGCACTCTG
```

Estimate the number of substitutions that have occurred in this region since humans and sheep last shared a common ancestor, using the Jukes-Cantor model.

Answer: Inspection of the sequences reveals that there are six differences (underlined):

```
Human: ATGGCCCTGT GGATGCGCCT CCTGCCCCTG CTGGCGCTGC TGGCCCTCTG
Sheep: ATGGCCCTGT GGACACGCCT GGTGCCCCTG CTGGCCCTGC TGGCACTCTG
```

The Jukes-Cantor model estimates the number of substitutions per site (K) based on the fraction of nucleotides that are different between two sequences (p), taking into account that a particular site can undergo multiple changes. According to this model, $K = -(3/4) \ln[1 - (4/3)(p)]$. Here, $p = 6/50 = 0.12$, so $K = -3/4 \ln[1 - (4/3)(0.12)] = 0.1308$. Since $K = 0.1308$, the estimated number of substitutions is $50 \times 0.1308 = 6.54$.

24.3 Using the alignment in problem 24.2 and assuming that humans and sheep last shared a common ancestor 80 million years ago, estimate the rate at which the sequence of the first 50 nucleotides in their preproinsulin genes has been accumulating substitutions.

Answer: Since substitutions are assumed to accumulate simultaneously and independently in both sequences, the substitution rate (r) is obtained by dividing the number of substitutions between the homologous sequences by $2T$, where T = divergence time: $r = K/2T$. Here, $K = 0.1308$ and $T = 80 \times 10^6$, so $r = 0.1308/(2 \times 80 \times 10^6) = 8.175 \times 10^{-10}$ substitutions/year.

24.4 Would the mutation rate be greater or less than the observed substitution rate for a sequence of a gene such as the one shown in problem 24.2? Why?

Answer: The mutation rate reflects changes in DNA sequence due to errors in replication or repair processes, while the substitution rate reflects changes in DNA sequence (i.e., mutations) that have passed through the filter of natural selection. Since mutations that are not synonymous may be selected against, an estimate of the number of substitutions in a coding region, such as in this analysis, is likely to be lower than the mutation rate. Therefore, the mutation rate is likely to be greater than the observed substitution rate.

24.5 If the rate of nucleotide evolution along a lineage is 1.0 percent per million years, what is the rate of substitution per nucleotide per year? What would be the observed rate of divergence between two species evolving at that rate since they last shared a common ancestor?

Answer: Using the assumption of the Jukes-Cantor model that each nucleotide is just as likely to change into any other nucleotide, when the rate of change to any one of three alternative nucleotides is α, the overall rate of substitution for any given nucleotide is 3α. Here, if the rate of nucleotide evolution is interpreted to be the observed rate of substitution—mutations that have passed through the filter of selection—$/3\alpha = 0.01$ substitutions/10^6 years $= 1 \times 10^{-8}$ substitutions/year, so $\alpha = 3.3 \times 10^{-9}$ substitutions/year. If the observed rate of substitution within each lineage in two diverging species is 3.3×10^{-9}, then, since substitutions accumulate simultaneously and independently in two diverging sequences, the observed rate of divergence will be 6.6×10^{-9}.

24.6 The average synonymous substitution rate in mammalian mitochondrial genes is approximately ten times the average value for synonymous substitutions in nuclear genes. Why would it be better to use comparisons of mitochondrial sequences to study human migration patterns and nuclear genes when studying the phylogenetic relationships of mammalian species that diverged 80 million years ago?

Answer: There are two broad issues to consider here. First, sequence differences must be able to be observed within the time period being assessed, and there must be a sufficient number of them so that meaningful conclusions can be drawn from their analysis. For the analysis of human migration patterns, the time scale is on the order of tens of thousands of years, while for the analysis of mammalian phylogenetic relationships, the time scale is on the order of millions of years. The average synonymous substitution rate in the mitochondrial genome is high enough for differences between different human populations to be observed and provides for a sufficient number of differences for a meaningful analysis to be performed. The average synonymous substitution rate in the nuclear genome may be too low to identify a sufficient number of differences for a meaningful analysis to be performed.

Second, over time, synonymous nucleotide substitutions will change repeatedly and will include changes back to their ancestral state. If the substitution rate is too high relative to the time period being examined, it may be difficult to clearly identify relationships due to noise in the data from this phenomenon. The average synonymous substitution rate in nuclear genes is high enough for differences to be observed among mammalian species over periods of millions of years, yet not so high that multiple substitutions cloud their phylogenetic interpretation. The rate of synonymous mutations in mitochondrial genes is so high that multiple substitutions cloud their phylogenetic interpretation. In summary, the rapid and regular rate of accumulation of nucleotide sequence differences in mitochondrial genomes allows for the phylogenetic analysis of closely related lineages, but for more distant lineages, it is better to use less rapidly changing sequences, such as those in the nuclear genome.

24.7 Why might mitochondrial DNA sequences accumulate substitutions at a faster rate than nuclear genes in the same organism?

Answer: The increased rate of substitutions in animal mtDNA may result from a higher mutation rate due to error-prone replication and DNA repair, since mitochondrial DNA polymerases lack proofreading ability; from increased concentrations of mutagens, such as free radicals resulting from mitochondrial metabolic processes; and/or from relaxed selection pressure, since most cells have at least several dozen mitochondria, each of which has multiple copies of the mitochondrial genome.

24.8 Why might substitution rates differ from one species to another, and how would such differences depart from Zuckerkandl and Pauling's assumptions for molecular clocks?

Answer: Substitution rates may differ between species because their generation times are different, and so the number of germ-line DNA replications (which occur once per generation) are different; because the species are different in their average repair efficiency; because the species have different average exposure to mutagens; and/or because the species are exposed to different opportunities to adapt to new ecological niches and environments.

Zuckerkandl and Pauling's molecular clock hypothesis was based on initial data suggesting that substitution rates were constant in homologous proteins, that is, that there is a steady rate of change between two sequences. If this were generally true, the number of differences between two homologous proteins could be correlated with the amount of time since speciation caused them to diverge independently. If the hypothesis were correct, it would facilitate the determination of the phylogenetic relationships between species and the times of their divergence. However, the molecular clock is not uniform for all species, and so molecular divergence cannot always be used for phylogenetic analysis and to date the times recent common ancestors existed. Before making these types of inferences, it is necessary to demonstrate that the species being examined have a uniform molecular clock.

24.9 Suppose we examine the rates of nucleotide substitution in two nucleotide sequences isolated from humans. In the first sequence (sequence A), we find a nucleotide substitution rate of 4.88×10^{-9} substitutions per site per year. The substitution rate is the same for synonymous and nonsynonymous substitutions. In the second sequence (sequence B), we find a synonymous substitution rate of 4.66×10^{-9} substitutions per site per year and a nonsynonymous substitution rate of 0.70×10^{-9} substitutions per site per year. Referring to Table 24.1, what might you conclude about the possible roles of sequence A and sequence B?

Answer: Nonsynonymous substitutions are those that code for different amino acids, while synonymous substitutions code for the same amino acid. In sequence A, the finding of the same, relatively high rate for synonymous and nonsynonymous substitutions suggests that this sequence may not code for a functional protein and may derive from a pseudogene. The rate seen for synonymous and nonsynonymous substitutions in sequence A is similar to that seen for pseudogenes shown in Table 24.1.

The different rates seen for synonymous and nonsynonymous substitutions in sequence B (and in particular, the much lower rate of nonsynonymous substitutions) suggest that this sequence encodes a protein. A low nonsynonymous substitution rate would be expected if nonsynonymous changes resulted in changes in protein function that were detrimental to fitness. Since most nonsynonymous substitutions would not "improve" a protein's function, such mutations would be eliminated by the filter of natural selection. Synonymous substitutions would be tolerated and seen at a higher frequency, as they would not alter protein function.

24.10 What evolutionary process might explain a coding region in which the rate of amino acid replacement is greater than the rate of synonymous substitution?

Answer: If there is evolutionary pressure, or natural selection, for diversity, then the rate of amino acid replacement, driven by nonsynonymous substitution, may be greater than the rate of synonymous substitution. This is seen at the major histocompatibility complex (MHC) in mammals, which is involved in immune function where diversity favors fewer individuals vulnerable to an infection by any single virus, and in viruses, which utilize error-prone replication coupled with diversifying selection. In both cases, this serves as a response to pressure for rapid evolution.

24.11 Natural selection does not always act just at the level of amino acid sequences in proteins. Ribosomal RNAs, for instance, are functionally dependent on extensive and specific intramolecular secondary structures that form when complementary nucleotide sequences within a single rRNA interact. Would the regions involved in such pairing accumulate mutations at the same rate as unpaired regions? Why?

Answer: While the sequences of rRNA regions that interact and provide for ribosomal function by pairing will be subject to mutation at the same rates as sequences that do not pair, mutations that disrupt pairing will be selected against. Such mutations will alter ribosomal function, and this provides a basis for selection against them. Consequently, substitutions in paired regions will not accumulate unless there are two simultaneous, complementary mutations—one in each of the two sites that pair. A high degree of conservation of nucleotides at particular sites within the rRNA genes therefore indicates that these nucleotides are functionally important.

24.12 The following three-way alignment is of the nucleotide sequences from the beginning of the human, rabbit, and duck α-globin genes:

```
Human: ATGGTGCTCT CTCCTGCCGA CAAGACCAAC GTCAAGGCCG CCTGGGACAA
Rabbit: ATGGTGCTGT CTCCCGCTGA CAAGACCAAC ATCAAGACTG CCTGGGAAAA
Duck: ATGGTGCTGT CTGCGGCTGA CAAGACCAAC GTCAAGGGTG TCTTCTCCAA
```

Given that humans and rabbits are known to be more closely related to each other than they are to ducks (an outgroup), use the relative rate test to determine whether there has been a change in the rate of substitution in this region of the human and rabbit genomes since they last shared a common ancestor.

Answer: First identify how many differences exist between pairs of the sequences. In the annotation below, H represents human; R represents rabbit; D represents duck; an underlined base indicates a base that is different between the mammalian and duck sequence; and an x below the aligned sequences identifies the position of a base that is different between the indicated pair of species.

```
H:          ATGGTGCTCT CTCCTGCCGA CAAGACCAAC GTCAAGGCCG CCTGGGACAA
R:          ATGGTGCTGT CTCCCGCTGA CAAGACCAAC ATCAAGACTG CCTGGGAAAA
D:          ATGGTGCTGT CTGCGGCTGA CAAGACCAAC GTCAAGGGTG TCTTCTCCAA
H vs. D:                 x x  x                     xx  x  xxxx
R vs. D:                 x x                 x     xx    x xxxx
H vs. R:                   x  x              x     x x         x
```

Based on this analysis, there are 10 substitutions between humans and ducks, 11 substitutions between rabbits and ducks, and 6 substitutions between humans and rabbits. With ducks as an outgroup, we can draw a species tree as follows:

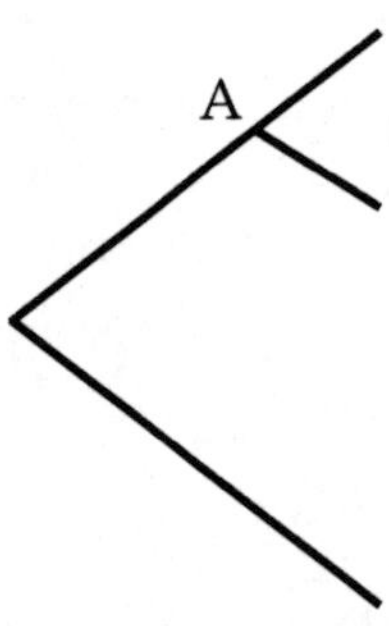

If the molecular clock hypothesis is correct, then $d_{\text{A-Humans}} = d_{\text{A-Rabbits}}$ (d in these equations represents distance; $d_{\text{A-Humans}}$ is the distance between the point of common ancestry (A) to humans). The path by which the distances are calculated should not matter if the molecular clocks throughout the tree are uniform. Check this by inserting the numbers of substitutions determined from the analysis of the sequences into equations that trace the paths in the species tree. These equations are presented in the text and reflect the relationships between distances in the species tree.

$$d_{\text{A-Humans}} = (d_{\text{Humans-Rabbits}} + d_{\text{Humans-Ducks}} - d_{\text{Rabbits-Ducks}})/2$$
$$= (6 + 10 - 11)/2$$
$$= 2.5$$

$$d_{\text{A-Rabbits}} = (d_{\text{Humans-Rabbits}} + d_{\text{Rabbits-Ducks}} - d_{\text{Humans-Ducks}})/2$$
$$= (6 + 11 - 10)/2$$
$$= 3.5$$

This analysis indicates that although $d_{\text{A-Humans}}$ and $d_{\text{A-Rabbits}}$ are similar, they are not identical. This suggests that there is a difference in the rate of substitution in this region of the rabbit and human genomes since they last shared a common ancestor.

24.13 What are some of the advantages of using DNA sequences to infer evolutionary relationships?

Answer: Using DNA sequence information to infer evolutionary relationships has several advantages. First, DNA sequences serve as historical records that can be unraveled to identify the dynamics of evolutionary history. Second, since they do not rely on phenotypes for comparison of genetic relatedness, they circumvent problems associated with phylogenetic classification in organisms based solely on phenotypes. These include problems in the identification of usable phenotypes in organisms that lack phenotypes that correlate with their degree of genetic relatedness or for which there is no fossil record evidence (e.g., eubacteria and archaeabacteria); problems in the identification of usable phenotypes in groups of organisms where few characters are shared (e.g., bacteria and mammals); and problems associated with convergent evolution, where distantly related organisms can share similar phenotypes. Third, they allow for an abundance of parameters that can be measured so that theories can be tested, as they provide highly accurate and reliable information, allow direct comparison of the genetic differences among organisms, are easily quantified, and can be used for all organisms.

24.14 As suggested by the popular movie *Jurassic Park,* organisms trapped in amber have proved to be a good source of DNA from tens and even hundreds of millions of years ago. However, when using such sequences in phylogenetic analyses, it is usually not possible to distinguish between samples that come from evolutionary dead ends and those that are the ancestors of organisms still alive today. Why would the former be no more useful than simply including the DNA sequence of another living species in an analysis?

Answer: An evolutionary dead end represents a taxon that is on a separate, no-longer-existing branch of the tree of life. In this sense, it serves just as another species for comparison, much as does a living species on a separate, existing branch of the tree of life. Phylogenetic analyses can not distinguish between existing and extinct species, since they only describe species relatedness.

24.15 In the phylogenetic analysis of a group of closely related organisms, the conclusions drawn from one locus were found to be at odds with those from several others. What might account for the discordant locus?

Answer: Sequence polymorphisms that result in divergence within genes often predate speciation events. If an ancestral polymorphism at a locus predates the split between two species, some members of a species may have retained the ancestral polymorphism, while others may have not retained it, having accumulated a newer polymorphism at the locus that occurred following speciation. Consequently, for certain loci, some members of a species will be more similar to other species than to their own or more closely related species. For these loci, phylogenetic trees made from a single gene will not always reflect the relationships between species. This is why it is important to use multiple loci when constructing species trees.

24.16 What is the chance of randomly picking the one rooted phylogenetic tree that describes the true relationship between a group of six organisms? Are the odds better or worse for randomly picking from among all the possible unrooted trees for those organisms?

Answer: Refer to Table 24.3 and the accompanying text. The number of possible rooted (N_R) and unrooted (N_U) trees for n taxa are given by the equations

$$N_R = (2n-3)!/2^{n-2}\,(n-2)! \quad \text{and} \quad N_U = (2n-5)!/2^{n-3}\,(n-3)!$$

For $n = 6$, $N_R = 9!/24\ 4! = 945$ and $N_U = 7!/2^3\ 3! = 105$. The chance of picking the one rooted tree that describes the true relationship between a group of six organisms is $1/945$ (0.11%). The odds are better in picking a random unrooted tree: $1/105$ (0.95%).

24.17 Increasing the amount of sequence information available for analysis usually has little effect on the length of time computer programs use to generate phylogenetic trees with the parsimony approach. Why doesn't the amount of sequence information affect the total number of possible rooted and unrooted trees?

Answer: The total number of possible rooted and unrooted trees is based on the number of taxa considered, not the amount of sequence information provided for each taxa. If n is the number of taxa considered, then the total number of rooted trees is $N_R = (2n-3)!/2^{n-2}$ $(n-2)!$ and the total number of unrooted trees is $N_U = (2n-5)!/2^{n-3}\,(n-3)!$ The amount of time required for computer generation of phylogenetic trees with the parsimony approach is not affected much by the amount of sequence information because in parsimony-based approaches, not all sites are used when considering molecular data. Rather, only informative sites—sites that have at least two different nucleotides present at least twice—are used. Simply increasing the amount of sequence information does not necessarily dramatically increase the number of informative sites, since there is no guarantee that the sequences that are added harbor informative sites. It is more important to identify regions that have informative sites and add these to the analysis. Once informative sites within an alignment are determined, the parsimony approach is used; the unrooted tree that invokes the fewest number of mutations for each of the sites is determined. Restricting the analysis to informative-only sites and invoking parsimony quickly eliminates many trees from further consideration.

24.18 When bootstrapping is used to assess the robustness of branching patterns in a tree of maximum parsimony, why is it more important to use sequences that have as many informative sites as possible than to simply use longer sequences?

Answer: Bootstrap tests draw repeatedly on different subsets of the original data and infer trees from the newly selected incomplete data set. Repeating this process hundreds or thousands of times identifies which portions of the inferred, maximum parsimony tree have the same groupings as many of the repetitions. These are the regions of the inferred tree that are especially well supported by the entire data set. When maximum parsimony is used, only informative sites—sites that have at least two different nucleotides present at least twice—are used for the analysis. Other sequence sites are not used in the analysis, as they are not considered biologically informative characters. Consequently, simply using longer sequences does not necessarily add any additional information to the analysis. It is more important to identify subsets of sequences that harbor a high density of informative sites and add these to the analysis. This will increase the size of the data set that can be selected from during the bootstrap analysis and therefore be more useful in assessing the robustness of the inferred tree.

24.19 What are the advantages of gene duplication (in whole or in part) in generating genes with new functions? Suggest an alternative way in which genes with new functions could arise.

Answer: Following a gene duplication event, two functional copies of a gene are present. Providing that one of them continues to provide a required function, the other may undergo changes in sequence, as it is free from the constraint of providing a required function. Not all of the changes, of course, will be desirable or lead to new functions; many may result in a loss of function and result in a pseudogene. However, some may result in a modified function that serves the organism. Subsequent misalignments between the altered, duplicated gene or a pseudogene and the retained functional gene provide an opportunity for recombination and gene conversion events to "repair" inactivated copies of a gene. Thus, a mutated, duplicated gene/pseudogene can have its function restored.

When only part of a gene is duplicated, there is a potential for domain shuffling. Domain shuffling is the duplication and rearrangement of domains in proteins that provide specific functions. This can lead to the assemblage of proteins with more complex domain arrangements, which can result in proteins with novel functions.

Gene duplications are therefore advantageous as a mechanism to generate genes with new functions because they provide a shortcut for modifying existing proteins. Rather than starting to build a gene "from scratch," duplications provide a template that already serves a function. The template can be tinkered with, repeatedly if recombination and gene conversion occur, in a manner that is largely free from selection. Of course, if the new function is deleterious to the organism (e.g., a dominant mutation resulting from a gain-of-function or new function), it will be selected against. Partial gene duplications are advantageous, since they allow combinations of existing functional domains to be reassembled into new arrangements and so effectively provide a set of building blocks for the assembly of new functions.

Additional mechanisms for the generation of genes with new functions could include the transposition, inversion, or translocation of chromosome segments that rearrange the transcriptional structure of an existing gene by placing it under the control of another gene's regulatory elements (e.g., altering the times or tissues it is expressed in during development and thereby providing "new functions" where none were previously); creating a novel fusion protein or introducing sites for mRNA processing that were previously not present; inserting copies of existing or foreign genes via viral retrotransposons, which could then, much like duplications, change over time; inserting and/or partially deleting of mobile repetitive elements into exonic or intronic regions that would introduce new regulatory and or RNA processing signals; and, of course, causing point, small deletion, and insertion mutations that could alter the function of an existing gene or modify the way existing RNA processing sites are used.